Residential Electrical Design Revised

by John E. Traister

Craftsman

Craftsman Book Company
6058 Corte del Cedro / P.O. Box 6500 / Carlsbad, CA / 92018

ISBN 0-934041-95-4

Cover: *Misty Bay at Silverlakes* built by Lowell Homes
Photo by *Robert Stein Photography*, Plantation, Florida

Contents

Chapter 5
Electric Services 49

Chapter 6
Wiring Methods 69

Chapter 7
Branch-Circuit Layout for Power 79

Chapter 8
Residential Lighting 95

Chapter 9
Outdoor Lighting 145

Chapter 10
Lighting Control 155

Chapter 1
Introduction

The use of electricity in houses began shortly after the opening of the California Electric Light Company in 1879 and Thomas Edison's Pearl Street Station in New York City in 1882. These two companies were the first to enter the business of producing and selling electric service to the public. In 1886, the Westinghouse Electric Company secured patents which resulted in the development and introduction of alternating current; this paved the way for rapid acceleration in the use of electricity.

The primary use of early home electrical systems was to provide interior lighting, but today's uses of electricity include:

- Heating and air conditioning
- Electrical appliances
- Electrical cooling
- Interior and exterior decorative lighting
- Communication systems
- Alarm systems

TYPES OF RESIDENTIAL CONSTRUCTION

Electricians and electrical designers must be able to visualize the building structure and its relation to the wiring in order to plan and coordinate the layout of the electrical system. Therefore, anyone involved with residential electrical systems should have, or acquire, a thorough knowledge of building construction for all types of building structures and should be able to interpret the drawing or plans in terms of the completed project with all of its necessary components.

In general, the following are the basic types of residential building construction:

- Wood frame
- Masonry
- Reinforced concrete
- Prefabricated structures

Usually, two or more basic types of construction are incorporated into one building.

Wood-Frame Structures

The most common form of residential building construction is the wood-frame type, as shown in Figure 1-1.

Masonry Structures

The masonry structure is constructed by placing clay bricks, stone, cement blocks, etc., one upon the other, and bonding them together with cement mortar. Except for basement floor slabs, the floor, the ceiling, and the roof construction usually have a wood frame. Figure 1-2 shows an example of masonry construction.

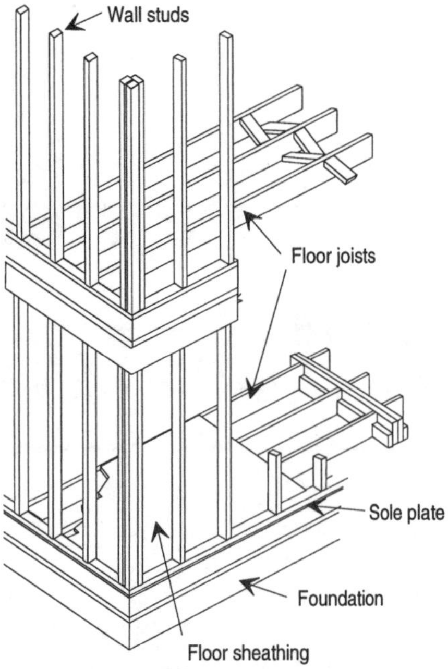
Wall studs
Floor joists
Sole plate
Foundation
Floor sheathing

Figure 1-1: Wood-frame construction

Reinforced Concrete

Reinforced-concrete construction is the type of construction sometimes used for residential foundations, although masonry structures of concrete blocks are more common today. Reinforced-concrete construction requires the building of forms, in which are placed the steel reinforcing bars necessary to form the foundation, the walls, the floor slabs, etc. Concrete is then poured into or onto the forms. When it has hardened sufficiently, the forms are stripped off. Figure 1-3 shows examples of reinforced-concrete construction.

Prefabricated Structures

Prefabricated structures have been used extensively over the past few decades for residential buildings. Such construction usually has a wood frame with plywood exterior sheathing and drywall interior sheathing. Sections of floors, walls, and roofs are constructed at a central construction factory and then shipped to the building site, where they are assembled by building-trade workers. In some instances, the electrical wiring is installed at

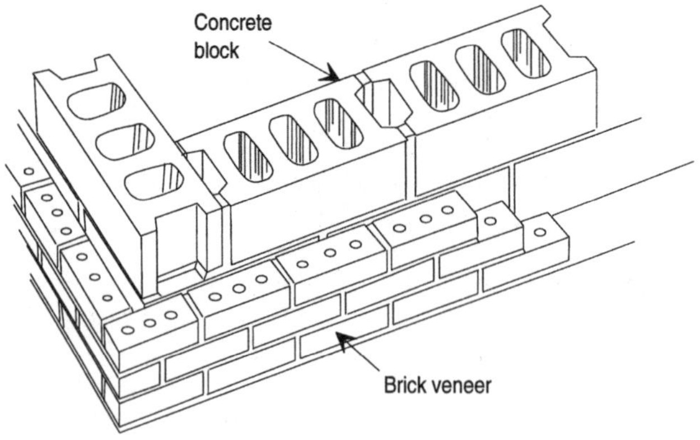
Concrete block
Brick veneer

Figure 1-2: Masonry construction

the factory also. In either case, the service-entrance — overheat or underground — must be installed once the structure is erected.

RESIDENTIAL WIRING SYSTEMS

The usual residential wiring system consists of the following sections:

- The electric service, which is the section of the electrical system extending from the point of attachment of the power company's conductors to the metering equipment.

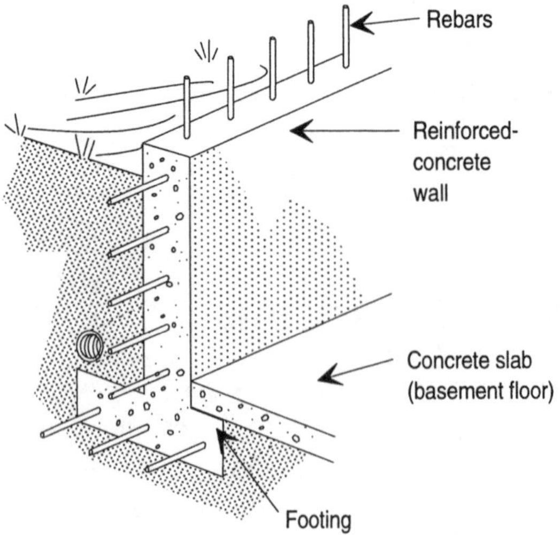
Rebars
Reinforced-concrete wall
Concrete slab (basement floor)
Footing

Figure 1-3: Reinforced-concrete construction

- The service-entrance equipment, which consists of the metering equipment and the main disconnecting means.
- The feeders that extend from the service entrance or other distribution equipment to subpanels or equipment.
- The branch-circuit wiring, which extends from the panelboard to various outlets (lights, receptacles, etc.).

These sections are usually found in all residential wiring systems, regardless of the size or type of the wiring system. Even large apartment buildings have these same sections; there are just more of them — one group for each apartment. Figure 1-4 illustrates these basic sections.

In planning any electrical system, there are certain general steps to be followed regardless of the type of construction. In planning a residential electrical system, the electrician must take certain factors into consideration. These include:

- Wiring method
- Overhead or underground electric service
- Type of building construction
- Grade of wiring devices and lighting fixtures
- Selection of lighting fixtures
- Type of heating and cooling system
- Control wiring for heating and cooling system
- Signal and alarm systems

The experienced electrician readily recognizes, within certain limits, the type of system that will be required. However, always check the local code requirements when selecting a wiring method. The *NE Code* gives minimum requirements for the practical safeguarding of persons and property from hazards arising from the use of electricity. These minimum requirements are not necessarily efficient, convenient, or adequate for good service or future expansion of electrical use. Some local building codes require electrical installations that surpass the requirements of the *NE Code*. For example, the *NE Code* requires that a certain type of service-entrance cable (called "SE cable") must be secured by means of cable straps every 2½ feet. Some electrical inspectors require these cable straps to be placed a minimum of 18 inches. In another area, the use of Type SE cable may not be allowed at all. Always check for any deviations from the *NE Code* in the area in which you are working. You'll save a lot of time and expense by doing so.

If more than one wiring method may be practical, you'll have to make a decision about which type of system to use prior to beginning the installation.

In a residential occupancy, the electrician should know that a 120/240-volt, single-phase service-entrance will invariably be provided by the local utility company. Furthermore, the service and feeders will be three-wire; the branch circuits will be either two- or three-wire, and the safety switches, service equipment, and panelboards or load centers will be three-wire, solid neutral. On each project, however, the electrician must determine where the point of service-drop attachment to the building will be located and how much of the service is to be provided as part of the electrical contract. The remaining portion will be installed by the utility company.

Summary

It is very important that anyone responsible for preparing electrical working drawings and specifications have a working knowledge of the elements of building construction, the ability to interpret architectural drawings — as well as electrical and mechanical drawings — and the ability to visualize the entire building structure and its relation to the electrical system. The same is true for workers who install the wiring. They must be able to interpret the electrical design laid out in the form of drawings and specifications.

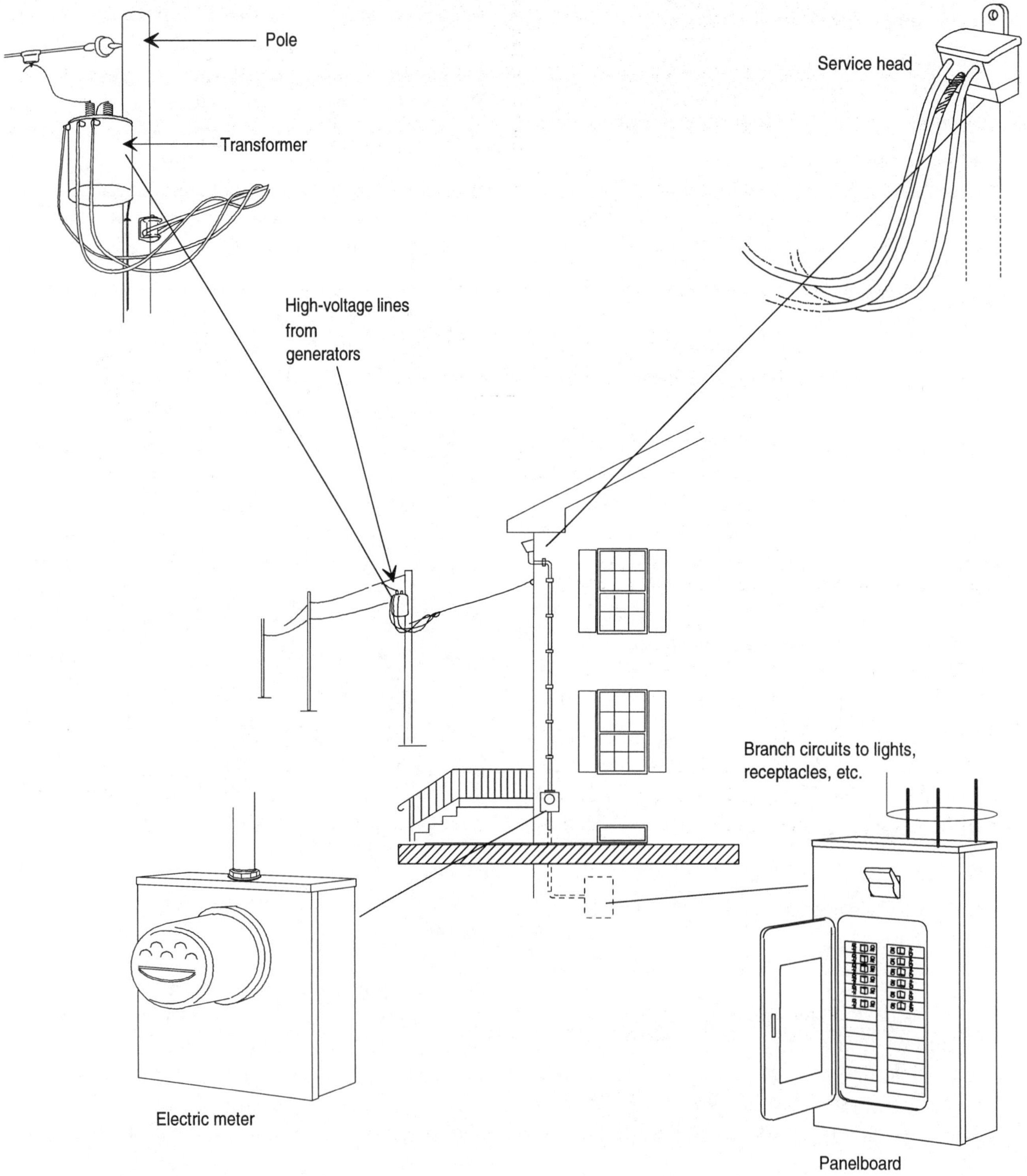
Pole
Service head
Transformer
High-voltage lines from generators
Branch circuits to lights, receptacles, etc.
Electric meter
Panelboard

Figure 1-4: Basic sections of a residential wiring system

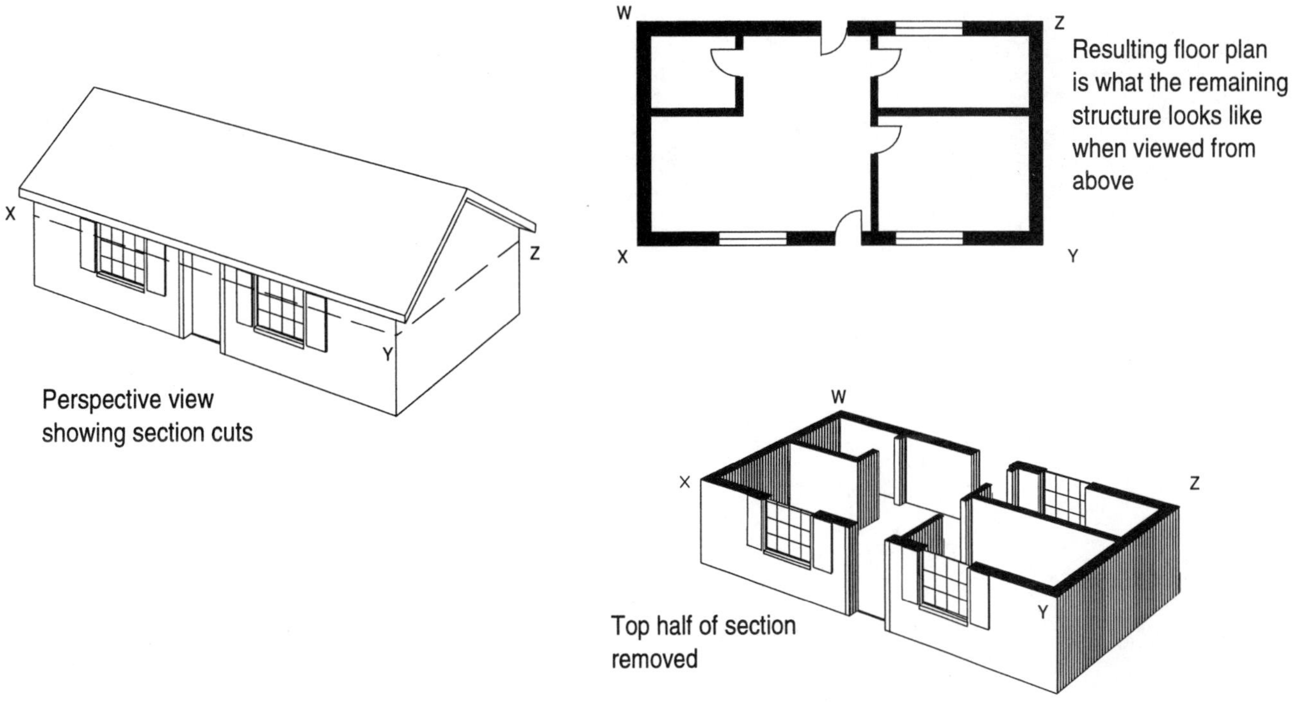
W
Z
Resulting floor plan is what the remaining structure looks like when viewed from above
X
Y
X
Z
Y
Perspective view showing section cuts
W
X
Z
Y
Top half of section removed

Figure 1-5: Principles of floor-plan development

Although blueprint reading will be covered in the chapters to follow, a description of a residential floor plan is in order here.

A floor plan is a drawing that shows the length and breadth of a building and the rooms which it contains. A separate plan is made for each floor.

Figure 1-5 shows how a floor plan is developed. An imaginary cut is made through the building as shown in the view on the left. The top half of this cut is removed (bottom view) and the resulting floor plan is what the remaining structure looks like when viewed directly from above.

Chapter 2

The *National Electrical Code®*

There are certain rules and regulations that govern the installation of residential wiring systems. First, and foremost, you will have to abide by the *National Electrical Code*. Second, some localities have their own codes or ordinances that sometimes surpass the requirements of the *NE Code*. Then, local power companies (utilities) have certain requirements that must be met — usually dealing with the service-entrance equipment.

Besides installation rules, you will also have to be concerned with the type and quality of materials that are used in residential wiring systems. Nationally recognized testing laboratories (Underwriters' Laboratories, Inc. is one) are product safety certification laboratories. They establish and operate product safety certification programs to make sure that items produced under the service are safeguarded against reasonably foreseeable risks. Some of these organizations maintain a worldwide network of field representatives who make unannounced visits to manufacturing facilities to countercheck products bearing their "seal of approval." See Figure 2-1 on page 14.

However, proper selection, overall functional performance and reliability of a product are factors that are not within the basic scope of U.L. activities.

INTRODUCTION TO THE *NATIONAL ELECTRICAL CODE®*

The *National Electrical Code (NE Code)* is one of the most important tools for the electrician. When used together with the electrical code for your local area, the *NE Code* provides the minimum requirements for the installation of electrical systems. You should always use the latest edition of the code as your on-the-job reference. It specifies the minimum provisions necessary for protecting people and property from electrical hazards.

Purpose of the *NE Code*

The primary purpose of the *National Electrical Code* is the practical safeguarding of persons and property from hazards arising from the use of electricity (*NE Code* 90-1(a)).

The *NE Code* is probably the most widely used and generally accepted code in the world. It has been translated into several languages. It is used as an electrical installation, safety, and reference guide in the United States. Many other parts of the world use it as well. Compliance with *NE Code* standards increases the safety of electrical installations — the reason the *NE Code* is so widely used.

Figure 2-1: The U.L. label is recognized nationwide as one "seal of approval"

The Layout of the *NE Code*

Figure 2-2 shows how the *NE Code* is organized. The main body of the text begins with an introduction, also titled Article 90. This introduction gives you an overview of the *National Electrical Code.* Items included in this section are:

- Purpose of the *NE Code* (90-1)
- Scope of the Codebook (90-2)
- Code Arrangement (90-3)
- Code Enforcement (90-4)
- Mandatory Rules and Explanatory Material (90-5)
- Formal Interpretation (90-6)
- Examination of Equipment for Safety (90-7)
- Wiring Planning (90-8)
- Metric Units of Measurement (90-9)

The Body of the *National Electrical Code*

The remainder of the *NE Code* is organized into nine chapters. Chapter 1 contains a complete list of definitions used in the *NE Code*. These definitions are referred to as Article 100. Article 110, also included in this chapter, gives the general requirements for electrical installations. It is important for you to be familiar with this general information and the definitions.

Article 100 Definitions: There are many definitions included in Article 100. You should become familiar with the definitions. Since a copy of the lastest *NE Code* is compulsory for any type of electrical wiring, there is no need to duplicate them here. However, here are two definitions that you should become especially familiar with:

- Labeled - Equipment or materials to which has been attached a label, symbol or other identifying mark of an organization acceptable to the authority having jurisdiction and concerned with product evaluation, that maintains periodic inspection of production of labeled equipment or materials, and by whose labeling the manufacturer indicates compliance with appropriate standards or performance in a specified manner.
- Listed - Equipment or materials included in a list published by an organization acceptable to the authority having jurisdiction and concerned with product evaluation, that maintains periodic inspection of production of listed equipment or materials, and whose listing states either that the equipment or material meets appropriate designated standards or has been tested and found suitable for use in a specified manner.

Chapters 2 through 8 each contain numerous articles and sections.

Each chapter focuses on a general category of electrical application, such as wiring and protection. Each article emphasizes a more specific segment of that category, such as branch circuits. Each section gives examples of a specific application of the code, such as multi-wire branch circuits.

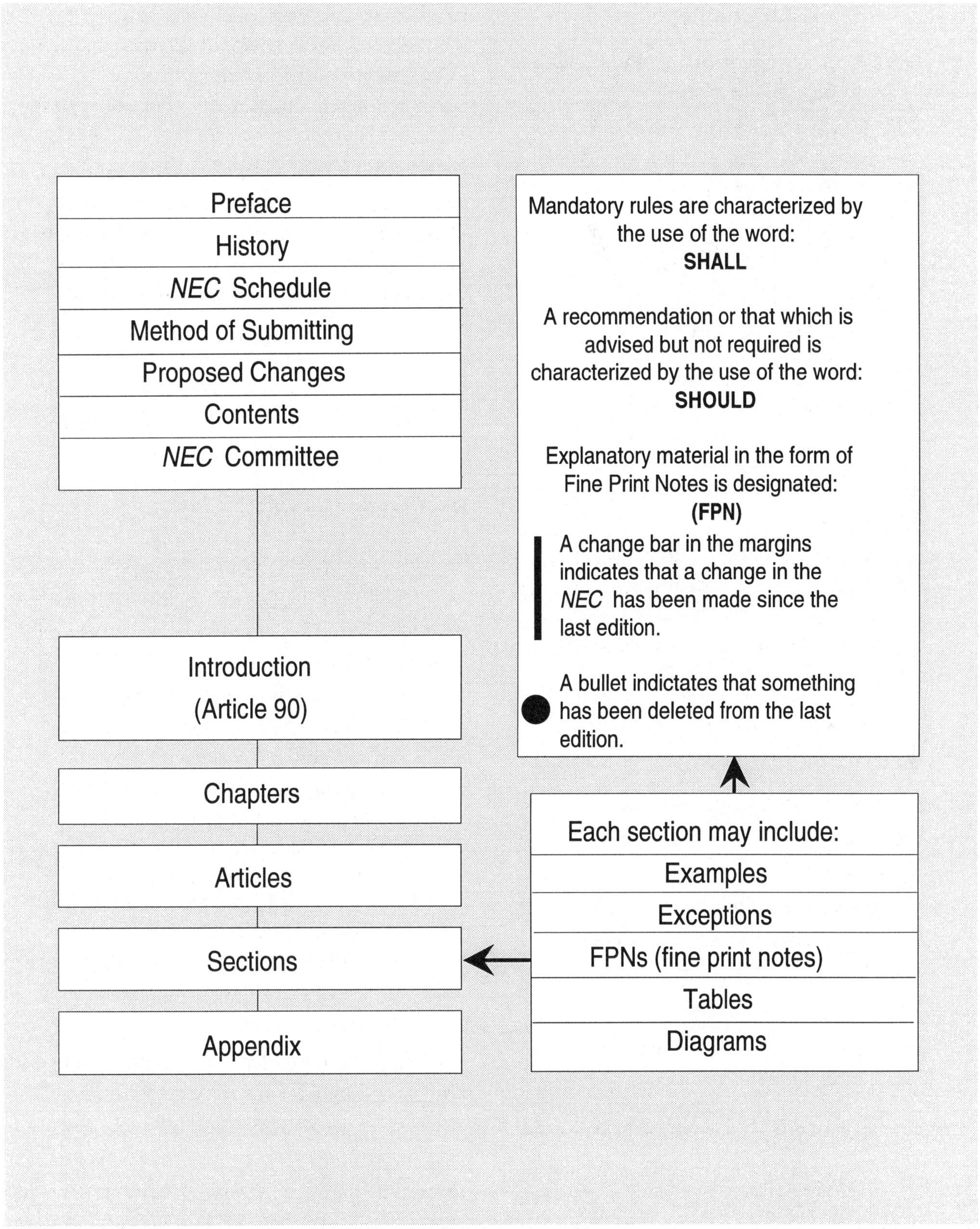
Preface
History
NEC Schedule
Method of Submitting
Proposed Changes
Contents
NEC Committee
Introduction (Article 90)
Chapters
Articles
Sections
Appendix
Mandatory rules are characterized by the use of the word:
SHALL
A recommendation or that which is advised but not required is characterized by the use of the word:
SHOULD
Explanatory material in the form of Fine Print Notes is designated:
(FPN)
A change bar in the margins indicates that a change in the NEC has been made since the last edition.
A bullet indictates that something has been deleted from the last edition.
Each section may include:
Examples
Exceptions
FPNs (fine print notes)
Tables
Diagrams

Figure 2-2: Layout of the *National Electrical Code*

Organization of the Chapters

The chapters of the *National Electrical Code* are organized into three major categories:

- Chapters 1, 2, 3, and 4 - The first four chapters present the rules for the design and installation of electrical wiring systems.
- Chapters 5, 6, and 7 - These chapters are concerned with special occupancies, equipment, and conditions. Rules in these chapters may modify or amend those in the first four chapters.
- Chapter 8 - This chapter covers communication systems, such as the telephone and telegraph systems, as well as radio and television receiving equipment.
- Chapter 9 - This chapter contains tables which specify the properties of conductors and rules for the use of conduit to enclose the conductors. The examples of Chapter 9 demonstrate the use of the rules for design given in the first four chapters.

What You Will See in the *National Electrical Code* Text

As you open the code book, you will notice several different types of text used. See Figure 2-2. Here is an explanation of each type of text:

- Black letters: Basic definitions and explanations of the code.
- Bold black letters: Headings for each code application.
- Exceptions: These explain the times when a specific code does not apply. Exceptions are written in *italics* under the code to which they pertain.
- Fine print notes: These are defined in the text by the abbreviation (FPN) placed before a paragraph printed in fine print. These are notes that explain something in an application, suggest other sections to read about the application, or give tips about the application.
- Tables: Tables are often included when there is more than one possible application of a code. You would use a table to look up the specifications of your job to find the proper conditions of the code.
- Diagrams: These may be included with code explanations to give you a picture of what your application may look like.

Navigating the *NEC*

In order to locate code information for a particular procedure you are performing, you should use the following steps:

Step 1: Familiarize yourself with Article 90 and Chapter 1 so you will have an understanding of the material covered in the code and the definitions used in it.

Step 2: Turn to the contents at the beginning of the code book.

Step 3: Locate the chapter which focuses on the category under which you are working.

Step 4: Find the article pertaining to your specific application.

Step 5: Turn to the page indicated. Each application will begin with an easily readable bold heading.

Note: An index is provided at the end of the code. The index lists specific topics and provides a reference to the location of the material within the code. The index is helpful when you're looking for something very specific.

An Example of Navigating

Suppose you are installing Type SE (service-entrance) cable on the side of a home. You know that this cable must be secured, but you aren't sure of the spacing between cable clamps. To find out this information, use the following procedure:

Step 1: Look in the *NE Code* Table of Contents and follow down the list until you find an appropriate category.

Step 2: Article 230 under Chapter 3 will probably catch your eye first, so turn to the page where Article 230 begins in the *NE Code*.

Step 3: Glance down the section numbers, 230-1, Scope, 230-2, Number of Services, etc. until you come to Section 230-51, Mounting Supports. Upon reading this section, you will find in paragraph (a) — Service -Entrance Cables — that "Service-entrance cable shall be supported by straps or other approved means within 12 inches (305 mm) of every service head, gooseneck, or connection to a raceway or enclosure and at intervals not exceeding 30 inches (762 mm)."

After reading this section, you will know that a cable strap is required within 12 inches of the service head and within 12 inches of the meter base. Furthermore, the cable must be secured in between these two termination points at intervals not exceeding 30 inches.

Another method of finding items in the *NE Code* is to refer to the Index in the back of the *NE Code* book.

NRTL, NEMA, and NFPA

To fully understand the *NE Code*, it is important to understand the organizations which govern it.

NRTL (Nationally Recognized Testing Laboratory)

Nationally Recognized Testing Laboratories are product safety certification laboratories. They establish and operate product safety certification programs to make sure that items produced under the service are safeguarded against reasonably foreseeable risks. NRTL maintains a worldwide network of field representatives who make unannounced visits to factories to countercheck products bearing the safety mark.

NEMA (National Electrical Manufacturers Association)

The National Electrical Manufacturers Association was founded in 1926. It is made up of companies that manufacture equipment used for generation, transmission, distribution, control, and utilization of electric power. The objectives of NEMA are to maintain and improve the quality and reliability of products; to ensure safety standards in the manufacture and use of products; to develop product standards covering such matters as naming, ratings, performance, testing, and dimensions. NEMA participates in developing the *NE Code* and the National Electrical Safety Code and advocates their acceptance by state and local authorities.

NFPA (National Fire Protection Association)

The NFPA was founded in 1896. Its membership is drawn from the fire service, business and industry, health care, educational and other institutions, and individuals in the fields of insurance, government, architecture, and engineering. The duties of the NFPA include:

- Developing, publishing, and distributing standards prepared by approximately 175 technical committees. These standards are intended to minimize the possibility and effects of fire and explosion.

- Conducting fire safety education programs for the general public.
- Providing information on fire protection, prevention, and suppression.
- Compiling annual statistics on causes and occupancies of fires, large-loss fires (over 1 million dollars), fire deaths, and firefighter casualties.
- Providing field service by specialists on electricity, flammable liquids and gases, and marine fire problems.
- Conducting research projects that apply statistical methods and operations research to develop computer modes and data management systems.

The Role of Testing Laboratories

Testing laboratories are an integral part of the development of the code. The NFPA, NEMA, and NRTL all provide testing laboratories to conduct research into electrical equipment and its safety. These laboratories perform extensive testing of new products to make sure they are built to code standards for electrical and fire safety. These organizations receive statistics and reports from agencies all over the United States concerning electrical shocks and fires and their causes. Upon seeing trends developing concerning association of certain equipment and dangerous situations or circumstances, this equipment will be specifically targeted for research.

Summary

The *National Electrical Code* specifies the minimum provisions necessary for protecting people and property from hazards arising from the use of electricity and electrical equipment. Anyone involved in any phase of the electrical industry must be aware of how to use and apply the code on your job. Using the code will help you to safely install and maintain the electrical equipment and systems that you come into contact with.

In the United States, cable wiring — Type AC (BX) and Type NM (Romex) — is probably the most common wiring method for residential-type construction. Local restrictions may limit the extent of its use, but cable wiring in some form is found in almost every place where electrical power is available.

Regardless of the type of cable specified in an electrical design, the *National Electrical Code* must be followed. Here are a few *NE Code* requirements that should be remembered and included either on the drawings or in the written specifications or during the actual installation:

- When armored cable or nonmetallic sheathed cable is used, the *NE Code* requires that a box be installed at each outlet.
- Insulating bushings are required at the termination of the armor on armored cable for the protection of the insulation between conductors and the armor.
- Nonmetallic sheathed cable must be secured in place by approved staples, straps, or similar fittings at intervals not exceeding 4½ feet, and within 12 inches from every outlet box.
- At least 6 inches of free conductor must be left at each outlet and switch point for the making of joints or the connection of fixtures or devices.
- Junction boxes must be so installed that the wiring contained in them is accessible without removing any fixed part of the building, sidewalk, or paving.

In the chapters to follow, additional *NE Code* requirements will be explained; that is, the *NE Code* regulations which apply to each step of the design are discussed in detail in their respective chapter. However, as pointed out previously, certain modifications may be necessary because of local and/or state electrical code requirements.

Chapter 3
Electrical Drawings

The drawings or plans used in the residential electrical field are mainly location plans and diagrams. The location plan, or floor plan as it is more commonly called, shows the outline of a building, area, or site where the electrical equipment is to be installed. Such plans usually are drawn to scale, and important dimensions are sometimes indicated by dimension lines and notations. The plan will also show exterior walls, interior partitions, doors, and windows in the building. In addition, it will indicate the placement of electrical equipment by the use of lines, symbols, and notations for lighting fixtures, convenience outlets, special-purpose outlets, motors, and other electrical devices.

Figure 3-1 shows an outline plan of a single-level residence without the electrical system drawn within the building. Figure 3-2 shows the same floor plan with the receptacles and related circuits

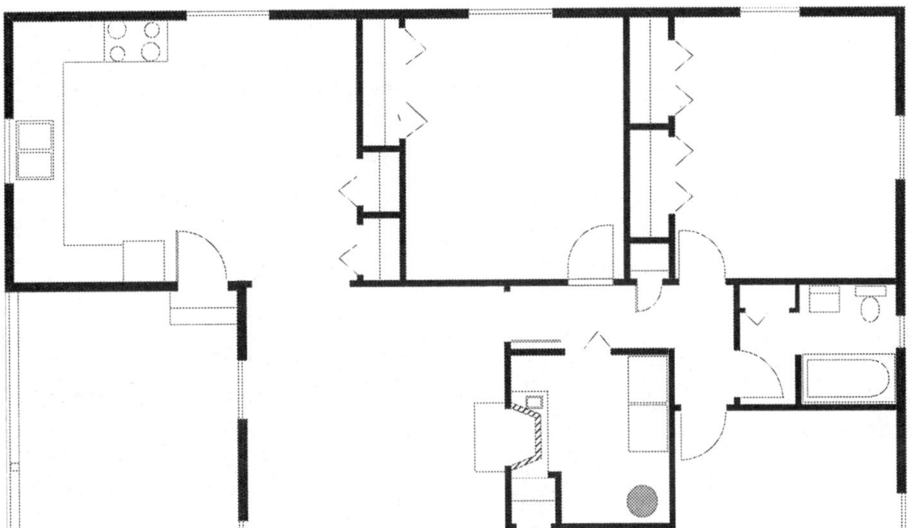

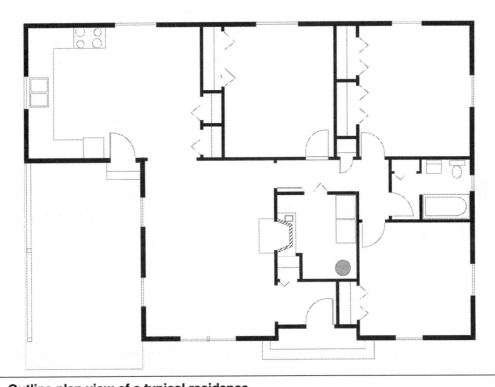

Figure 3-1: Outline plan view of a typical residence

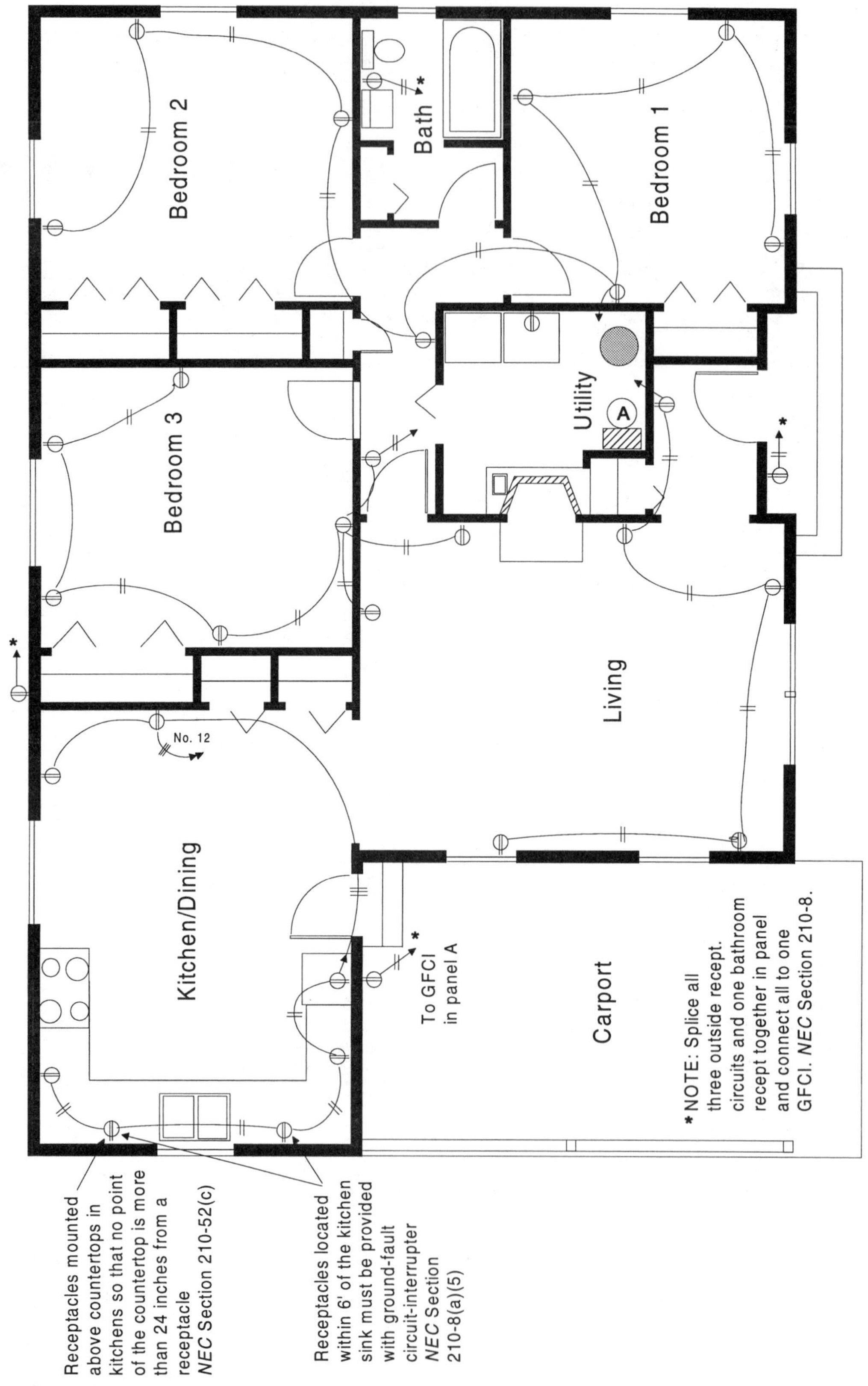
Bedroom 2
Bath
Bedroom 1
Bedroom 3
Utility
A
Living
No. 12
Kitchen/Dining
To GFCI in panel A
Carport
* NOTE: Splice all three outside recept. circuits and one bathroom recept together in panel and connect all to one GFCI. NEC Section 210-8.
Receptacles mounted above countertops in kitchens so that no point of the countertop is more than 24 inches from a receptacle NEC Section 210-52(c)
Receptacles located within 6' of the kitchen sink must be provided with ground-fault circuit-interrupter NEC Section 210-8(a)(5)

Figure 3-2: Residential floor plan with receptacles and related circuitry shown

drawn. The main load center or power panel is also shown. This type of plan is normally referred to as an *electrical power plan.*

Architectural plans also use symbols to indicate the various details of construction, such as the materials used in the construction of the partitions. The details also show the interior finish of the walls, door and window schedule numbers, etc. All walls and partitions are dimensioned for clarification. The outline plan in Figure 3-1, however, only shows those details that are considered necessary for the installation of the electrical system.

All that need be shown on the electrical drawings are a building outline drawn to scale, partitions, door and window locations, stairwells, etc. If the electrician needs to know certain other elements of the construction of the building in order to install his electrical wiring, he can refer to the architectural drawings. In other words, there is no need to repeat any architectural details on the electrical drawings — only details pertinent to the electrical system.

ELECTRICAL DIAGRAMS

Electrical diagrams are drawings that indicate the basic scheme, or plan, according to which the electrical equipment will be connected and the specific purpose of the equipment. Electrical diagrams are seldom, if ever, drawn to scale, and they are normally referred to as elementary block or schematic diagrams.

Elementary block diagrams, such as the one shown in Figure 3-3, use rectangular or square blocks to represent the various pieces of electrical equipment. In Figure 3-3, the larger blocks also

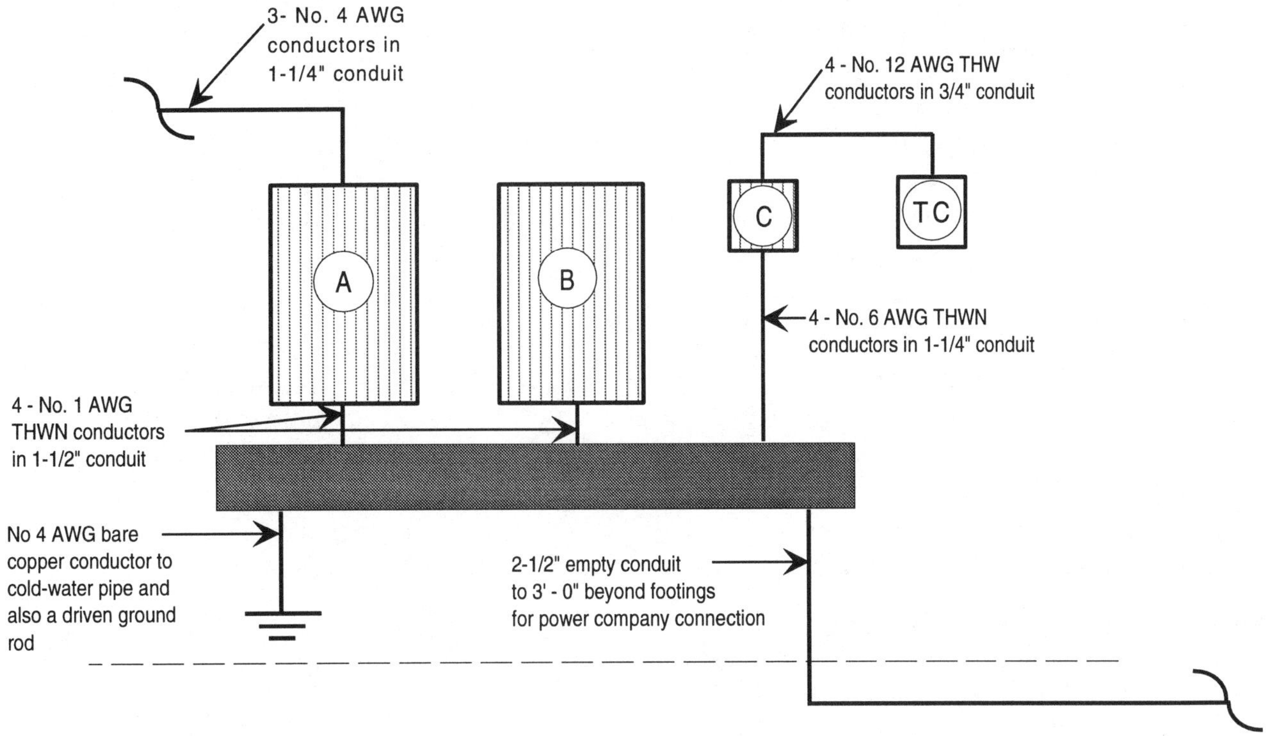

3- No. 4 AWG conductors in 1-1/4" conduit
4 - No. 12 AWG THW conductors in 3/4" conduit
A
B
C
TC
4 - No. 6 AWG THWN conductors in 1-1/4" conduit
4 - No. 1 AWG THWN conductors in 1-1/2" conduit
No 4 AWG bare copper conductor to cold-water pipe and also a driven ground rod
2-1/2" empty conduit to 3' - 0" beyond footings for power company connection

Figure 3-3: Elementary block diagram (power-riser diagram) used to show electric-service details

represent electrical panelboards (Panel A, B, etc.), and the smaller block represents a time clock (TC) to control the outside lighting. The various lines represent electrical feeders or conduit raceways containing conductors. Notes on the drawing indicate such items as conduit and wire size, catalog numbers of electrical equipment, etc. Notice that this diagram is not drawn to scale and should be identified as such. A block diagram is useful for indicating the overall structure of an electrical system.

Schematic diagrams use symbols to represent electrical devices and to show their connections by both solid and broken lines. According to the number of lines used to connect the various devices, the diagrams are classified as single-line, three-line, or complete diagrams.

The single-line diagram uses a single line to indicate the circuits, regardless of the actual number of wires in the circuit. For example, the duplex receptacles shown in Figure 3-2 are connected by solid lines. Only one single line is shown connecting the outlets. However, two wires are used in some of the circuits and three wires in the others. These are indicated by the slash marks through the lines; that is, two slash marks indicate two wires and three slash marks indicate three wires. If a circuit had four wires, there would be four slash marks, etc. In many cases, however, no slash marks indicate a two-wire circuit. Only more than two wires is indicated by the slash marks.

The solid lines connecting the outlets indicate that the circuits are installed in the ceiling or walls. If the outlets were connected by broken lines, the lines would indicate that the circuits were to be run into the floor.

In general, electrical drawings for residential electrical systems will indicate all electrical outlets, branch circuits, feeders, panelboards, service-entrance equipment, and other similar electrical details. The drawings will also include various schedules, wiring diagrams, and detail drawings of electrical equipment when deemed necessary. The number of drawings will vary depending upon the size and complexity of the job. On very small jobs, the architect may merely show the electrical outlets on the architectural floor plan, make an allowance (in dollars) for lighting fixtures to be chosen by the owners, and indicate by notation that "all electrical work will be installed according to provisions set forth in the *National Electrical Code* and local ordinances . . ."

When preparing electrical drawings, an outline of the architectural floor plan(s) is first laid out. The following are the steps in this procedure:

Step 1: Clean the drawing board and all drawing equipment. This is necessary in order to have a clean and neatly finished drawing.

Step 2: Take the print (usually blue line) of the architectural floor plan and lay it on the drawing board. If the print has been folded, rub out the creases as much as possible because they will cause smudge marks on the tracing paper. Now line up the horizontal line on the print with the T-square, parallel bar, or drafting machine. When the print is lined up perfectly, secure it to the board with small pieces of drafting tape.

Step 3: Place a sheet of tracing paper over the print, line it up, and secure this to the drawing board — again with drafting tape.

Step 4: With the use of a straightedge, trace all necessary horizontal lines on the floor plan of the building. Next, trace all vertical lines. This drawing should show just an outline of the building, all partitions, doors, windows, etc. It should not show dimension lines, wall finishes, or any other architectural detail not necessary for the electrical work.

Step 5: A medium-hard lead (about 4H) and a light line should be used for drawing the outline of the floor plan. On architectural drawings, a heavier line is used for drawing the floor

Lighting-Fixture Schedule				
Symbol	Type	Manufacturer and Catalog Number	Mounting	Lamps
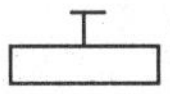	A			
	B			
	C			
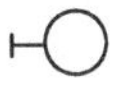	D			
	E			
Etc.	Etc.			

Figure 3-4: Lighting-fixture schedule

plan because the building construction lines are more important on architectural than on electrical drawings. On electrical drawings, however, the outlets, circuits, and similar electrical details are the most important. Therefore, the building lines are sometimes drawn in lightly as a background and all of the electrical outlets, etc., are drawn with heavy or dark lines.

Step 6: Once the tracing is completed, check it over very carefully; make certain that no doors, windows, partitions, etc., are omitted. Now remove the tape holding the tracing paper in place and remove the tracing. Again, check the tracing against the print for any errors or omissions.

The electrical designer takes the architect's drawings and designs a suitable electrical system for the building. This involves calculating the service-entrance size, the number and size of the circuits, the size of the feeders, and the location of the outlets. It also involves determining if the lighting is sufficient and which lighting fixtures will blend with the basic style of the home. Usually the designer's layout is then sketched roughly on prints of the building outline and given to an electrical draftsman to complete. It is both the designer's and the draftsman's responsibility to provide detailed and accurate drawings of the electrical system, and the final result should indicate exactly what is required.

ELECTRICAL SCHEDULES

A schedule is a systematic method of presenting notes or lists of equipment in tabular form on a drawing. When properly organized and thoroughly understood, schedules are not only powerful timesaving methods for the electrical designer, but also save the draftsman, the specification writer, and the workman on the job much valuable time.

For example, the lighting-fixture schedule shown in Figure 3-4 has provisions to list the manufacturer and catalog number of each fixture;

Lighting-Fixture Schedule				
Symbol	Type	Manufacturer and Catalog Number	Mounting	Lamps
	A	LIGHTOLIER 10234	WALL	2-40W T-12WWX
	B	LIGHTOLIER 10420	SURFACE	2-40W T-12WWX
	C	ALKCO RPC-210-6E	POST	2-8W T-5
	D	P7S AL 2936	WALL	1-100W A
	E	P7S 110	SURFACE	1-100W A

Figure 3-5: Lighting-fixture schedule with data filled in

the number, type, and wattage of the lamps in the fixture; the voltage of the circuit supplying current to the fixture; and the mounting characteristics of the fixture. A column for remarks is sometimes included to facilitate the installation of the fixture. The second column from the left marked "Type" gives the designated fixture number as shown on the floor plans for proper identification.

Sometimes all of the same information can be found in the written specifications for the project, but the electricians do not always have access to the specifications, whereas they usually do have constant access to the working drawings. Therefore, the schedule (drawn on the plans) is an excellent means of providing essential information in a clear, concise, and accurate manner, allowing all concerned with the project to carry out their assignments in the least amount of time.

Figure 3-5 shows the same type of lighting-fixture schedule with the data filled in the appropriate spaces. This schedule, combined with the project's lighting floor plans, should enable any qualified electrician to install the lighting fixtures and their related wiring exactly as specified.

Electric-Heat Schedules

Each day, more and more existing homes are converting to flameless electric heat, and the use of electric heat in new homes is increasing even more. Electric baseboard heaters along with various other types of "unit" heaters usually are the responsibility of the electrical contractor, whereas electric forced-air systems utilizing ductwork usually are the responsibility of the mechanical contractor. Therefore, this book will cover only resistance-type electric unit heaters.

The schedule shown in Figure 3-6 is excellent for use on electrical drawings; it quickly and clearly lists the required unit heaters for any given project. Like the lighting-fixture schedule in Figures 3-4 and 3-5, this schedule provides spaces for

Electric-Heat Schedule					
H'TR Type	Manufacturer's Description	Dimensions	Volts	Mounting	Wattage/Remarks
A	Nutone Cat. No. 438	8	240	Surface	2000 Watts
B	Nutone Cat. No. 416	6	240	Surface	1500 Watts
C	Nutone Cat. No. 402	2	240	Surface	500 Watts

Figure 3-6: Electric-heat schedule

the manufacturer and catalog number of the unit. It lists the total wattage of the unit and gives physical dimensions, voltage, etc., to facilitate installation. The example given was taken from the drawings for a summer cabin, and this schedule and the floor plans give all the information necessary for a correct installation.

More examples of electric heat schedules can be found later in this book.

Panelboard Schedules

Panelboard schedules are used to indicate the electric service equipment specified for a given residence. Such schedules should indicate the required ampere rating of the equipment, accepted manufacturers, number of fuses or circuit breakers within the panel, and similar data that will indicate, beyond any question of a doubt, exactly what is required of the electrical contractor.

The panelboard schedule in Figure 3-7 is called the short form and is used for smaller residences. While sufficient for its purpose, it does not give any circuit-load data such as the schedule shown in Figure 3-8.

The short-form panelboard schedule is again shown in Figure 3-9 — this time with data filled in for a certain project. The panel number (A) is indicated in the left space of the schedule. For locating purposes, the panel designation is also

Panelboard Schedule										
Panel No.	Type Cabinet	Panel Mains			Branches					Item Fed or Remarks
		Amps	Volts	Phase	1P	2P	3P	Prot.	Frame	

Figure 3-7: Panelboard schedule (short form)

shown on the plans of the residence. The panel cabinet is to be mounted on the surface in this case, but many panelboards are flush mounted, in which case the schedule would indicate "flush" instead of "surface." The schedule also tells that the panelboard is to be rated for a total load of 100 amperes at 120/240 volts, single-phase. Under this data is the manufacturer's name and the type of panelboard; that is Square D, Type QO.

The columns under the heading "Branches" are provided to list the overcurrent protection devices (in this instance they are circuit breakers). The single-pole (1-P) 15-A circuit breakers are provided for the lighting and convenience outlet circuits; the 1-P, 20-A breakers are for circuits feeding the kitchen and laundry room outlets; and the 2-P breakers are for heavy-power-consumption devices or equipment such as air conditioners, water heaters, electric ranges, clothes dryers, etc. The last column, "Item Fed or Remarks," lists the items fed on each row of circuit breakers.

The panelboard schedule in Figure 3-10 shows the same loads on the long-form panelboard schedule. In this schedule the panelboard and circuit data are the same as in Figure 3-9 except that they are more detailed; that is, circuit numbers, as well as the total estimated load for each circuit, are given.

Kitchen-Equipment Schedule

Sometimes when a residential kitchen is fully equipped with several appliances, such

Panelboard Schedule

Panel____________ 1 ϕ3 Wire________Mounted_________Ampere Main

Location___________________________________ _____ Ampere Bus

CCT No.	volt-amperes ϕ A	volt-amperes ϕ B	Description	CB Pole	CB AIC	Phase A	Phase B	CB AIC	CB Pole	Description	volt-amperes ϕ A	volt-amperes ϕ B	CCT No.
1						●							2
3							●						4
5						●							6
7							●						8
9						●							10
11							●						12
13						●							14
15							●						16
17						●							18
19							●						20

	Total VA/ ϕ A
	Total VA/ ϕ B
	Total VA
	Line Amperes

Figure 3-8: Panelboard schedule (long form)

Panelboard Schedule

Panel No.	Type Cabinet	Panel Mains			Branches					Item Fed or Remarks
		Amps	Volts	Phase	1-P	2-P	3-P	Prot.	Frame	
A	Surf.	100A	120/240	1	8	-	-	15A	70A	Lights & Recepts.
Sq. D Type QO with 100-A Main Circuit Breaker					3	-	-	20A		Appliance Circuits
					-	1	-	30A		Water Heater
					-	1	-	40A		Elect. Range
					5	-	-	15A		Spares

Figure 3-9: Application of short-form panelboard schedule

as dishwasher, cooktop, wall ovens, electric barbecue, garbage disposal, built-in mixer center, etc., the electrical designer will use a kitchen-equipment schedule to specify the required installation.

A typical kitchen-equipment schedule is shown in Figure 3-11 on page 30. This schedule provides spaces for the name or description of each piece of equipment, the horsepower or kilowatts, the voltage, the wire and conduit sizes, and the overcurrent protection. The "Furnished by" column designates who will supply the piece of equipment, and the remaining column is for remarks. Such remarks may give the mounting height of the outlet or they may designate whether a receptacle is needed or if a direct connection should be made.

Figure 3-12 is a floor plan of a large residential kitchen. Although the equipment numbers are shown and each piece of equipment is circuited, several details needed for proper installation are missing. However, if the complete kitchen-equipment schedule is referred to (Figure 3-11), all details necessary for correct installation will be found.

COMPUTER-AIDED DESIGN

The use of computers is a cost effective method of increasing design and drafting productivity in residential electrical applications. Computer-aided design (CAD) can increase productivity up to ten or more times with some of the CAD and graphic systems now currently available.

A CAD system generates drawings from computer programs and the advantages are many:

- CAD can be learned and used easily by drafters and designers.
- Changes can be made quickly and easily.
- Commonly-used symbols can be easily retrieved.
- Drawings can be stored on computer diskettes and retrieved at any time — usually by merely pressing a couple of keys on the computer keyboard.
- Any number of copies may quickly and easily be printed on a plotter or printer.

Panelboard Schedule

Panel A 120/240V 1 ϕ3 Wire Surface Mounted 100 Ampere Main

Location Utility Room 125A Ampere Bus

CCT No.	volt-amperes ϕ A	volt-amperes ϕ B	Description	CB Pole	CB AIC	Phase A	Phase B	CB AIC	CB Pole	Description	volt-amperes ϕ A	volt-amperes ϕ B	CCT No.
1	800	-	Lighting	1		●			1	Lighting	1050		2
3		1200	Lighting	1			●		1	Receptacles		1200	4
5	1200		Receptacles	1		●			1	Receptacles	1200		6
7		1200	Receptacles	1			●		1	Receptacles		1200	8
9	1500		Appliance Cir.	1		●			1	Appliance Cir.	1500		10
11		1200	Spare				●		1	Appliance Cir.		1500	12
13	2250		Water Heater	2		●			2	Elect. Range	4000		14
15		2250	↓				●			↓		4000	16
17	1200		Spare			●				Spare	1200		18
19		1200	Spare				●			Spare		1200	20

15,900	Total VA/ ϕ A	32,050/240 = 133.54 connected amperes
16,150	Total VA/ ϕ B	21,292.5/240 = 88.71 demand amperes
32,050	Total VA	
88.71	Line Amperes	

Figure 3-10: Application of long-form panelboard schedule

ELECTRICAL SYMBOLS

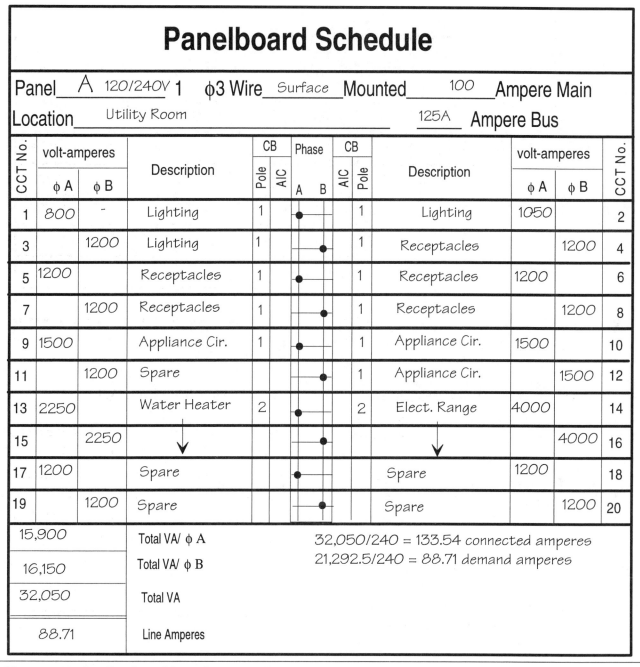

It was previously explained that the purpose of an electrical drawing was to show the location of various electrical circuits, outlets, and equipment; the size of electrical service and feeders; and other details of construction necessary to properly install the electrical system. In the preparation of such drawings, symbols are used to simplify the draftsman's work; this, in turn, saves both the designer and the client/owner a great deal of time and money. For example, the pictorial drawing in Figure 3-13 on page 31 shows a ceiling-mounted smoke detector. If drafters had to draw this picture

Kitchen-Equipment Schedule								
Equip. No.	Description	HP or Kw	Volts	Connection			Furnished By	Remarks
				Wire	Conduit	Prot.		
1	Cooktop	4.5	240	#8	Cable	30A		
2	Oven	4.5	240	#8	↓	30A		
3	Garbage disposal	1.0	120	#12	↓	15A		
4	Refrigerator	0.9	120	#12	↓	15A		
5	Clothes dryer	4.5	240	#12	↓	15A		
6	Washer	1.0	120	#8	↓	30A		

Figure 3-11: Kitchen-equipment schedule

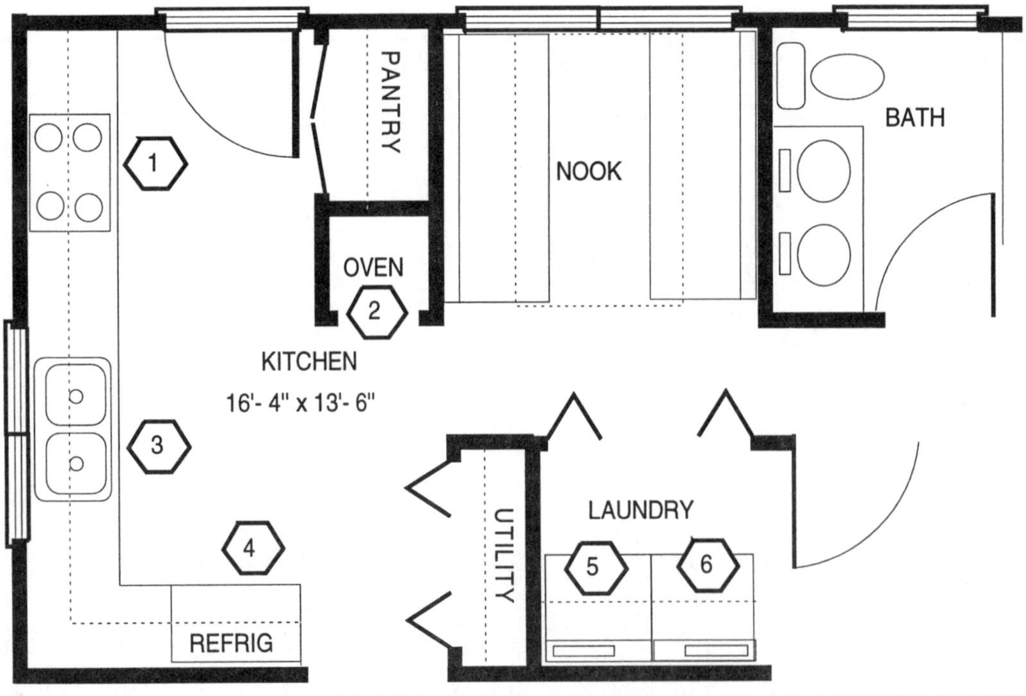
PANTRY
NOOK
BATH
1
OVEN
2
KITCHEN
16'- 4" x 13'- 6"
3
UTILITY
LAUNDRY
5
6
4
REFRIG

Figure 3-12: Kitchen floor plan

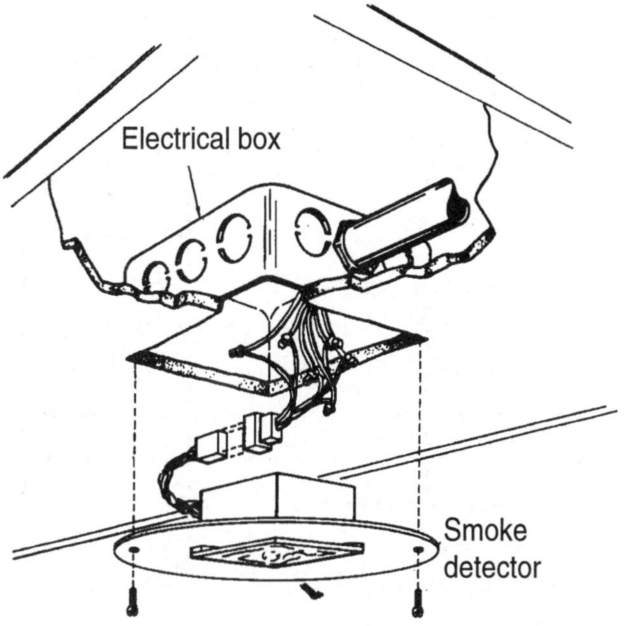
Electrical box
Smoke detector

Figure 3-13: Pictorial drawing of a smoke detector

on the drawings every time a smoke detector was located, it could conceivably take days to finish one floor-plan layout. But, if a symbol could be used to represent the smoke detector, several detectors could be located and drawn in only a few minutes.

In drawing electrical plans, most engineers, designers, and drafters use electrical symbols adopted by the American National Standards Institute (ANSI).

Note that all the outlet symbols in Figure 3-14 have the same basic form — a circle; however, the addition of a line or other mark gives each an individual meaning. It is also apparent that the difference in meaning can be indicated by the addition of letters or an abbreviation to the symbol. A good practice to follow in learning electrical symbols is to first learn the basic form and then apply the variations for obtaining different meanings.

The following lists of electrical symbols (Figure 3-15) were prepared jointly by the Consulting Engineers Council/US and the Construction Specifications Institute, Inc. The drawing symbols in Figure 3-16 on page 39 were modified especially for residential and multidwelling buildings, by a consulting engineering firm. Symbols from both lists will be used throughout this book whenever examples of working drawings are given.

A variety of special templates have been designed to simplify the job of drafters when they are drawing various electrical symbols. These templates are made of thin pieces of transparent plastic out of which different shapes have been punched. The draftsman uses the templates to trace these shapes onto the drawing. Such templates minimize many tedious and time-consuming operations.

The remaining chapters in this book are designed to provide the reader with the knowledge necessary to prepare electrical designs of high quality.

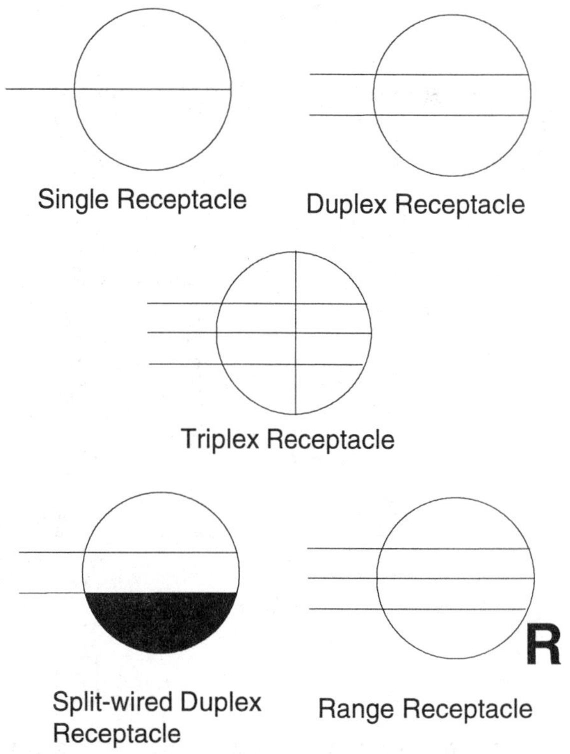
Single Receptacle
Duplex Receptacle
Triplex Receptacle
Split-wired Duplex Receptacle
R
Range Receptacle

Figure 3-14: Typical outlet symbols appearing on electrical drawings

SWITCH OUTLETS

Single Pole Switch	S
Double Pole Switch	S_2
Three-Way Switch	S_3
Four-Way Switch	S_4
Key-Operated Switch	S_K
Switch and Fusestat Holder	$S_{F}H$
Switch and Polit Lamp	S_P
Fan Switch	S_F
Switch for Low-Voltage Switching System	S_L
Master Switch for Low-Voltage Switching System	S_{LM}
Switch and Single Receptacle	S
Switch and Duplex Receptacle	S
Door Switch	S_D
Time Switch	S_T
Momentary Contact Switch	S_{MC}
Ceiling Pull Switch	(S)
"Hand-Off-Auto" Control Switch	HOA
Multi-Speed Control Switch	M
Push Button	•

RECEPTACLE OUTLETS

Where weatherproof, explosionproof, or other specific types of devices are to be required, use the upper-case subscript letters to specify. For example, weatherproof single or duplex receptales would have the upper case WP subscript letters noted alongside of the symbol. All outlets must be grounded.

Single Receptacle Outlet	
Duplex Receptacle Outlet	
Triplex Receptacle Outlet	
Quadruplex Receptacle Outlet	
Duplex Receptacle Outlet Split Wired	
Triplex Receptacle Outlet Split Wired	
250 Volt Receptacle Single Phase Use Subscript Letter to Indicate Function (DW - Dishwasher, RA - Range) or Numerals (with explanation in symbols schedule)	
250 Volt Receptacle Three Phase	
Clock Receptacle	(C)
Fan Receptacle	(F)
Floor Single Receptacle Outlet	
Floor Duplex Receptacle Outlet	
Floor Special-Purpose Outlet	*
Floor Telephone Outlet - Public	
Floor Telephone Outlet - Private	

** Use numeral keyed explanation of symbol usage*

Figure 3-15: Electrical drawing symbols

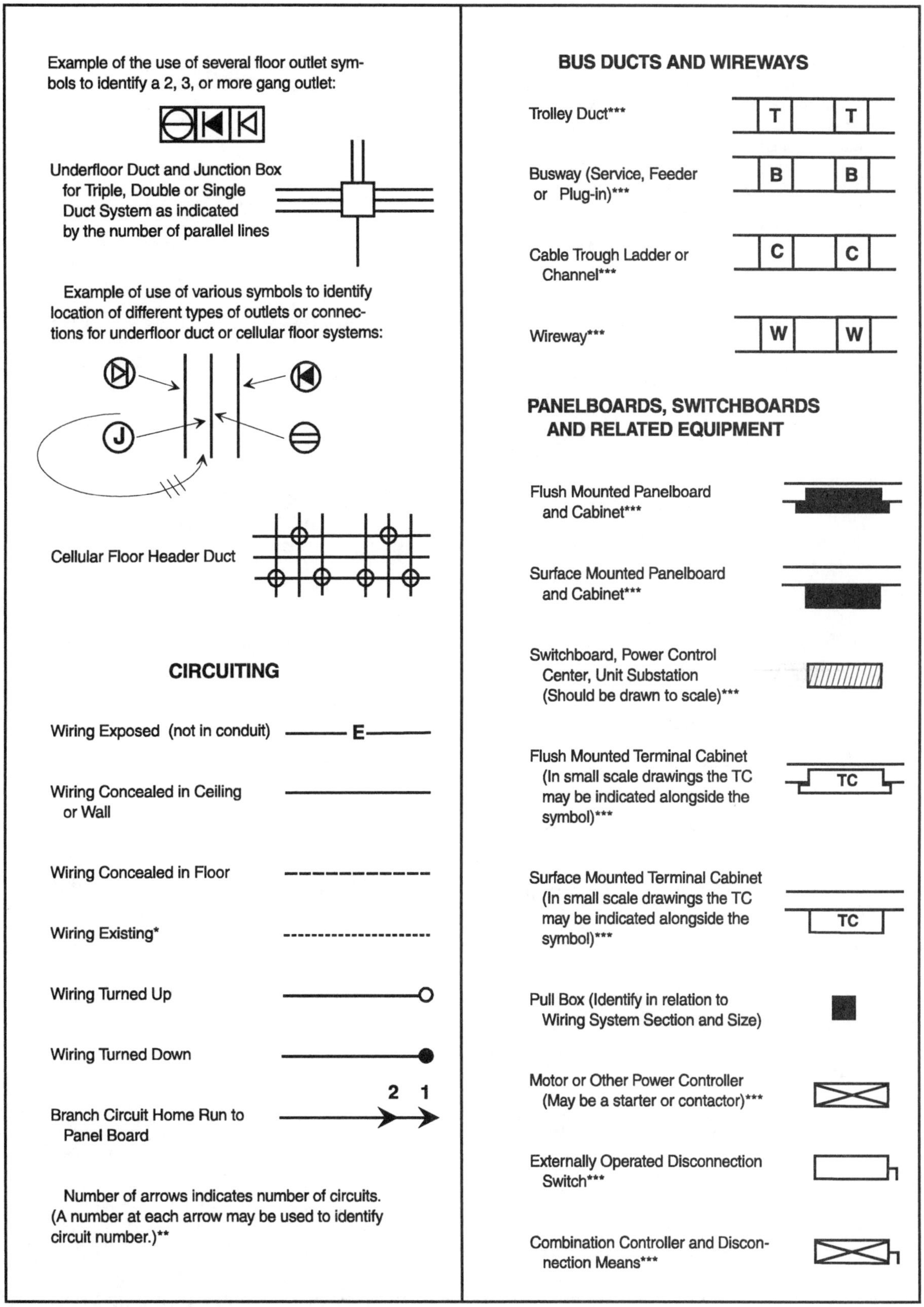
Example of the use of several floor outlet symbols to identify a 2, 3, or more gang outlet:
Underfloor Duct and Junction Box for Triple, Double or Single Duct System as indicated by the number of parallel lines
Example of use of various symbols to identify location of different types of outlets or connections for underfloor duct or cellular floor systems:
J
Cellular Floor Header Duct
CIRCUITING
Wiring Exposed (not in conduit)
E
Wiring Concealed in Ceiling or Wall
Wiring Concealed in Floor
Wiring Existing*
Wiring Turned Up
Wiring Turned Down
Branch Circuit Home Run to Panel Board
2 1
Number of arrows indicates number of circuits. (A number at each arrow may be used to identify circuit number.)**
BUS DUCTS AND WIREWAYS
Trolley Duct***
T T
Busway (Service, Feeder or Plug-in)***
B B
Cable Trough Ladder or Channel***
C C
Wireway***
W W
PANELBOARDS, SWITCHBOARDS AND RELATED EQUIPMENT
Flush Mounted Panelboard and Cabinet***
Surface Mounted Panelboard and Cabinet***
Switchboard, Power Control Center, Unit Substation (Should be drawn to scale)***
Flush Mounted Terminal Cabinet (In small scale drawings the TC may be indicated alongside the symbol)***
TC
Surface Mounted Terminal Cabinet (In small scale drawings the TC may be indicated alongside the symbol)***
TC
Pull Box (Identify in relation to Wiring System Section and Size)
Motor or Other Power Controller (May be a starter or contactor)***
Externally Operated Disconnection Switch***
Combination Controller and Disconnection Means***

Figure 3-15: Electrical drawing symbols *(Cont.)*

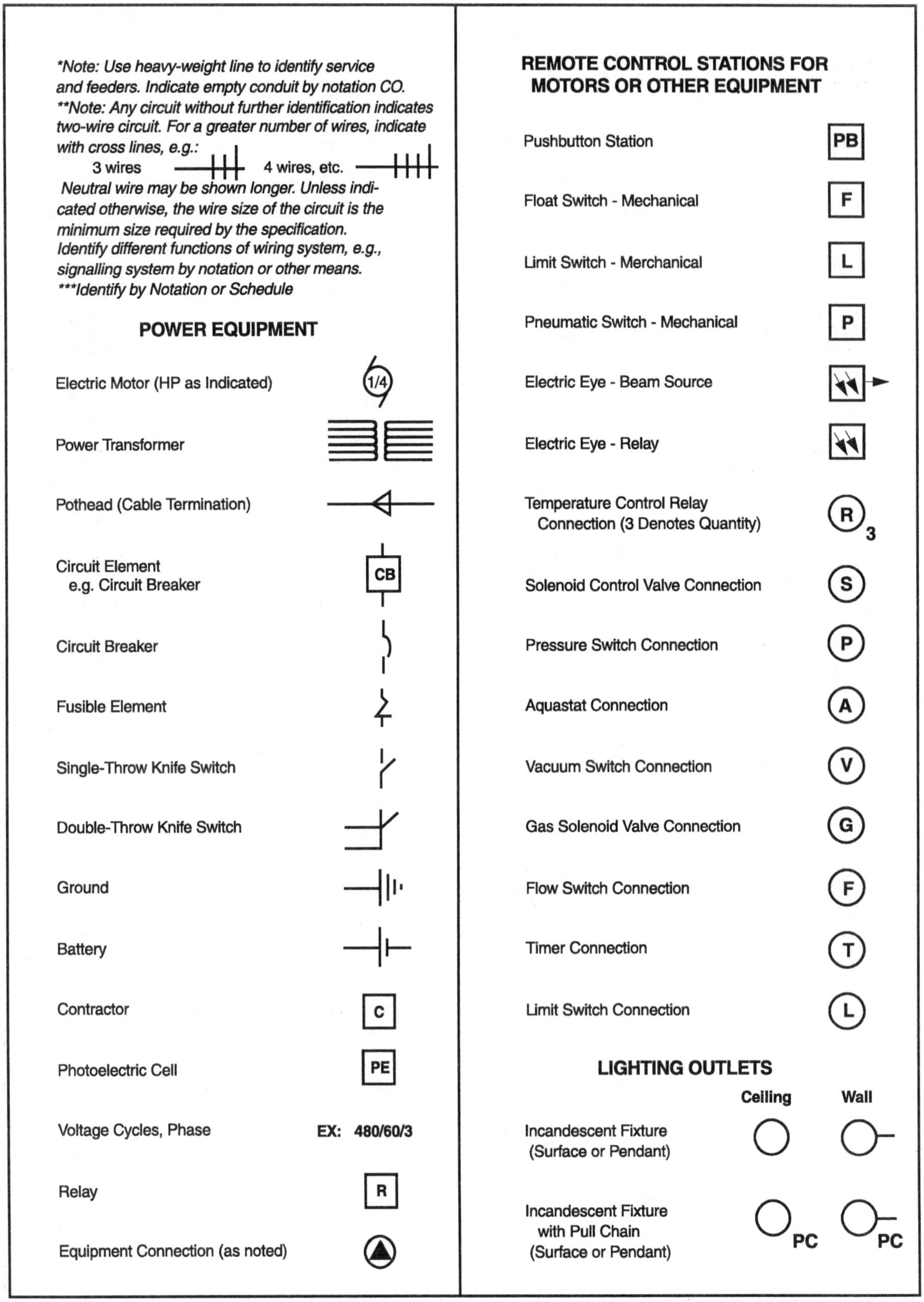
*Note: Use heavy-weight line to identify service and feeders. Indicate empty conduit by notation CO.
**Note: Any circuit without further identification indicates two-wire circuit. For a greater number of wires, indicate with cross lines, e.g.:
3 wires
4 wires, etc.
Neutral wire may be shown longer. Unless indicated otherwise, the wire size of the circuit is the minimum size required by the specification.
Identify different functions of wiring system, e.g., signalling system by notation or other means.
***Identify by Notation or Schedule
POWER EQUIPMENT
Electric Motor (HP as Indicated)
1/4
Power Transformer
Pothead (Cable Termination)
Circuit Element
e.g. Circuit Breaker
CB
Circuit Breaker
Fusible Element
Single-Throw Knife Switch
Double-Throw Knife Switch
Ground
Battery
Contractor
C
Photoelectric Cell
PE
Voltage Cycles, Phase
EX: 480/60/3
Relay
R
Equipment Connection (as noted)
REMOTE CONTROL STATIONS FOR MOTORS OR OTHER EQUIPMENT
Pushbutton Station
PB
Float Switch - Mechanical
F
Limit Switch - Merchanical
L
Pneumatic Switch - Mechanical
P
Electric Eye - Beam Source
Electric Eye - Relay
Temperature Control Relay Connection (3 Denotes Quantity)
R
3
Solenoid Control Valve Connection
S
Pressure Switch Connection
P
Aquastat Connection
A
Vacuum Switch Connection
V
Gas Solenoid Valve Connection
G
Flow Switch Connection
F
Timer Connection
T
Limit Switch Connection
L
LIGHTING OUTLETS
Ceiling
Wall
Incandescent Fixture (Surface or Pendant)
Incandescent Fixture with Pull Chain (Surface or Pendant)
PC
PC

Figure 3-15: Electrical drawing symbols *(Cont.)*

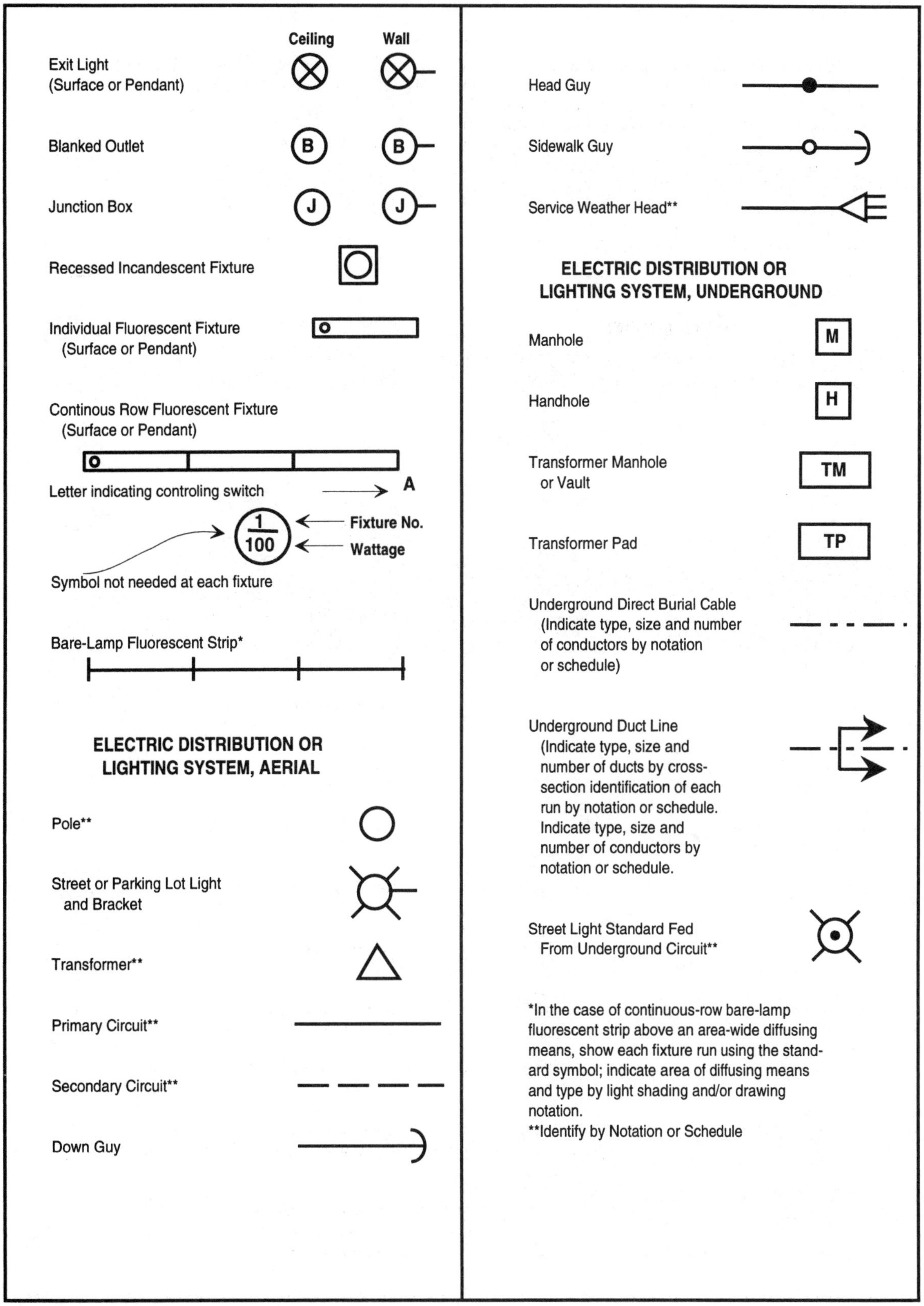
Ceiling
Wall
Exit Light
(Surface or Pendant)
Blanked Outlet
B
Junction Box
J
Recessed Incandescent Fixture
Individual Fluorescent Fixture
(Surface or Pendant)
Continous Row Fluorescent Fixture
(Surface or Pendant)
Letter indicating controling switch
A
1
100
Fixture No.
Wattage
Symbol not needed at each fixture
Bare-Lamp Fluorescent Strip*
ELECTRIC DISTRIBUTION OR
LIGHTING SYSTEM, AERIAL
Pole**
Street or Parking Lot Light
and Bracket
Transformer**
Primary Circuit**
Secondary Circuit**
Down Guy
Head Guy
Sidewalk Guy
Service Weather Head**
ELECTRIC DISTRIBUTION OR
LIGHTING SYSTEM, UNDERGROUND
Manhole
M
Handhole
H
Transformer Manhole
or Vault
TM
Transformer Pad
TP
Underground Direct Burial Cable
(Indicate type, size and number
of conductors by notation
or schedule)
Underground Duct Line
(Indicate type, size and
number of ducts by cross-
section identification of each
run by notation or schedule.
Indicate type, size and
number of conductors by
notation or schedule.
Street Light Standard Fed
From Underground Circuit**
*In the case of continuous-row bare-lamp fluorescent strip above an area-wide diffusing means, show each fixture run using the standard symbol; indicate area of diffusing means and type by light shading and/or drawing notation.
**Identify by Notation or Schedule

Figure 3-15: Electrical drawing symbols *(Cont.)*

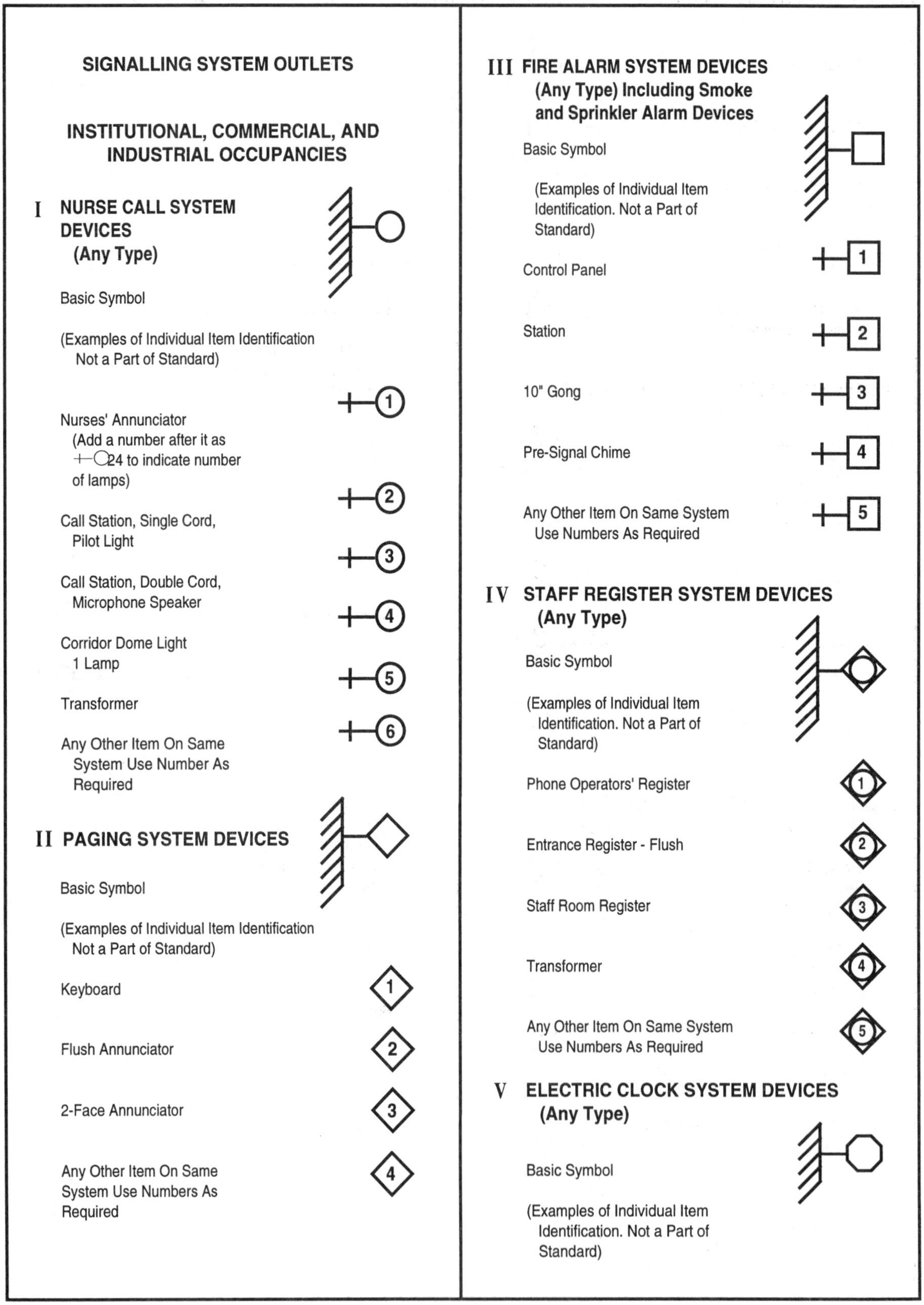
SIGNALLING SYSTEM OUTLETS
INSTITUTIONAL, COMMERCIAL, AND INDUSTRIAL OCCUPANCIES
I NURSE CALL SYSTEM DEVICES (Any Type)
Basic Symbol
(Examples of Individual Item Identification Not a Part of Standard)
1
Nurses' Annunciator (Add a number after it as 24 to indicate number of lamps)
2
Call Station, Single Cord, Pilot Light
3
Call Station, Double Cord, Microphone Speaker
4
Corridor Dome Light 1 Lamp
5
Transformer
6
Any Other Item On Same System Use Number As Required
II PAGING SYSTEM DEVICES
Basic Symbol
(Examples of Individual Item Identification Not a Part of Standard)
Keyboard
1
Flush Annunciator
2
2-Face Annunciator
3
Any Other Item On Same System Use Numbers As Required
4
III FIRE ALARM SYSTEM DEVICES (Any Type) Including Smoke and Sprinkler Alarm Devices
Basic Symbol
(Examples of Individual Item Identification. Not a Part of Standard)
Control Panel
1
Station
2
10" Gong
3
Pre-Signal Chime
4
Any Other Item On Same System Use Numbers As Required
5
IV STAFF REGISTER SYSTEM DEVICES (Any Type)
Basic Symbol
(Examples of Individual Item Identification. Not a Part of Standard)
Phone Operators' Register
1
Entrance Register - Flush
2
Staff Room Register
3
Transformer
4
Any Other Item On Same System Use Numbers As Required
5
V ELECTRIC CLOCK SYSTEM DEVICES (Any Type)
Basic Symbol
(Examples of Individual Item Identification. Not a Part of Standard)

Figure 3-15: Electrical drawing symbols *(Cont.)*

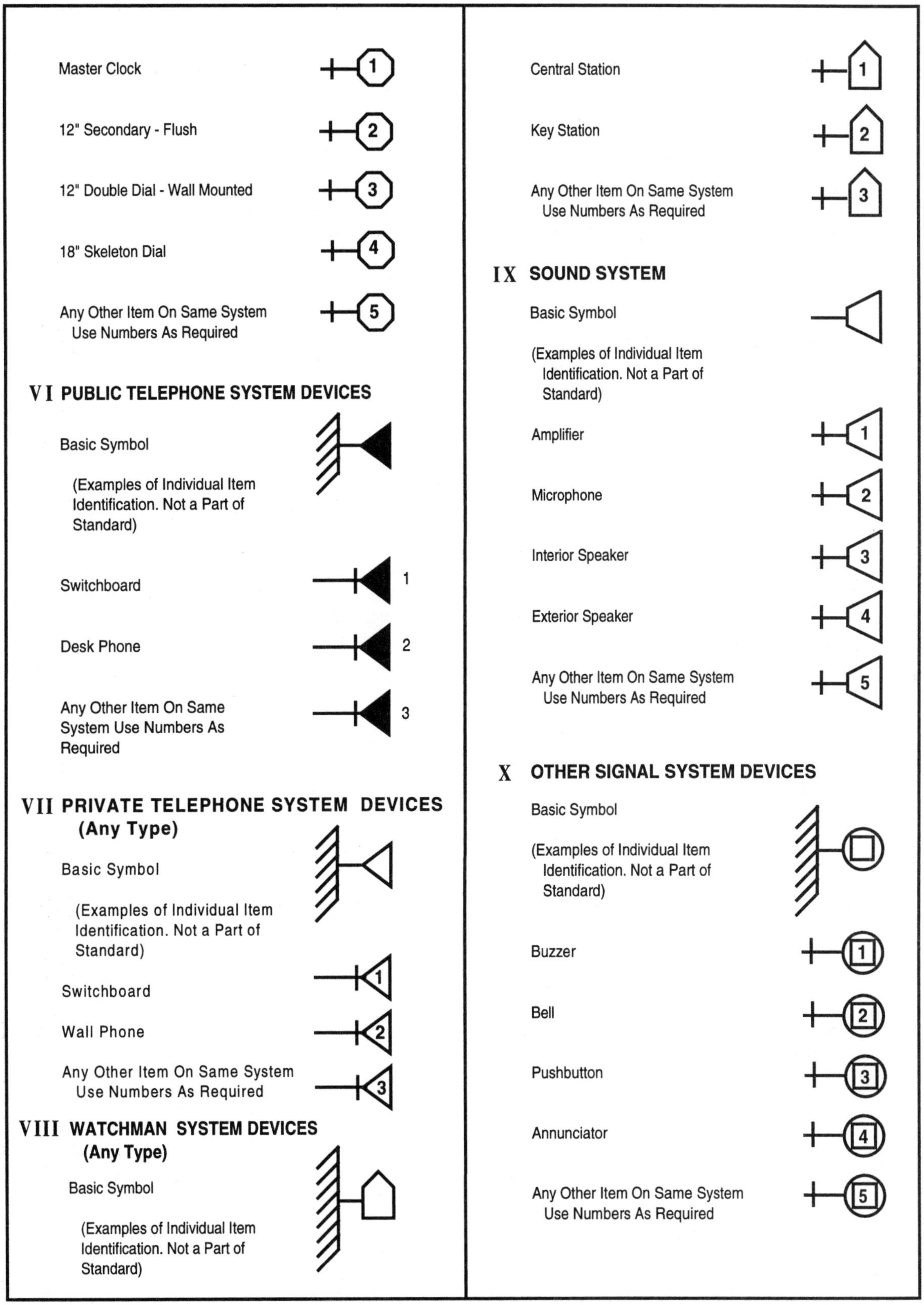
Master Clock
12" Secondary - Flush
12" Double Dial - Wall Mounted
18" Skeleton Dial
Any Other Item On Same System Use Numbers As Required
VI PUBLIC TELEPHONE SYSTEM DEVICES
Basic Symbol
(Examples of Individual Item Identification. Not a Part of Standard)
Switchboard
1
Desk Phone
2
Any Other Item On Same System Use Numbers As Required
3
VII PRIVATE TELEPHONE SYSTEM DEVICES (Any Type)
Basic Symbol
(Examples of Individual Item Identification. Not a Part of Standard)
Switchboard
Wall Phone
Any Other Item On Same System Use Numbers As Required
VIII WATCHMAN SYSTEM DEVICES (Any Type)
Basic Symbol
(Examples of Individual Item Identification. Not a Part of Standard)
Central Station
Key Station
Any Other Item On Same System Use Numbers As Required
IX SOUND SYSTEM
Basic Symbol
(Examples of Individual Item Identification. Not a Part of Standard)
Amplifier
Microphone
Interior Speaker
Exterior Speaker
Any Other Item On Same System Use Numbers As Required
X OTHER SIGNAL SYSTEM DEVICES
Basic Symbol
(Examples of Individual Item Identification. Not a Part of Standard)
Buzzer
Bell
Pushbutton
Annunciator
Any Other Item On Same System Use Numbers As Required

Figure 3-15: Electrical drawing symbols *(Cont.)*

RESIDENTIAL OCCUPANCIES

Signalling system symbols for use in identifying standardized residential-type signal system items on residential drawings where a descriptive symbol list is not included on the drawing. When other signal system items are to be identified, use the above basic symbols for such items together with a descriptive symbol list.

Pushbutton	
Buzzer	
Bell	
Combination Bell - Buzzer	
Chime	CH
Annunciator	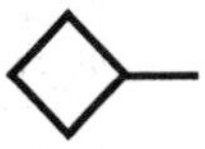
Electric Door Opener	D
Maid's Signal Plug	M
Interconnection Box	
Bell-Ringing Transformer	BT
Outside Telephone	
Interconnecting Telephone	
Television Outlet	TV

Figure 3-15: Electrical drawing symbols *(Cont.)*

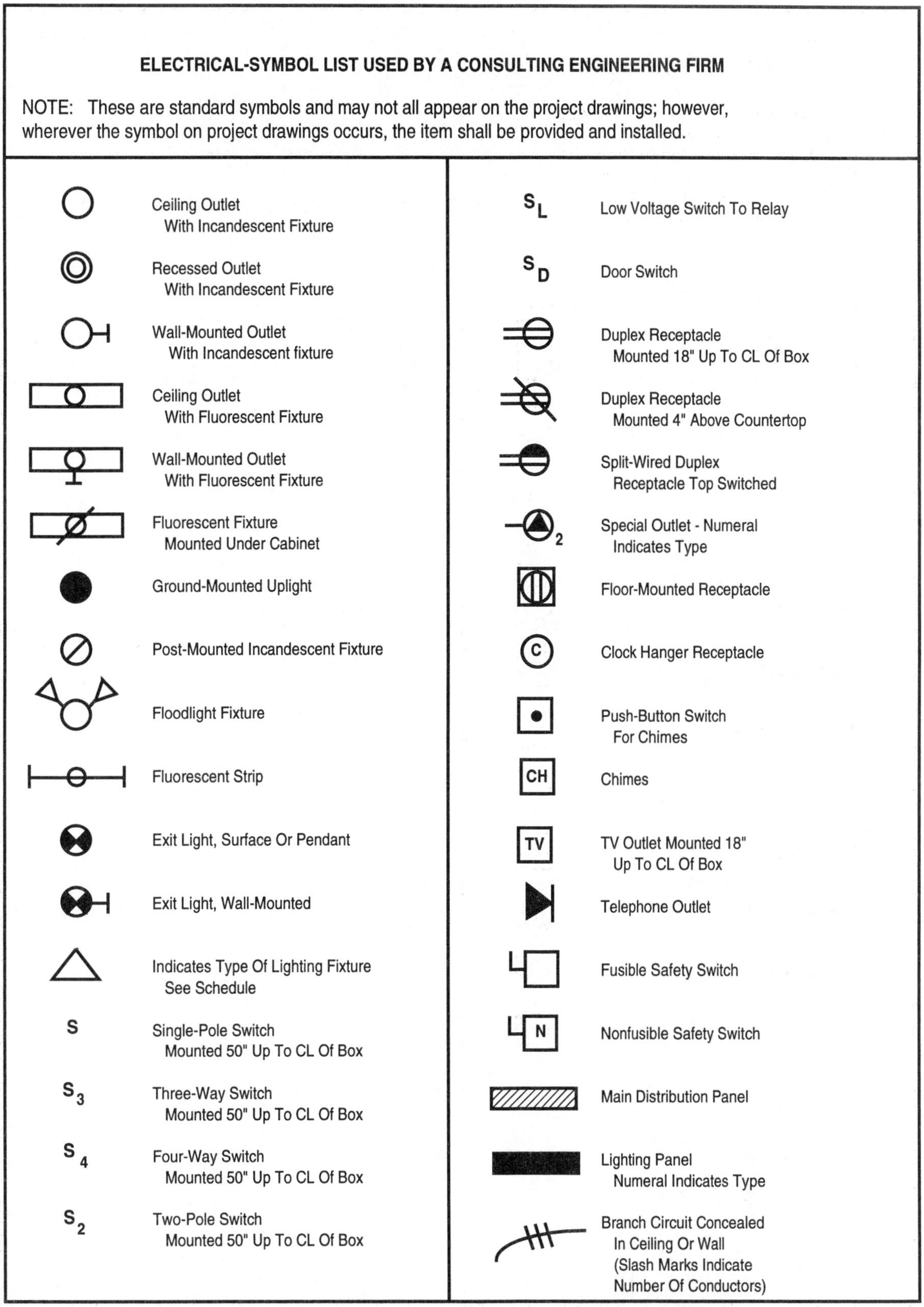
ELECTRICAL-SYMBOL LIST USED BY A CONSULTING ENGINEERING FIRM
NOTE: These are standard symbols and may not all appear on the project drawings; however, wherever the symbol on project drawings occurs, the item shall be provided and installed.
Ceiling Outlet With Incandescent Fixture
Recessed Outlet With Incandescent Fixture
Wall-Mounted Outlet With Incandescent fixture
Ceiling Outlet With Fluorescent Fixture
Wall-Mounted Outlet With Fluorescent Fixture
Fluorescent Fixture Mounted Under Cabinet
Ground-Mounted Uplight
Post-Mounted Incandescent Fixture
Floodlight Fixture
Fluorescent Strip
Exit Light, Surface Or Pendant
Exit Light, Wall-Mounted
Indicates Type Of Lighting Fixture See Schedule
S
Single-Pole Switch Mounted 50" Up To CL Of Box
S3
Three-Way Switch Mounted 50" Up To CL Of Box
S4
Four-Way Switch Mounted 50" Up To CL Of Box
S2
Two-Pole Switch Mounted 50" Up To CL Of Box
SL
Low Voltage Switch To Relay
SD
Door Switch
Duplex Receptacle Mounted 18" Up To CL Of Box
Duplex Receptacle Mounted 4" Above Countertop
Split-Wired Duplex Receptacle Top Switched
2
Special Outlet - Numeral Indicates Type
Floor-Mounted Receptacle
C
Clock Hanger Receptacle
Push-Button Switch For Chimes
CH
Chimes
TV
TV Outlet Mounted 18" Up To CL Of Box
Telephone Outlet
Fusible Safety Switch
N
Nonfusible Safety Switch
Main Distribution Panel
Lighting Panel Numeral Indicates Type
Branch Circuit Concealed In Ceiling Or Wall (Slash Marks Indicate Number Of Conductors)

Figure 3-16: Electrical drawing symbols modified by a consulting engineering firm

ELECTRICAL-SYMBOL LIST USED BY A CONSULTING ENGINEERING FIRM

NOTE: These are standard symbols and may not all appear on the project drawings; however, wherever the symbol on project drawings occurs, the item shall be provided and installed.

Symbol	Description
	Branch Circuit Conceiled In Floor Or Ceiling Below
	Low - Voltage Cable
5	Indicates Type Of Heater (See Schedule)
	Indicates Homerun To Panelboard (Number Of Arrows Indicates Circuits)
WP	Weatherproof
M 1/2	Motor Outlet (Numeral Indicates HP)
J	Junction Box
D	Dimmer Control For Light Fixture
	Electric Baseboard Heater
	Flush-Mounted Electric Floor Heater
	Ceiling Electric Panel Heater
	Infrared Electric Heater Ceiling Mounted
T	Double-Pole Thermostat For Electric Heat
F	Fire Alarm Striking Station
G	Fire Alarm Gong
D	Fire Detector
SD	Smoke Detector
B	Program Bell
Y	Yard Gong
M	Microphone, Wall-Mounted
M	Microphone, Floor-Mounted
S	Speaker, Wall-Mounted
S	Speaker, Recessed

Figure 3-16: Electrical drawing symbols modified by a consulting engineering firm *(Cont.)*

Chapter 4

Planning an Electrical Design

GENERAL

In planning *any* electrical system, there are certain general steps to be followed regardless of the type of construction. In planning a residential electrical system the designer must take certain factors into consideration. These are:

1. Wiring method.
2. Overhead or underground electrical service.
3. Type of building construction.
4. Type of service entrance and equipment.
5. Grade of wiring devices and lighting fixtures.
6. Selection of lighting fixtures.
7. Type of heating and cooling system.
8. Control wiring for heating and cooling system.
9. Signal and alarm systems.

The experienced designer readily recognizes, within certain limits, the type of system that will be required. However, the designer should always check the local code requirements when selecting a wiring method. If more than one wiring method may be practical, he must decide which type of system he will provide prior to the preparation of the design.

In a residential occupancy, the designer will know that 120/240 volt, single-phase service will invariably be provided by the utility company. The designer also knows that the service and feeders will be three-wire, that the branch circuits will be either two- or three-wire, and that the safety switches, service equipment, and panelboards will be three-wire, solid neutral. On each project, however, the designer must determine where the point of service drop attachment to the building will be located and whether the service is to be provided as part of the contract or by the utility company.

LIGHTING OUTLETS

Lighting fixtures should be carefully selected and located to afford the best distribution of light, yet they should fit in with the architectural scheme of the structure and interior decor. Often the designer will provide a fixture allowance in the bid and then let the contractor or owner make the selection. However, it seems that the most satisfactory method is for the electrical designer to locate and select the various lighting fixtures on a preliminary sketch and then meet with the owner to discuss this preliminary design. During the meeting, the designer should have illustrations of the various lighting fixtures that he has selected. At this time, modifications can be made.

PRELIMINARY CONFERENCES

Whenever an owner contemplates having a new home constructed, he usually commissions an architect to prepare the complete working drawings and specifications. The architect, in turn, begins his initial sketches and finally develops these into *preliminary drawings.* It is at this point that the electrical designer is called in to discuss the electrical system within the building.

A conference is arranged for the owner to discuss his desires with the architect and the electrical designer. During this conference, the electrical designer should determine the owner's style of living, the architectural style of the house, and the electrical conveniences desired by the owner. At this time the electrical designer should be prepared to make recommendations to the owner about all the electrical conveniences that are available. During this preliminary conference much time can be saved if a standard form is used to collect the necessary data. Such a form should include the following:

1. Temporary electric facilities: Who shall furnish the temporary electric facilities, and who will have use of them? Who will pay for energy consumed during the construction period?
2. Service entrance: Overhead or underground?
3. Service switch and panel: Circuit breakers or switches and fuses? This same information should be determined for load centers.
4. Branch-circuit wiring: Does the owner or architect have any preference? This, of course, must be in accordance with the *NE Code* and local ordinances; the designer should be prepared to make recommendations.
5. Wiring devices: Type and color of each?
6. Wiring-device plates: Metal or composition? Color and finish?
7. Low-voltage, remote-control switches: The designer should determine if the project warrants discussion of low-voltage switching of lighting circuits. If so, the designer should be prepared to discuss the advantages of this system with the architect and owners.
8. Air conditioning: What type is wanted; that is, central or through-wall room units? The mechanical designer will normally handle this phase of the construction, but the electrical designer must know the type to be used in order to plan the wiring system, feeders, etc. He must also find out other necessary data about the air-conditioning system, but he can do this later at another conference with the mechanical engineer.
9. Electric heating: If electric baseboard heating or any other type of individual room-controlled heating is desired, the electrical designer should indicate this on the electrical drawings. However, a conflict may arise with the mechanical designer. If a forced-air duct system is planned, it will definitely be the mechanical designer's responsibility.
10. Signal and communications: The preliminary conference should be used to discuss the desired location of the telephone outlet and any other electrical conveniences chosen by the owner, such as intercom system, fire-alarm system, sound system, central vacuum system, etc. The electrical designer should be prepared to discuss the benefits of such systems along with approximate costs.

DESIGNING THE ELECTRICAL SYSTEM

In some cases, especially in residential renovation projects, an architect will not be used. Rather, the homeowner will hire a remodeling contractor to do the work and bypass the architect. If an electrical layout or design is desired, the electrical designer will have to obtain existing drawings (if available) from the owner. If such drawings are not available, the electrical designer will have to measure the existing building and make working drawings of his own to enable working drawings of the electrical system to be made. Otherwise, most of the previously mentioned steps will remain the same.

After the preliminary drawings have been approved by the owner, the architect begins on the actual working drawings and specifications. The electrical designer usually will have to wait a while for reference drawings to be made available before beginning the electrical-design drawings. During this waiting period, the designer can obtain other necessary data for the design. This may include checking with the local utility company to determine the best possible route for the service drop, verifying local codes and ordinances, and determining any shortage of materials from suppliers that may delay the work once the project is under construction.

Once the architectural drawings are received, the designer makes outline drawings of the house floor plans on tracing paper; these drawings will eventually comprise the electrical drawings. Then the designer makes sketches, locating all electrical outlets; that is, convenience outlets, special-purpose outlets, lighting outlets, etc. After laying out the circuitry, he gives the sketches to an electrical draftsman who does the finished drawings.

Ideally, the electrical designer should have a floor plan of all levels of the building (Figure 4-1); elevation drawings of the building (Figures 4-2 and 4-3), and a plot plan (Figure 4-4) showing all outside utilities.

Note that the floor plan in Figure 4-1 has dimensions indicated on the drawing. Unless the placement of a piece of electrical equipment is critical, these dimensions will normally be omitted on the electrical drawing — mainly because they serve no practical purpose for the electrical contractor or for the electrician wiring the house. If certain dimensions are needed, the electrician may refer to the architectural drawings.

The method of selecting lighting fixtures was discussed earlier in this chapter. If they are selected by the designer, each type should be indicated in a lighting fixture schedule located on the drawings. Data should include the manufacturer's catalog number, type and wattage of each lamp, and similar information.

Next comes the calculation of all feeders and service entrance equipment. Later this is transferred to the working drawings in the form of physical location on the floor plans, wire and conduit sizes in a power-riser diagram, and complete information about the service equipment and load centers in the panelboard schedules.

The plumbing and mechanical drawings must be checked to determine which equipment furnished by the mechanical contractor requires electrical service and to determine the required size of each feeder for that equipment.

Once the drawings are completed, the written specifications are prepared in order to further describe the materials and the methods to be used in the construction of the electrical system for the residence.

After the electrical drawings are completed, the electrical designer should carefully check them over to make certain that there will be no conflicts in the architectural, structural, or mechanical work and to ensure that there are no errors or omissions.

The architect will, at times, request from the electrical designers an estimate of the electrical-system cost to aid him in determining the probable cost of the building or project, in advance of the actual request for formal bids by contractors.

DESIGNER'S RESPONSIBILITIES

The electrical designer will act as liaison between the architect and the electrical contractor, handling the details of the electrical construction from the time the work is started, through the bidding and construction sequences, to the final approval and acceptance of the finished job. Usually the architect will depend on the judgment of the electrical designer in making decisions on problems arising from the work.

In nearly all installations, the contract specifications will require the inspection or approval of the various items of material and the wiring method by the electrical designer. The electrical designer should therefore make periodic inspections of the job as it progresses in case a mistake is made (on either the designer's or the contractor's part). The expense will be less if the error is caught early.

It is a rare electrical job, indeed, that does not have one or more changes in the original contract before it is completed. These electrical changes are

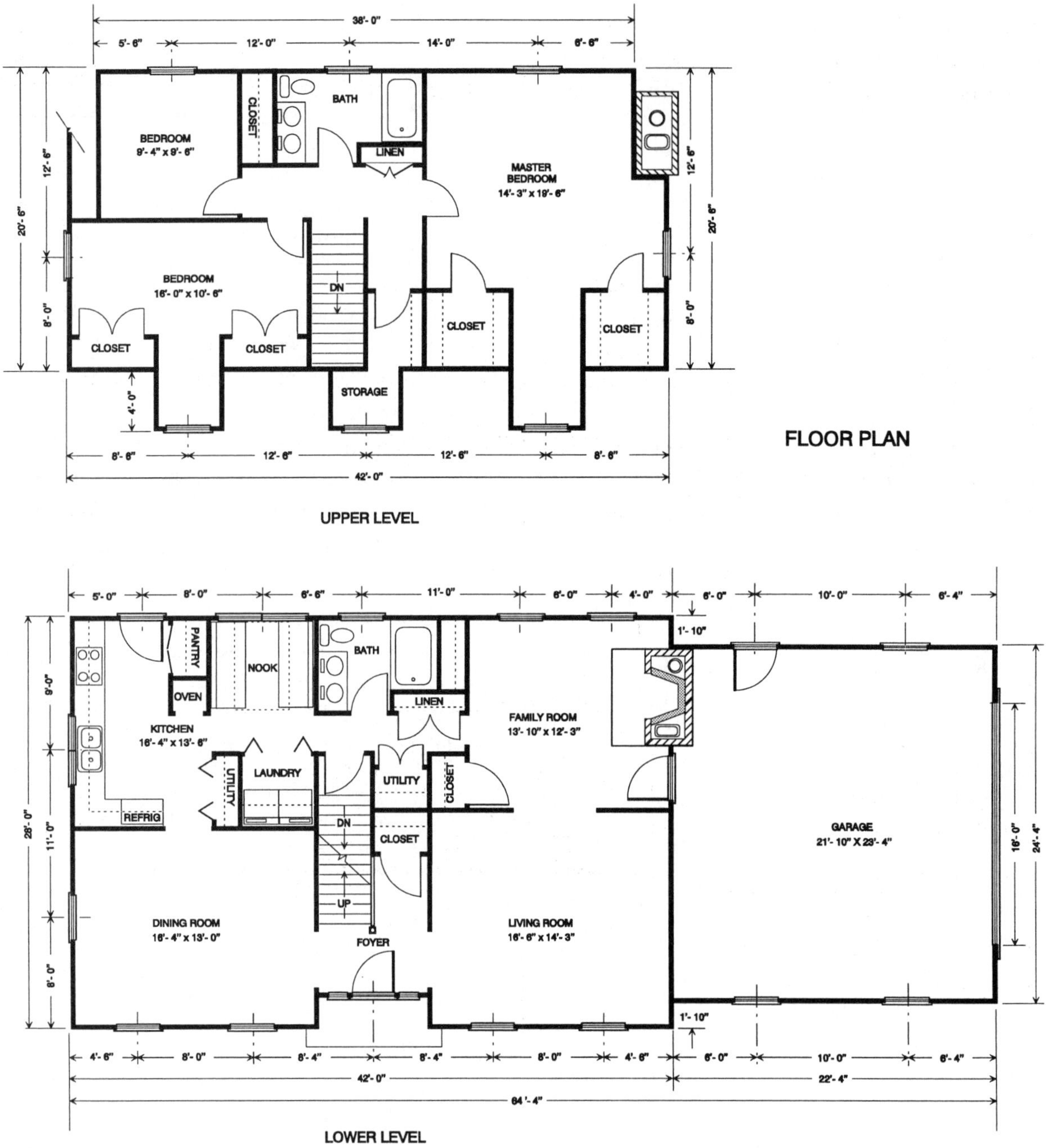

FLOOR PLAN
UPPER LEVEL
BEDROOM
9'- 4" x 9'- 6"
CLOSET
BATH
LINEN
MASTER BEDROOM
14'- 3" x 19'- 6"
BEDROOM
16'- 0" x 10'- 6"
DN
CLOSET
CLOSET
CLOSET
CLOSET
STORAGE
38'- 0"
42'- 0"
LOWER LEVEL
PANTRY
NOOK
BATH
OVEN
LINEN
KITCHEN
16'- 4" x 13'- 6"
FAMILY ROOM
13'- 10" x 12'- 3"
LAUNDRY
UTILITY
UTILITY
CLOSET
REFRIG
DN
CLOSET
UP
GARAGE
21'- 10" X 23'- 4"
DINING ROOM
16'- 4" x 13'- 0"
FOYER
LIVING ROOM
16'- 6" x 14'- 3"
42'- 0"
22'- 4"
64'- 4"

Figure 4-1: Typical architectural floor plan from which the electrical designer will work

usually due to changes in building design, changes in type of equipment, or changes in the owner's requirements, At such times, the electrical designer should call upon the contractor to prepare an estimate of the cost of the change in order that a proper additional charge may be approved or a credit given.

The electrical designer's specification should state that no changes or additions are to be made without written approval and instructions from the proper authority. The reason for this is that the owners will sometimes come on the job during construction and decide that they want an additional outlet here or there, assuming that this little change will cost nothing or next to nothing. However, when the final contractor's bill comes in for approval, it is found that these changes here and there have increased the original bid price by several hundred dollars. This is certain to cause hard feelings all around, and many times a court hearing is necessary to resolve the problem. Such action will cost all concerned regardless of who wins.

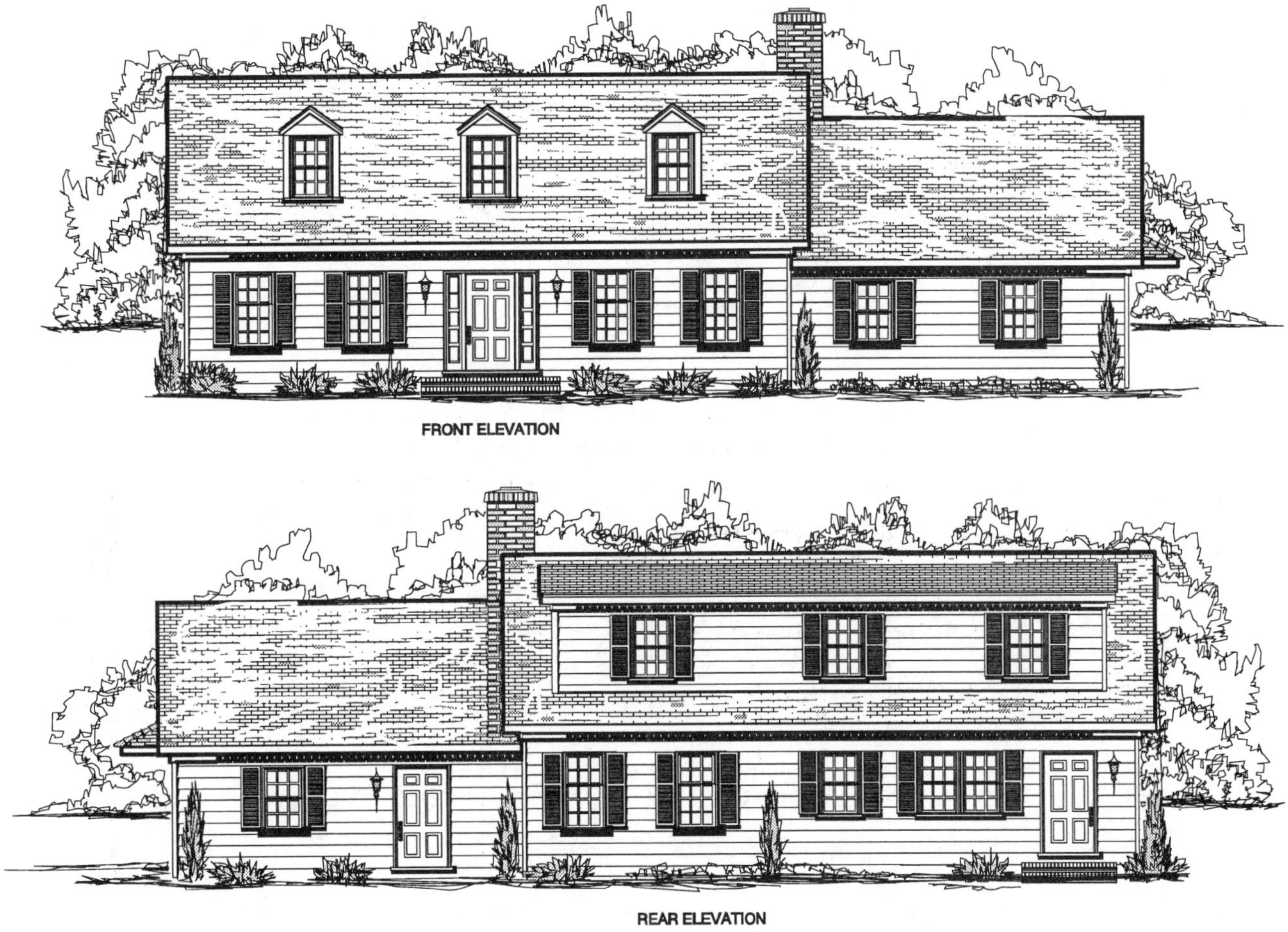
FRONT ELEVATION
REAR ELEVATION

Figure 4-2: Front elevation of the residence in Figure 4-1

LEFT ELEVATION

RIGHT ELEVATION

Figure 4-3: Right and left elevations for the residence shown in Figure 4-1

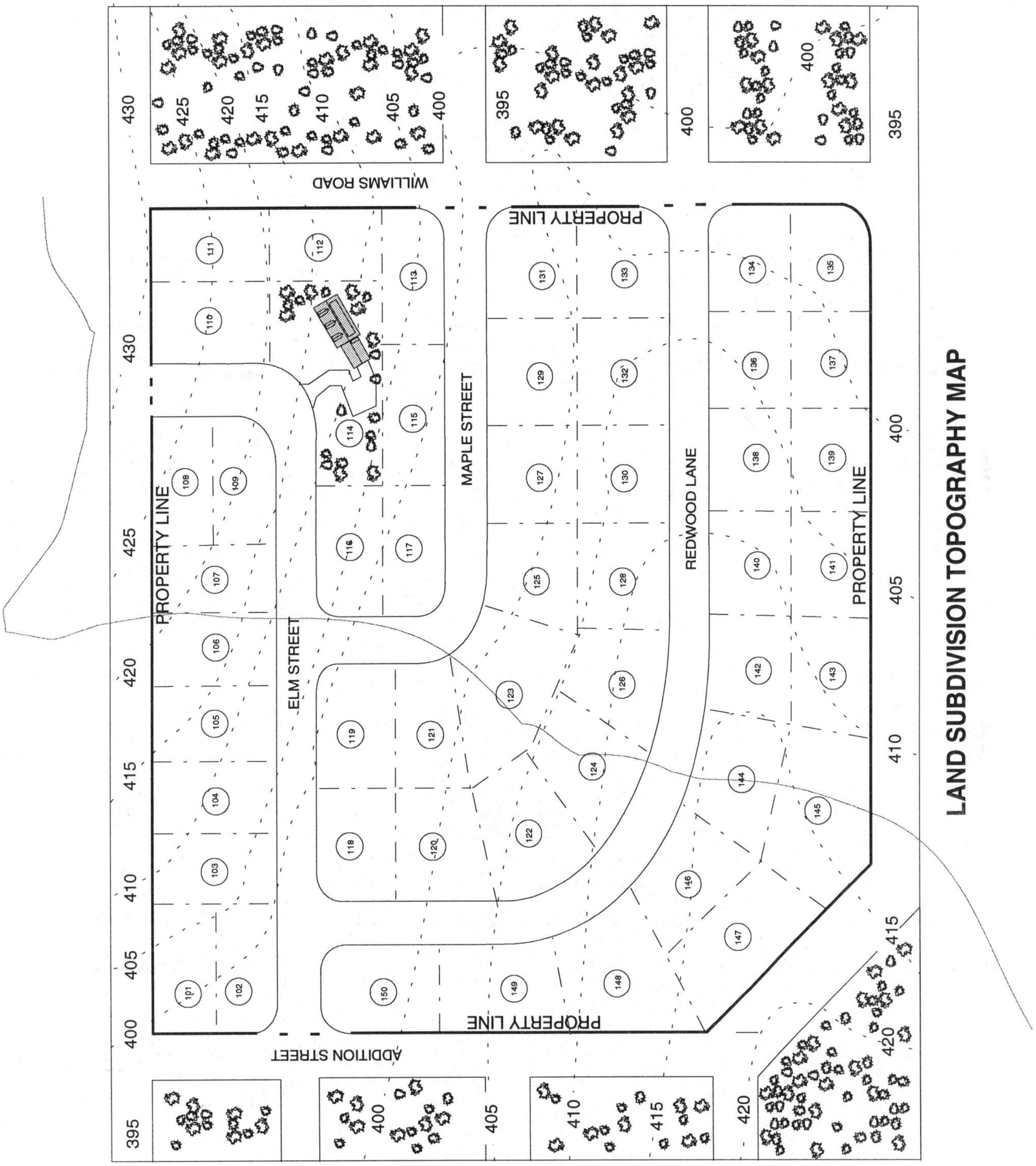
LAND SUBDIVISION TOPOGRAPHY MAP
WILLIAMS ROAD
MAPLE STREET
REDWOOD LANE
ELM STREET
ADDITION STREET
PROPERTY LINE
PROPERTY LINE
PROPERTY LINE
PROPERTY LINE

Figure 4-4: Plot plan showing the location of the building and related outside utilities

Consequently, a great effort should be made to avoid such disputes.

It is very important that the electrical designer understand the responsibilities of his or her position and reflect this knowledge in his or her everyday work. Doing so will benefit all concerned with the building construction project, and make the work go easier.

Chapter 5
Electric Services

Electric services can range in size from a small 120-volt, single-phase,15-ampere service (the minimum allowed by *NE Code* Section 230-79(a) for a roadside vegetable stand to huge industrial installations involving substations dealing with thousands of volts and amperes. Regardless of the size, all electric services are provided for the same purpose: for delivering electrical energy from the supply system to the wiring system on the premises served. Consequently, all establishments containing equipment that utilizes electricity require an electric service.

Figure 5-1 shows how electric power is transmitted from generating plants to points of utilization, while Figure 5-2 shows the basic sections of a typical residential electric service. In this latter illustration, note that the high-voltage lines terminate on a power pole near the building being served. A bank of transformers is mounted on the pole to reduce the transmission voltage to a usable level (120/240V, single-phase, three-wire in this case). The remaining sections are described as follows:

- Service drop: The overhead conductors, through which electrical service is supplied, between the last power company pole and the point of their connection to the service facilities located at the building or other support used for the purpose.
- Service entrance: All components between the point of termination of the overhead service drop or underground service lateral and the building's main disconnecting device, except for metering equipment.
- Service-entrance conductors: The conductors between the point of termination of the overhead service drop or underground service lateral and the main disconnecting device in the building or on the premises.
- Service-entrance equipment: Provides overcurrent protection to the feeder and service conductors, a means of disconnecting the feeders from energized service conductors, and a means of measuring the energy used by the use of metering equipment.

When the service conductors to the building are routed underground, as shown in Figure 5-3, these conductors are known as the service lateral, defined as follows:

- Service lateral: The underground conductors through which service is supplied between the power company's distribution facilities and the first point of their connection to the building or area service facilities.

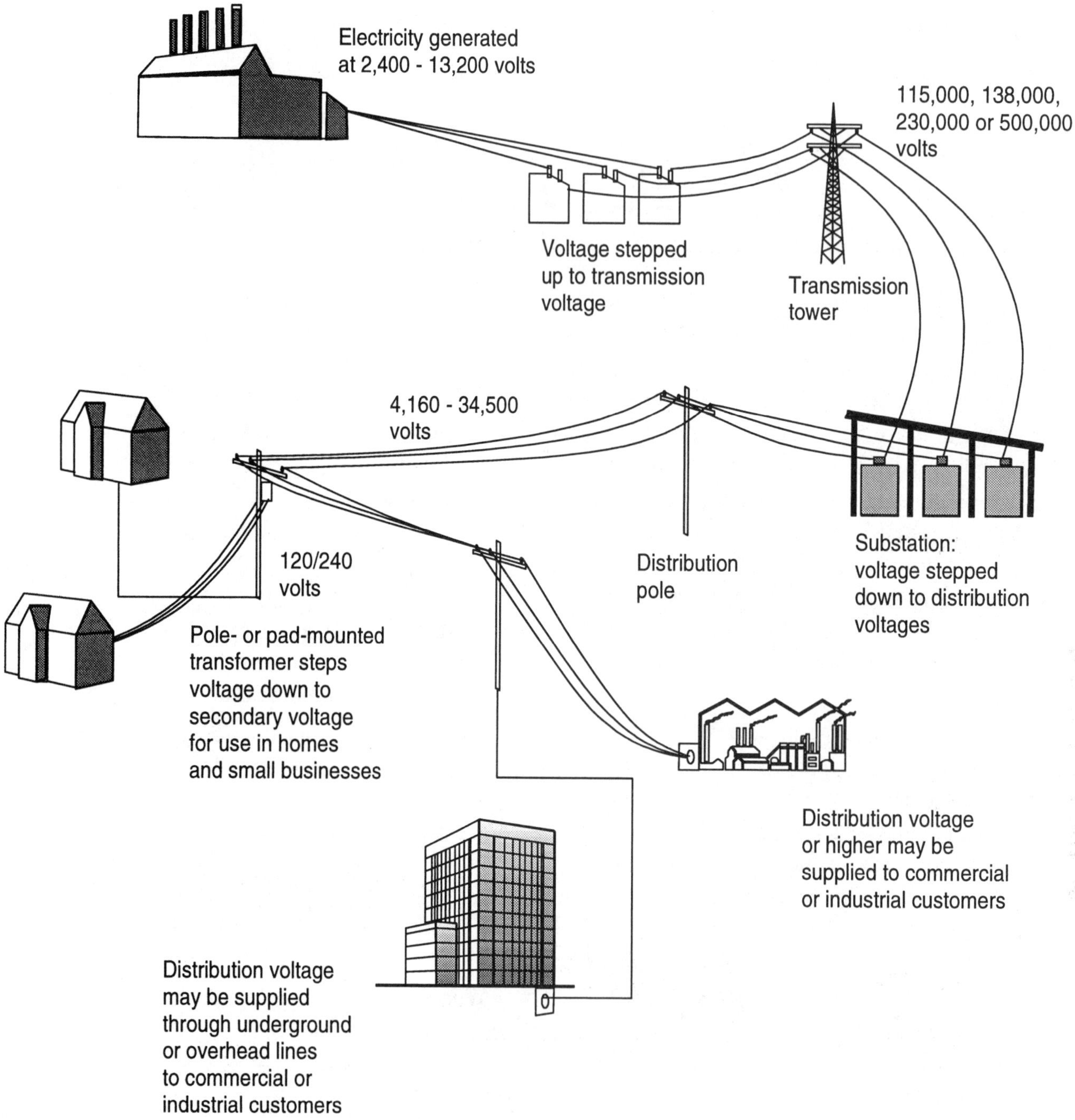
Electricity generated
at 2,400 - 13,200 volts
115,000, 138,000,
230,000 or 500,000
volts
Voltage stepped
up to transmission
voltage
Transmission
tower
4,160 - 34,500
volts
120/240
volts
Distribution
pole
Substation:
voltage stepped
down to distribution
voltages
Pole- or pad-mounted
transformer steps
voltage down to
secondary voltage
for use in homes
and small businesses
Distribution voltage
or higher may be
supplied to commercial
or industrial customers
Distribution voltage
may be supplied
through underground
or overhead lines
to commercial or
industrial customers

Figure 5-1: Basic sections of an electrical distribution system — from generators to points of utilization

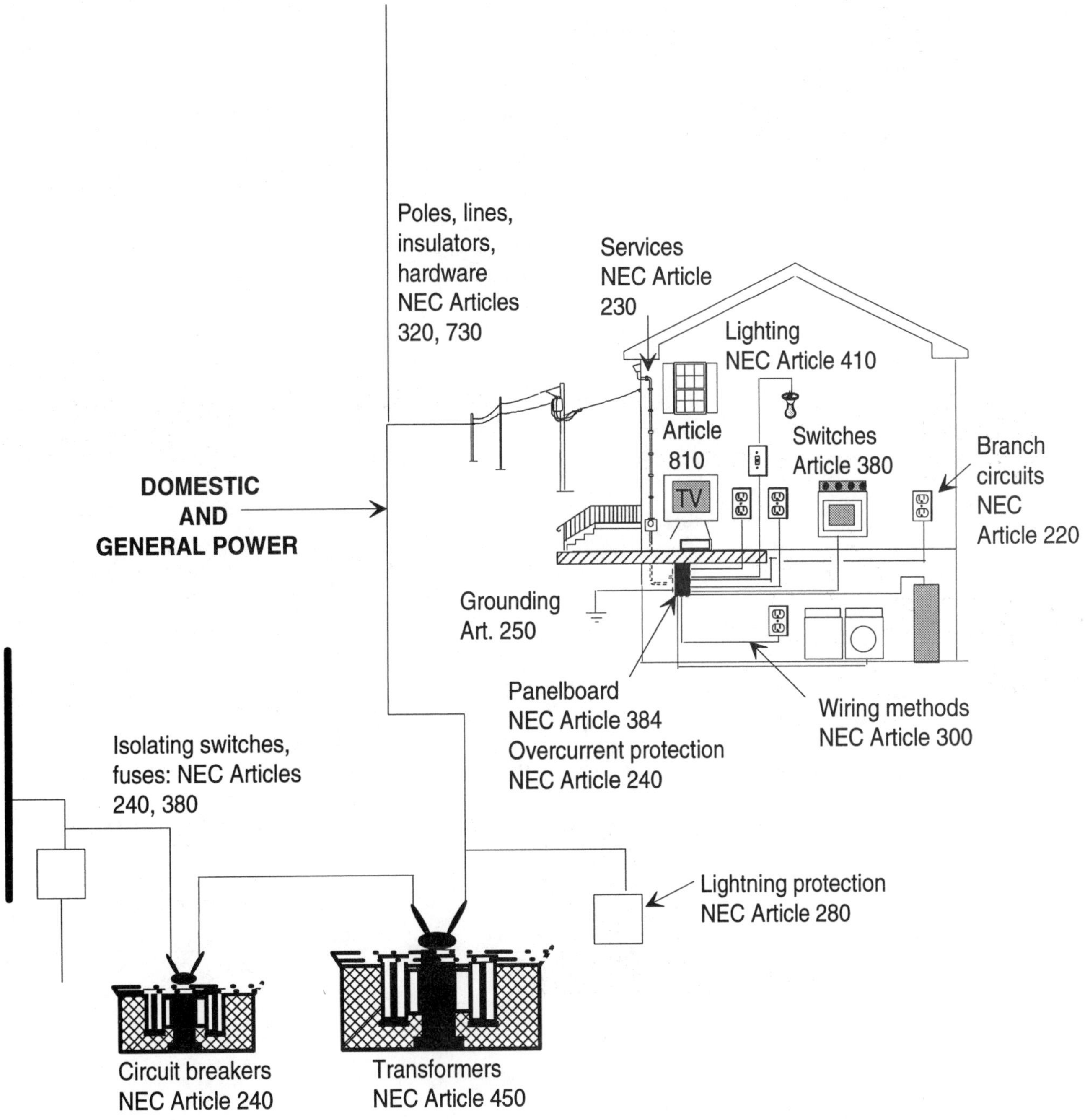
Poles, lines, insulators, hardware NEC Articles 320, 730
Services NEC Article 230
Lighting NEC Article 410
Article 810
TV
Switches Article 380
Branch circuits NEC Article 220
DOMESTIC AND GENERAL POWER
Grounding Art. 250
Panelboard NEC Article 384 Overcurrent protection NEC Article 240
Wiring methods NEC Article 300
Isolating switches, fuses: NEC Articles 240, 380
Lightning protection NEC Article 280
Circuit breakers NEC Article 240
Transformers NEC Article 450

Figure 5-2: Parts of a typical residential electric service

SIZING THE ELECTRIC SERVICE

Sometimes it is confusing just which comes first; the layout of the outlets, or the sizing of the electric service. In many cases, the service size (size of main disconnect, panelboard, service conductors, etc.) can be sized using *National Electrical Code (NE Code)* procedures before the outlets are actually located. In other cases, the outlets will have to be laid out first. However, in either case, the service-entrance and panelboard locations will have to be determined before the circuits can be installed — so the electrician will know in which direction (and to what points) the circuit homeruns will terminate. Let's take an actual residence and size the electric service according to the latest edition of the *NE Code*.

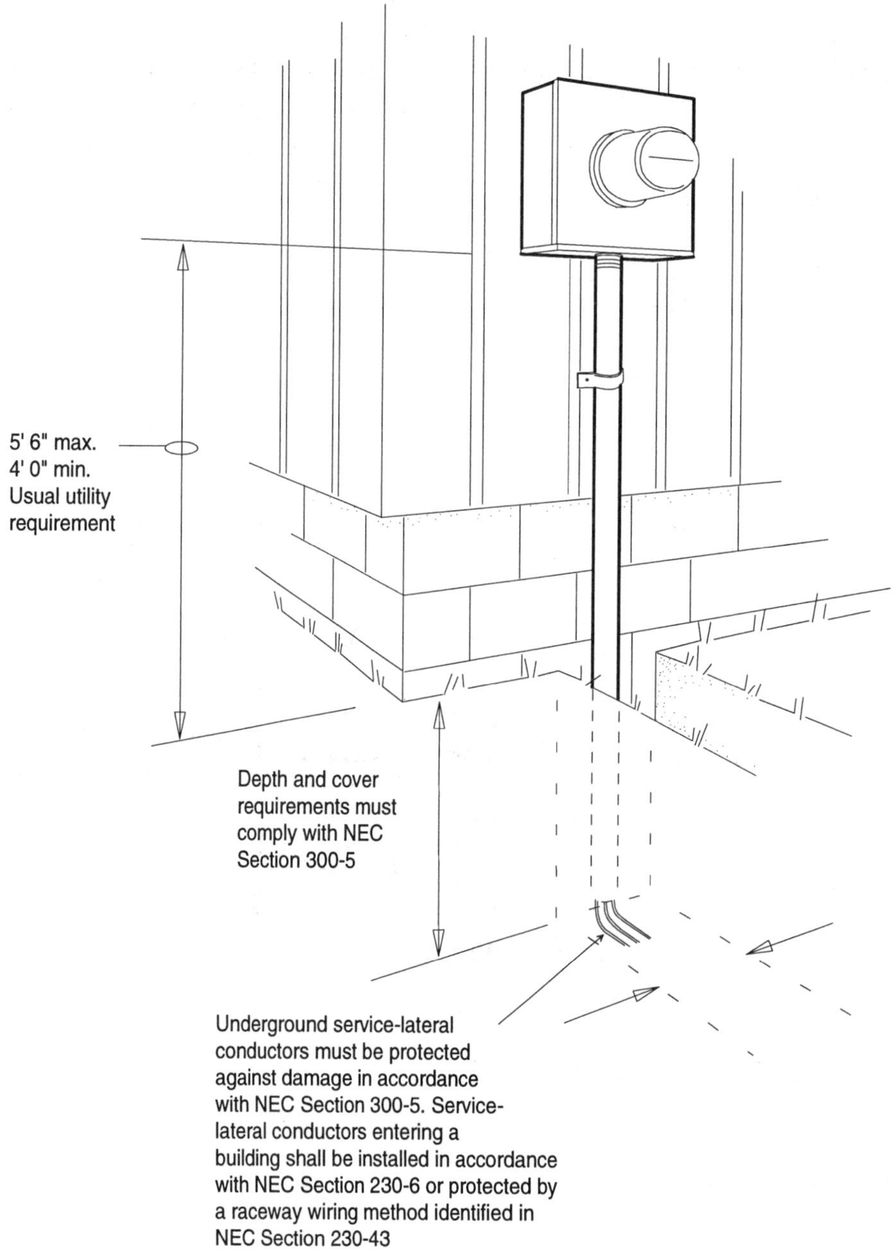
5' 6" max.
4' 0" min.
Usual utility requirement
Depth and cover requirements must comply with NEC Section 300-5
Underground service-lateral conductors must be protected against damage in accordance with NEC Section 300-5. Service-lateral conductors entering a building shall be installed in accordance with NEC Section 230-6 or protected by a raceway wiring method identified in NEC Section 230-43

Figure 5-3: Underground service lateral

A floor plan of a small residence is shown in Figure 5-4. This building is constructed on a concrete slab with no basement or crawl space. There is an unfinished attic above the living area, and an open carport just outside the kitchen entrance. Appliances include a 12 kVA (12,000 volt-amperes or 12 kilovolt-amperes) electric range and a 4.5 kVA water heater. There is also a washer/dryer (rated at 5.5 kVA) in the utility room. Gas heaters are installed in each room with no electrical requirements. What size service-entrance should be provided for this residence if no other information is specified?

GENERAL LIGHTING LOADS

General lighting loads are calculated on the basis of *NE Code* Table 220-3(b). For all residential occupancies, 3 volt-amperes (watts) per square foot of living space is the figure to use. This includes non-appliance duplex receptacles into which table lights, television, etc. may be connected. Therefore, the area of the building must be calculated first. If the building is under construction, the dimensions can be determined by scaling the working drawings used by the builder. If the residence is an existing building, with no drawings, actual measurements will have to be made on the site.

Using the floor plan of the residence in Figure 5-4 as a guide, an architect's scale is used to measure the longest width of the building (using outside dimensions) which is 33 feet. The longest length of the building is 48 feet. These two measurements multiplied together give ($33' \times 48'$ =) 1584 square feet of living area. However, there is

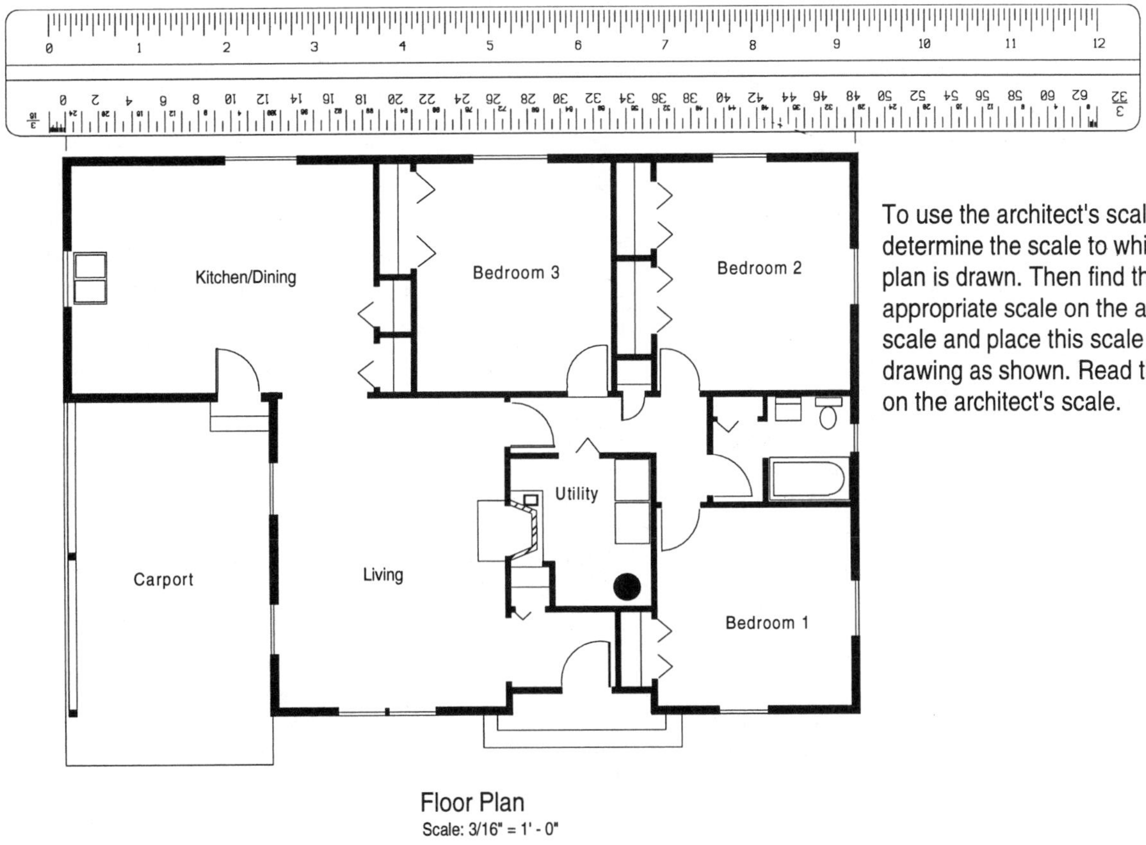
Kitchen/Dining
Bedroom 3
Bedroom 2
Utility
Carport
Living
Bedroom 1
Floor Plan
Scale: 3/16" = 1' - 0"

To use the architect's scale, first determine the scale to which the plan is drawn. Then find the appropriate scale on the architect's scale and place this scale onto the drawing as shown. Read the dimensions on the architect's scale.

Figure 5-4: Floor plan of a small residence

an open carport on the lower left of the drawing. This carport area will have to be calculated and then deducted from the 1584 square-foot figure above to give a true amount of living space. This open area (carport) is 12 feet wide by 19.5 feet long. So, 12′ by 19.5′ = 234 square feet. Consequently, the carport area deducted from 1584 square feet leaves (1584 – 234 =) 1350 square feet of living area.

When using the square-foot method to determine lighting loads for buildings, *NE Code* Section 220-3(b) requires the floor area for each floor to be computed from the outside dimensions. When calculating lighting loads for residences, the computed floor area must not include open porches, carports, garages, or unused or unfinished spaces not adaptable for future use.

CALCULATING ELECTRIC LOAD

Figure 5-5 shows a standard calculation worksheet for a single-family dwelling. This form contains numbered blank spaces to be filled in while making the service calculation. Using this worksheet as a guide, we have previously determined the total area of our sample dwelling to be 1350 square feet of living space. This figure is entered in the appropriate space (1) on the form and multiplied by 3 VA for a total general lighting load of 4050 volt-amperes (2).

Small Appliance Loads

NE Code Section 220-4(b) requires at least two, 120-volt, 20-ampere small appliance circuits to be installed for small appliance loads in the kitchen, dining area, breakfast nook, and similar areas where toasters, coffee makers, etc. will be used. *NE Code* Section 220-16 gives further requirements for residential small appliance circuits; that is, each must be rated at 1500 volt-amperes. Since two such circuits are used in our sample residence, the number "2" is entered in the appropriate space (3) and then multiplied for a total small appliance load of 3000 volt-amperes (4).

Laundry Circuit

NE Code Section 220-4(c) requires an additional 20-ampere branch circuit to be provided for the exclusive use of the laundry area (5). This circuit must not have any other outlets connected except for the laundry receptacle(s) as required by *NE Code* Section 210-52(f). Therefore, enter 1500 (volt-amperes) in space (6) in the form.

Thus far we have enough information to complete the first portion of the service-calculation form.

General Lighting .... 4050 VA (2)
Small Appliance Load........... 3000 VA (4)
Laundry Load........................ 1500 VA (6)
Total General Lighting &
Appliance Loads................ 8550 VA (7)

Demand Factors

All residential electrical outlets are never used at one time. There may be a rare instance where all the lighting could be on for a short time every night, but if so, all the small appliances, all burners on the electric range, water heater, furnace, dryer, washer, and the numerous receptacles throughout the house will never be used simultaneously. Knowing this, the *NE Code* allows a diversity or demand factor in sizing electric services. Our calculation continues:

First 3000 VA is rated at 100% — 3000 VA (8)

The remaining 5550 watts (VA) may be rated at 35% (the allowable demand factor). Therefore:

$$5550 \times .35 = 1942.5 \text{ VA } (10)$$

Therefore, the net General Lighting & Small Appliance Load equals 4942.5 VA (11). Also enter this number in Box 12 in the form.

The electric range, water heater, and clothes dryer must now be considered in the service calculation. Although we previously learned that the nameplate rating of the electric range is 12 kVA, seldom will every burner be on high at once. Nor

I. GENERAL LIGHTING LOADS

TYPE OF LOAD	CALCULATION	TOTAL VA	*NE CODE* REFERENCE
Lighting load	(1)__________sq. ft. x 3 VA =	(2)__________VA	Table 220-3(b)
Small appliance loads	(3)__________circuits x 1500VA=	(4)__________VA	Section 220-16(a)
Laundry load	(5)__________circuits x 1500VA=	(6)__________VA	Section 220-16(b)
Lighting, small appliance, laundry	Total VA =	(7)__________VA	

LOAD	**CALCULATION**	**DEMAND FACTOR**	**TOTAL VA**
Lighting, small appliance, laundry	First 3000VA x	**100%=**	(8)3000VA
Lighting, small appliance, laundry	(9)Remaining VA__________x	**35% =**	(10)__________VA
Lighting, small appliance, laundry	Add #8 & #10 above	=	(11)__________VA

TOTAL CALCULATED LOAD FOR LIGHTING, SMALL APPLIANCES, & LAUNDRY CIRCUIT(S) (Enter Item #11 above in box #12)

12

II. LARGE APPLIANCE LOADS

TYPE OF LOAD	NAMEPLATE RATING	DEMAND FACTOR	TOTAL VA	*NE CODE* REFERENCE
Electric range	Not over 12 kVA	Use 8 kVA	(13) 8,000 VA	Table 220-19
Clothes dryer	(14)________VA	100%	(15)________VA	Table 220-18
Water heater	(16)________VA	100%	(17)________VA	
Other appliances	(18)________VA	100%	(19)________VA	

TOTAL CALCULATED LOAD FOR LARGE APPLIANCES (Add items #15, #17, and #19 above). ENTER TOTAL IN BOX #20

TOTAL CALCULATED LOAD (Add boxes #12 & #20) ENTER TOTAL IN BOX #21

20

III. CONVERT VA TO AMPERES

21

$$\frac{Total\ VA\ in\ Box\ \#21}{240\ volts} = amperes \qquad \frac{VA}{240\ (volts)} = (\#22)\_\_\_\_\_\_\_\_ amperes$$

IV. UNGROUNDED AND GROUNDED (NEUTRAL) CONDUCTOR SIZE—*NE CODE* ARTICLE 310

Notes to Ampacity Tables: No. 3

MINIMUM UNGROUNDED CONDUCTOR SIZE		MINIMUM GROUNDED (NEUTRAL) CONDUCTOR SIZE		
COPPER AWG SIZE	**ALUMINUM OR COPPER-CLAD AL**	**RATING IN AMPERES**	**COPPER AWG SIZE**	**ALUMINUM OR COPPER-CLAD AL**
4	2	100	8	4
3	1	110	6	3
2	1/0	125	4	2
1	2/0	150	3	1
1/0	3/0	175	2	1/0
2/0	4/0	200	1	2/0

Figure 5-5: Calculation worksheet for residential service requirements

will the oven remain on all the time during cooking. When the oven reaches the temperature set on the oven controls, the thermostat shuts off the power until it cools down. Again, the *NE Code* allows a diversity or demand factor.

When one electric range is installed and the nameplate rating is not over 12 kVA, *NE Code* Table 220-19 allows a demand factor resulting in a total rating of 8 kVA. Therefore, 8 kVA may be used in the service calculations instead of the nameplate rating of 12 kVA. The electric clothes dryer and water heater, however, must be calculated at 100% when using this method to calculate residential electric services. The total large appliance load is entered in Box 20 in the form. Here is the service calculation thus far:

Net General Lighting & Small Appliance
Load ....4942.5 VA (11) & (12)
Electric Range
(using demand factor) ....8000 VA (13)
Clothes Dryer ....5500 VA (14) & (15)
Water Heater ....4500 VA (16) & (17)

Total load ....22,942.5 VA (21)

Required Service Size

The conventional electric service for residential use is 120/240-volt, 3-wire, single-phase. Services are sized in amperes and when the volt-amperes are known on single-phase services, amperes may be found by dividing the (highest) voltage into the total volt-amperes.

$$22942.5(VA) \div 240(volts) = 95.6 \text{ amperes} \quad (22)$$

When the net computed load exceeds 10 kVA, or there are six or more two-wire branch circuits, the minimum size service conductors and panelboard must be 100 amperes as required in *NE Code* Section 230-42.

SIZING RESIDENTIAL NEUTRAL CONDUCTORS

The neutral conductor in a 3-wire, single-phase service carries only the unbalanced load between the two "hot" legs. Since there are several 240-volt loads in the above calculations, these 240-volt loads will be balanced and therefore reduce the load on the service neutral conductor. Consequently, in most cases, the service neutral does not have to be as large as the ungrounded ("hot") conductors.

In the above example, the water heater does not have to be included in the neutral-conductor calculation, since it is strictly 240 volts with no 120-volt loads. The clothes dryer and electric range, however, have 120-volt lights that will unbalance the current between phases. The *NE Code* allows a demand factor of 70% for these two appliances. Using this information, the neutral conductor may be sized accordingly:

General Lighting and
Appliance Load .... 4942.5 VA
Electric Range
(8000 VA × .70) .... 5600 VA
Clothes Dryer
(5500 VA × .70) .... 3850 VA

Total .... 14,392.5 VA

To find the total phase-to-phase amperes, divide the total volt-amperes by the voltage between phases.

$$14,392.5 \div 240 \text{ V} = 59.96 \text{ or } 60 \text{ amperes}$$

The service-entrance conductors have now been calculated and must be rated at 100 amperes with a neutral conductor rated for at least 60 amperes. See Figure 5-6 for a completed calculation form for the residence in question.

In *NE Code* "Notes to Ampacity Tables (No. 3) . . ." that follow *NE Code* Table 310-19, special consideration is given 120/240 volts, single-phase

SERVICE LOAD CALCULATION ONE-FAMILY DWELLING—STANDARD CALCULATION

I. GENERAL LIGHTING LOADS

TYPE OF LOAD	CALCULATION	TOTAL VA	*NE CODE* REFERENCE
Lighting load	(1) 1350 sq. ft. x 3 VA =	(2) 4050 VA	Table 220-3(b)
Small appliance loads	(3) 2 circuits x 1500VA=	(4) 3000 VA	Section 220-16(a)
Laundry load	(5) 1 circuits x 1500VA=	(6) 1500 VA	Section 220-16(b)
Lighting, small appliance, laundry	Total VA =	(7) 8550 VA	

LOAD	CALCULATION	DEMAND FACTOR	TOTAL VA
Lighting, small appliance, laundry	First 3000VA x	100%=	(8) 3000VA
Lighting, small appliance, laundry	(9)Remaining VA 5550 x	35% =	(10) 1942.5 VA
Lighting, small appliance, laundry	Add #8 & #10 above	=	(11) 4942.5 VA

TOTAL CALCULATED LOAD FOR LIGHTING, SMALL APPLIANCES, & LAUNDRY CIRCUIT(S) (Enter Item #11 above in box #12)

12	4942.5

II. LARGE APPLIANCE LOADS

TYPE OF LOAD	NAMEPLATE RATING	DEMAND FACTOR	TOTAL VA	*NE CODE* REFERENCE
Electric range	Not over 12 kVA	Use 8 kVA	(13) 8,000 VA	Table 220-19
Clothes dryer	(14) 5500 VA	100%	(15) 5500 VA	Table 220-18
Water heater	(16) 4500 VA	100%	(17) 4500 VA	
Other appliances	(18)____ VA	100%	(19)____ VA	

TOTAL CALCULATED LOAD FOR LARGE APPLIANCES (Add items #15, #17, and #19 above). ENTER TOTAL IN BOX #20

20	18,000 VA

TOTAL CALCULATED LOAD (Add boxes #12 & #20) ENTER TOTAL IN BOX #21

21	22,942.5

III. CONVERT VA TO AMPERES

$$\frac{\textit{Total VA in Box \#21}}{240\ (\textit{volts})} = \textit{amperes} \qquad \frac{22{,}942.5\ \ VA}{240\ (\textit{volts})} = (\#22)\ \underline{95.6}\ \textit{amperes}$$

IV. UNGROUNDED AND GROUNDED (NEUTRAL) CONDUCTOR SIZE—*NE CODE* ARTICLE 310

Notes to Ampacity Tables: No. 3

MINIMUM UNGROUNDED CONDUCTOR SIZE			MINIMUM GROUNDED (NEUTRAL) CONDUCTOR SIZE	
COPPER AWG SIZE	ALUMINUM OR COPPER-CLAD AL	RATING IN AMPERES	COPPER AWG SIZE	ALUMINUM OR COPPER-CLAD AL
4	2	100	8	4
3	1	110	6	3
2	1/0	125	4	2
1	2/0	150	3	1
1/0	3/0	175	2	1/0
2/0	4/0	200	1	2/0

Figure 5-6: Completed service-entrance calculation worksheet

residen tial services. Conductor sizes are shown in the table that follows these *NE Code* notes. This same table is also shown in Figures 5-5 and 5-6 of this chapter. Reference to this table shows that the *NE Code* allows a No. 4 AWG copper or No. 2 AWG aluminum or copper-clad aluminum for a 100 ampere service. Furthermore, the *NE Code* states that the grounded or neutral conductors must never be more than two wire sizes smaller than the ungrounded conductors. Therefore, the grounded conductor may not be smaller than No. 8 AWG copper or No. 4 aluminum.

When sizing the grounded conductor for services, provisions stated in *NE Code* Sections 215-2, 220-22, and 230-42 must be met, along with other applicable sections.

SIZING THE LOAD CENTER

Each ungrounded conductor in all circuits must be provided with overcurrent protection—either in the form of fuses or circuit breakers. If more than six such devices are used, a means of disconnecting the entire service must be provided—using either a main disconnect switch or a main circuit breaker.

To calculate the number of fuse holders or circuit breakers required in our sample residence, let's look at the general lighting load first. Since there is a total general lighting load of 4050 volt-amperes, this figure can be divided by 120 volts (Ohm's Law states amperes = VA/V) which equals:

$$4050 \div 120 = 33.75 \text{ amperes}$$

Either 15- or 20-ampere (amp) circuits may be used for the lighting load. Two 20-ampere circuits (2 × 20) equals 40 amperes, so two 20-ampere circuits would be adequate for the lighting. However, two 15-ampere circuits total only 30 amperes and 33.75 amperes are needed. Therefore, if 15-ampere circuits are used, three are required for the total lighting load.

In addition to the lighting circuits, the sample residence will require a minimum of two 20-ampere circuits for the small-appliance load and one 20-ampere circuit for the laundry. Thus far, we can count the following branch circuits.

General Lighting
Load .... Three 15-ampere circuits
Small Appliance
Load .... Two 20-ampere circuits
Laundry Load........ One 20-ampere circuit
Total Six SP circuit breakers

Most load centers and panelboards are provided with an even number of circuit breaker (cb) spaces or fuse holders; that is, 4, 6, 8, 10 etc. But before the panelboard can be selected, space must be provided for the 240-volt loads—each requiring a 2-pole circuit breaker or fuse holder. The rating of overcurrent-protection devices is sized by dividing the demand volt-amperes by the voltage. For example, since the demand load of the electric range is 8 kVA, 8000 volt-amperes divided by 240 volts equals 33.3 amperes. The closest standard overcurrent-protection device is 40 amperes; this will be the size used—a 40-amp, 2-pole circuit breaker. In some existing installations, you might find a 2-pole fuse block containing two 40-amp cartridge fuses being used to feed a residential electric range. The remaining 240-volt circuits are calculated in a similar fashion, which results in the following:

- Electric Range—One 40-amp, 2-pole circuit breaker
- Clothes Dryer—One 30-amp, 2-pole circuit breaker
- Water Heater—One 30-amp, 2-pole circuit breaker

These three appliances will therefore require an additional (2 poles × three appliances =) 6 spaces in the load center or panelboard. Adding these six spaces to the six required for the general lighting and small appliance loads, requires at least a 12-space load center to handle the circuits in our sample residence.

GROUND-FAULT CIRCUIT-INTERRUPTERS

Although the placement of the various outlets has not been done at this point, the experienced electrician knows that circuits providing power to certain areas of the home require ground-fault circuit-interrupters (GFCIs) to be installed for additional protection of people using these circuits. Such areas include:

- All outside receptacles
- Receptacles used in bathrooms
- Receptacles located in residential garages
- Receptacles located in unfinished basements
- Receptacles located in crawl spaces
- Receptacles installed within six feet of a kitchen or bar sink

Since there is no basement or crawl space in our sample residence, these two areas do not apply to this project. However, there is a bathroom and kitchen. Furthermore, outside receptacles will be provided. Therefore, at least one small-appliance circuit will be provided with GFCI protection along with one circuit supplying outdoor receptacles. The bathroom receptacles can be connected to the GFCI circuit supplying the outdoor receptacles, or a GFCI receptacle can be used. This brings the total number to at least two circuits that will require GFCI protection.

GFCI circuit breakers require one space in the load center or panelboard—the same as a 1-pole circuit breaker. Consequently, at least two more spaces must be added to the 12 obtained in previous calculations; the panelboard spaces now total 14.

We could get by with a 14-space circuit-breaker load center with a 100-ampere main circuit breaker. However, it is always better to provide a few extra spaces for future use. Some local ordinances require 20% extra space in any residential load center for additional circuits that may be added later. The circuit breakers themselves do not have to be installed, but space should be provided for them. This puts the size of the load center for the house in question up to 18 spaces, providing four extra spaces for future use.

This first example given for sizing residential electrical services is very basic. Most modern homes have many more electrical conveniences for added comfort. Some of these include air conditioning, electric heat, garbage disposal, dishwasher, and perhaps a central vacuum system. Optional methods of service-entrance calculations, including these options, are presented later in this chapter.

GROUNDING ELECTRIC SERVICES

The grounding system is a major part of the electrical system. Its purpose is to protect life and equipment against the various electrical faults that can occur. It is sometimes possible for higher-than-normal voltages to appear at certain points in an electrical system or in the electrical equipment connected to the system. Proper grounding ensures that the high electrical charges that cause these high voltages are channeled to earth or ground before damaging equipment or causing danger to human life.

When we refer to *ground*, we are talking about ground potential or earth ground. If a conductor is connected to the earth or to some conducting body that serves in place of the earth, such as a driven ground rod (electrode) or cold-water pipe, the conductor is said to be grounded. The neutral conductor in a three- or four-wire service, for example, is intentionally grounded and therefore becomes a grounded conductor. However, a wire used to connect this neutral conductor to a grounding electrode or electrodes is referred to as a grounding conductor. Note the difference in the two meanings; one is *groundED,* while the other provides a means of *groundING.*

There are two general classifications of protective grounding:

- System grounding
- Equipment grounding

The system ground relates to the service-entrance equipment and its interrelated and bonded components. That is, system and circuit conductors are grounded to limit voltages due to lighting, line surges, or unintentional contact with higher voltage lines, and to stabilize the voltage to ground during normal operation.

Equipment grounding conductors are used to connect the noncurrent-carrying metal parts of equipment, conduit, outlet boxes, and other enclosures to the system grounded conductor, the grounding electrode conductor, or both, at the service equipment or at the source of a separately derived system. Equipment grounding conductors are bonded to the system grounded conductor to provide a low impedance path for fault current that will facilitate the operation of overcurrent devices under ground-fault conditions. Equipment grounding is covered later in this book.

Article 250 of the *NE Code* covers general requirements for grounding and bonding. Nearly 75 changes or additions have been made to this Article since the 1990 *NE Code* was printed; these changes are reflected in the new 1993 *NE Code*. This should be reason enough to carefully read all parts of this article over several times until you have a thorough understanding of its contents.

To better understand a complete grounding system, let's take a look at a conventional residential system beginning at the power company's high-voltage lines and transformer. A pole-mounted transformer is fed with a two-wire 7200-volt system which is transformed and stepped down to a 3-wire, 120/240-volt, single-phase electric service suitable for residential use. Figure 5-7 shows the voltage between phase A and phase B to be 240 volts. However, by connecting a third wire (neutral) on the secondary winding of the transformer — between the other two—the 240 volts are split in half, giving 120 volts between either phase A or phase B and the neutral conductor. Consequently, 240 volts are available for household appliances such as ranges, hot-water heaters, clothes dryers, and the like, while 120 volts are available for lights, small appliances, tvs, and the like.

Referring again to the diagram in Figure 5-7, conductors A and B are ungrounded conductors, while the neutral is a grounded conductor. If only

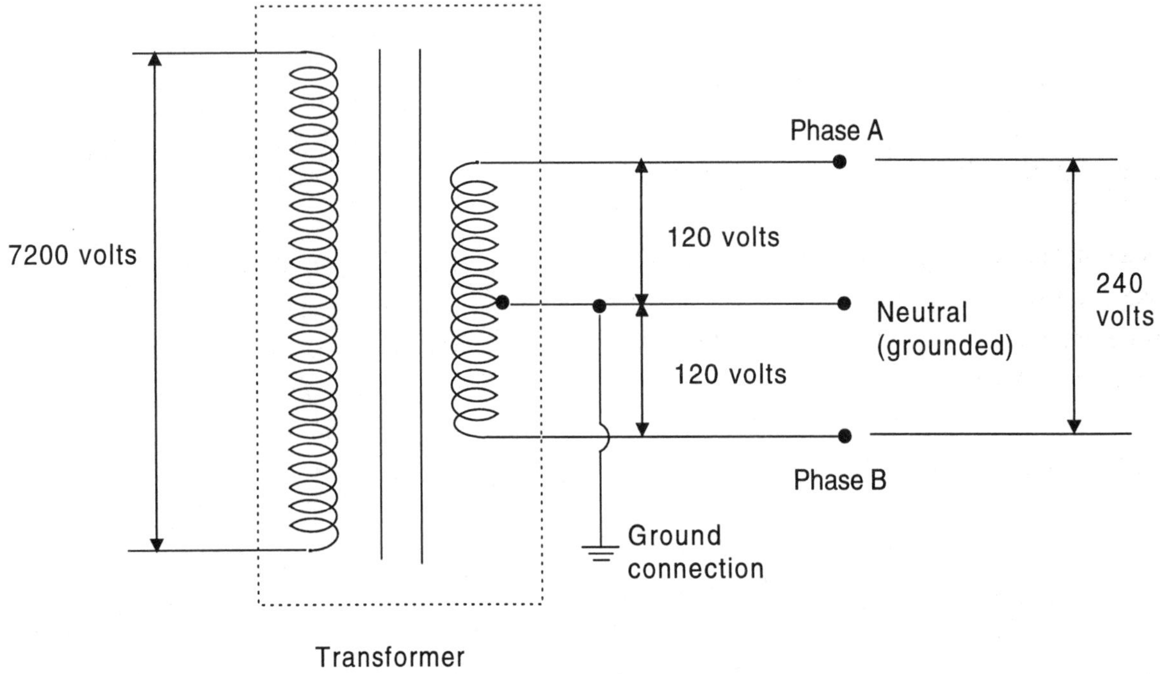
7200 volts
Phase A
120 volts
240
volts
Neutral
(grounded)
120 volts
Phase B
Ground
connection
Transformer

Figure 5-7: Wiring diagram of 7200-volt to 120/240-volt, single-phase transformer connection

the point where most systems are grounded — the neutral bus in the main panelboard. See Figure 5-8.

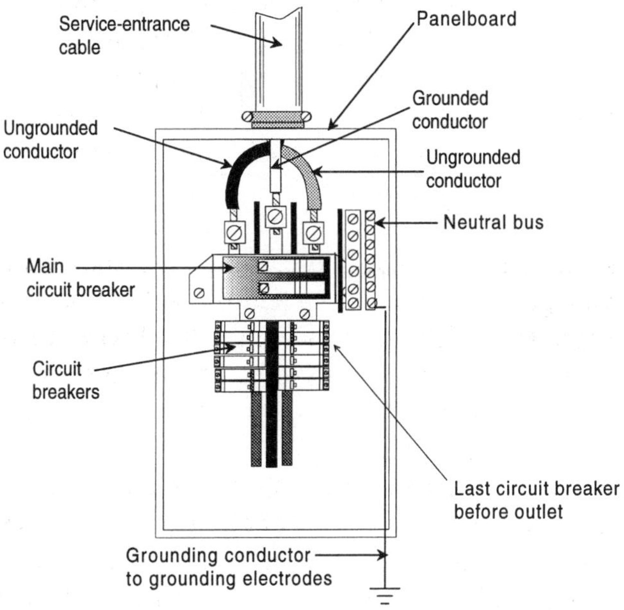
Service-entrance cable
Panelboard
Grounded conductor
Ungrounded conductor
Ungrounded conductor
Neutral bus
Main circuit breaker
Circuit breakers
Last circuit breaker before outlet
Grounding conductor to grounding electrodes

Figure 5-8: Interior view of panelboard

240-volt loads were connected, the neutral (grounded conductor) would carry no current. However, since 120-volt loads are present, the neutral will carry the unbalanced load and becomes a current-carrying conductor. For example, if phase A carries 60 amperes and phase B carries 50 amperes, the neutral conductor would carry only (60 - 50 =) 10 amperes. This is why the *NE Code* allows the neutral conductor in an electric service to be smaller than the ungrounded conductors.

The typical pole-mounted service-entrance is normally routed by messenger cable from a point on the pole to a point on the building being served, terminating in a meter housing. Another service conductor is installed between the meter housing and the main service switch or panelboard. This is

GROUNDING METHODS

Methods of grounding an electric service are covered in *NE Code* Section 250-81. In general, all of the following (if available) and any made electrodes must be bonded together to form the grounding electrode system:

- An underground water pipe in direct contact with the earth for no less than 10 feet.
- The metal frame of a building where effectively grounded.
- An electrode encased by at least 2 inches of concrete, located within and near the bottom of a concrete foundation or footing that is in direct contact with the earth. Furthermore, this electrode must be at least 20 feet long and must be made of electrically conductive coated steel reinforcing bars or rods of not less than ½-inch diameter, or consisting of at least 20 feet of bare copper conductor not smaller than No. 2 AWG wire size.
- A ground ring encircling the building or structure, in direct contact with the earth at a depth below grade not less than 2½ feet. This ring must consist of at least 20 feet of bare copper conductor not smaller than No. 2 AWG wire size.

In most residential structures, only the water pipe will be available, and this water pipe must be supplemented by an additional electrode as specified in *NE Code* Sections 250-81(a) and 250-83. With these facts in mind, let's take a look at a typical residential electric service, and the available grounding electrodes. See Figure 5-9.

Our sample residence has a metal underground water pipe that is in direct contact with the earth for more than 10 feet, so this is one valid grounding

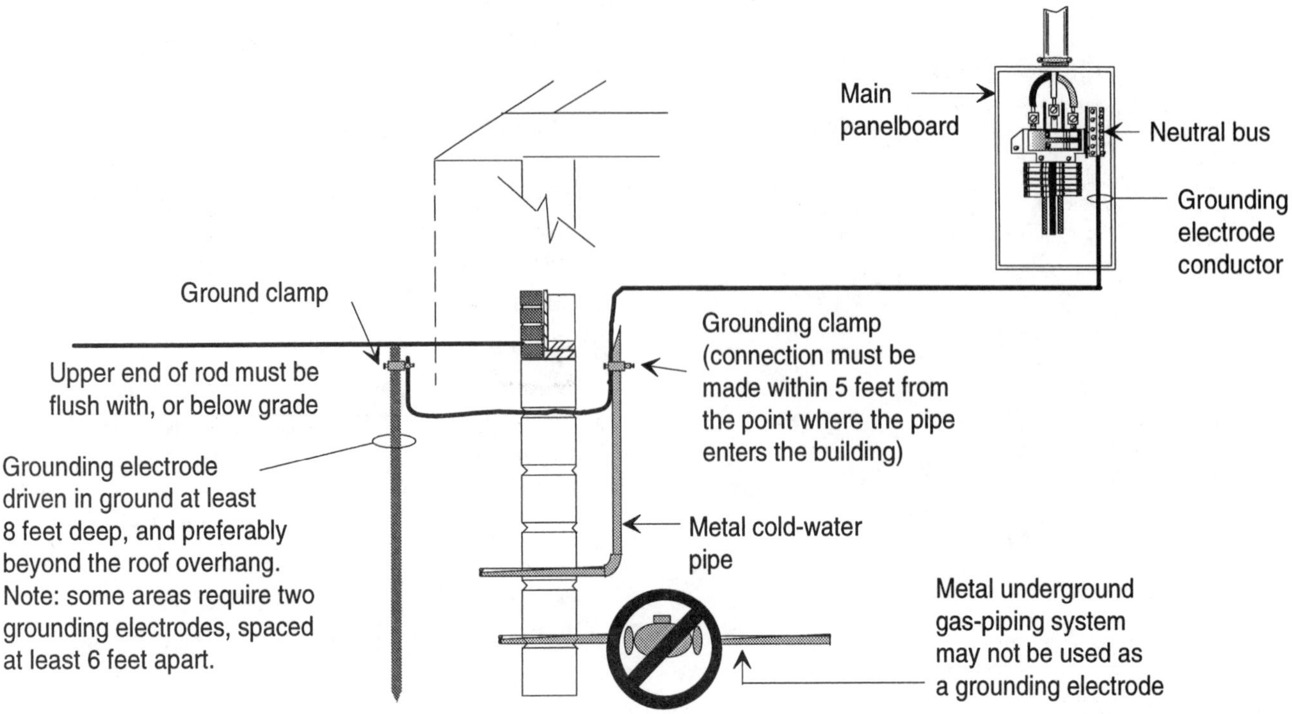
Main
panelboard
Neutral bus
Grounding
electrode
conductor
Ground clamp
Upper end of rod must be
flush with, or below grade
Grounding electrode
driven in ground at least
8 feet deep, and preferably
beyond the roof overhang.
Note: some areas require two
grounding electrodes, spaced
at least 6 feet apart.
Grounding clamp
(connection must be
made within 5 feet from
the point where the pipe
enters the building)
Metal cold-water
pipe
Metal underground
gas-piping system
may not be used as
a grounding electrode

Figure 5-9: Components of a residential service grounding system

source. The house also has a metal underground gas-piping system, but this may not be used as a grounding electrode (*NE Code* Section 250-83(a)). *NE Code* Section 250-81(a) further states that the underground water pipe must be supplemented by an additional electrode of a type specified in Section 250-81 or in Section 250-83. Since a grounded metal building frame, concrete-encased electrode, or a ground ring are not normally available for most residential applications, *NE Code* Section 250-83 — Made and Other Electrodes — must be used in determining the supplemental electrode. In most cases, this supplemental electrode will consist of either a driven rod or pipe electrode, specifications for which are shown in Figure 5-10.

An alternate method to the pipe or rod method is a plate electrode. Each plate electrode must expose not less than 2 square feet of surface to the surrounding earth. Plates made of iron or steel must be at least $\frac{1}{4}$-inch thick, while plates of nonferrous metal such as copper need only be .06 inch thick.

Either type of electrode must have a resistance to ground of 25 ohms or less. If not, they must be augmented by an additional electrode spaced not less than 6 feet from each other. In fact, many locations require two electrodes regardless of the resistance to ground. This, of course, is not an *NE Code* requirement, but is required by some power companies and local ordinances in some cities and counties. Always check with the local inspection authority for such rules that surpass the requirements of the *NE Code.*

Grounding Conductors

The grounding conductor, connecting the panelboard neutral bus to the water pipe and grounding electrodes, must be of either copper, aluminum, or copper-clad aluminum. Furthermore, the material selected must be resistant to any corrosive condition existing at the installation or it must be suitably

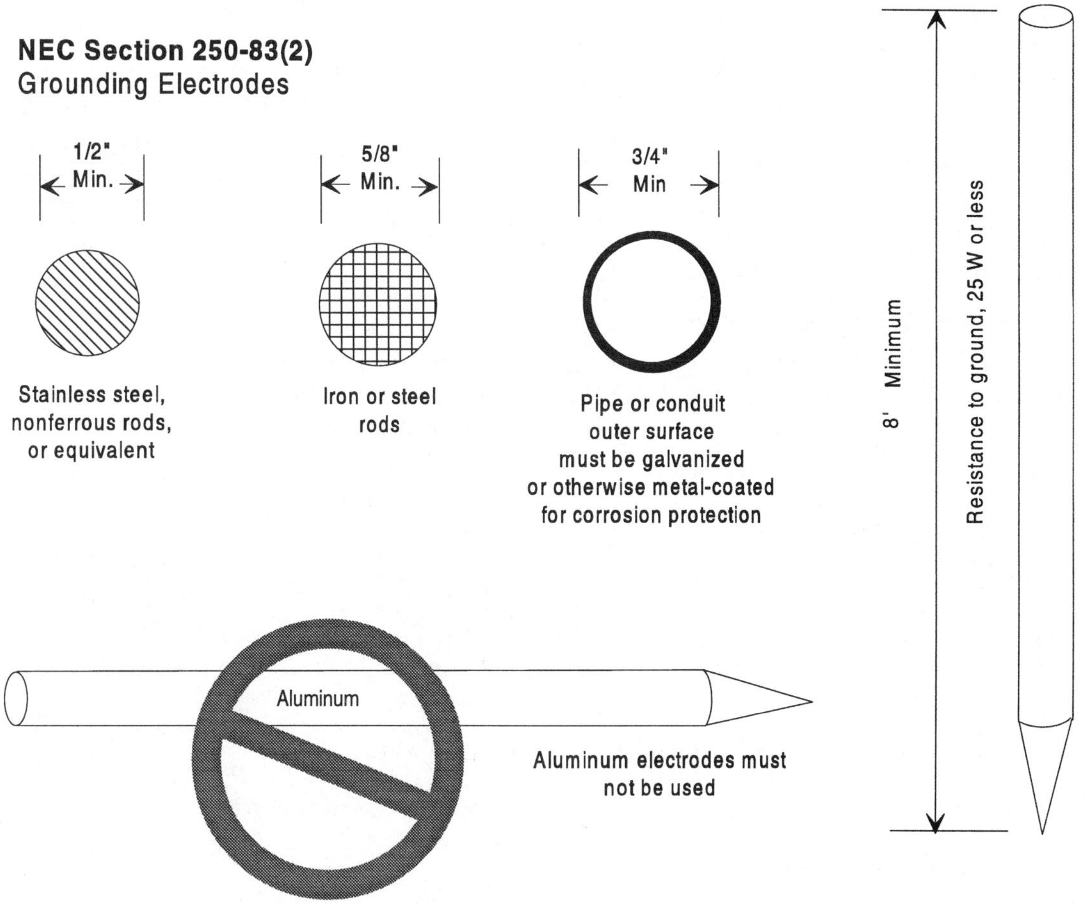
NEC Section 250-83(2)
Grounding Electrodes
1/2" Min.
5/8" Min.
3/4" Min
Stainless steel, nonferrous rods, or equivalent
Iron or steel rods
Pipe or conduit outer surface must be galvanized or otherwise metal-coated for corrosion protection
8' Minimum
Resistance to ground, 25 W or less
Aluminum
Aluminum electrodes must not be used

Figure 5-10: Specifications of rod and pipe grounding electrodes

protected against corrosion. The grounding conductor may be either solid or stranded, covered or bare, but it must be in one continuous length without a splice or joint—except for the following conditions:

- Splices in busbars are permitted.
- Where a service consists of more than one single enclosure, it is permissible to connect taps to the grounding electrode conductor provided the taps are made within the enclosures. They are not to be made outside of the enclosure.
- Grounding electrode conductors may also be spliced at any location by means of irreversible compression-type connectors listed for the purpose or the exothermic welding process. These methods prevent easy removal.

The size of grounding conductors depends on service-entrance size; that is, the size of the largest service-entrance conductor or equivalent for parallel conductors. The table in Figure 5-11 gives the proper sizes of grounding conductors for various sizes of electric services. Equipment grounding is thoroughly covered later on in this book.

INSTALLING THE SERVICE-ENTRANCE

In practical applications, the electric service is normally one of the last parts of an electrical system to be installed. However, it is one of the

Size of Largest Service-Entrance Conductor or Equivalent for Parallel Conductors		Size of Grounding Electrode Conductor	
Copper	Aluminum or Copper-Clad Aluminum	Copper	Aluminum or Copper-Clad Aluminum
2 or smaller	0 or smaller	8	6
1 or 2	2/0 or 3/0	6	4
2/0 or 3/0	4/0 or 250 kcmil	4	2
Over 3/0 through 350 kcmil	Over 250 kcmil through 500 kcmil	2	0
Over 350 kcmil through 600 kcmil	Over 500 kcmil through 900 kcmil	0	3/0
Over 600 kcmil through 1100 kcmil	Over 900 kcmil through 1750 kcmil	2/0	4/0
Over 1100 kcmil	Over 1750 kcmil	3/0	250 kcmil

Figure 5-11: Grounding electrode conductor sizes for service-entrance grounding

first considerations when "laying out" a residential electrical system. The reasons are many; a few of them follow:

- Electrician must know in which direction, and to what location, to route the circuit homeruns while roughing-in the electrical wiring.
- Provisions must be made for sleeves through footings and foundations in cases where underground systems (service laterals) are used.
- The local power company must be notified as to the approximate size of service required so they may plan the best way to furnish a service drop to the property.

SERVICE DROP LOCATIONS

The location of the service drop, electric meter and the load center should be considered first. It is always wise to consult the local power company to obtain their recommendations; where you want the service drop and where they want it may not coincide. A brief meeting with the power company about the location of the service drop can prevent much grief, confusion, and perhaps expense.

While considering the placement of the service drop, it must be routed so that the service-drop conductors are not readily accessible. They must have a clearance of not less than 3 feet from windows, doors, porches, fire escapes, or similar locations. Furthermore, when service-drop conductors pass over rooftops, driveways, yards, etc., they must have clearances as specified in the *NE Code* Section 230-24.

A plot plan — sometimes called *site plan* — is often available for new construction. The plot plan shows the entire property, with the building or buildings drawn in their proper location on the plot of land. It also shows sidewalks, driveways, streets, and existing utilities — both overhead and underground.

A plot plan of our sample residence is shown in Figure 5-12. In reviewing this drawing, we see that the closest power pole is located across a public street from the house in question. After consulting

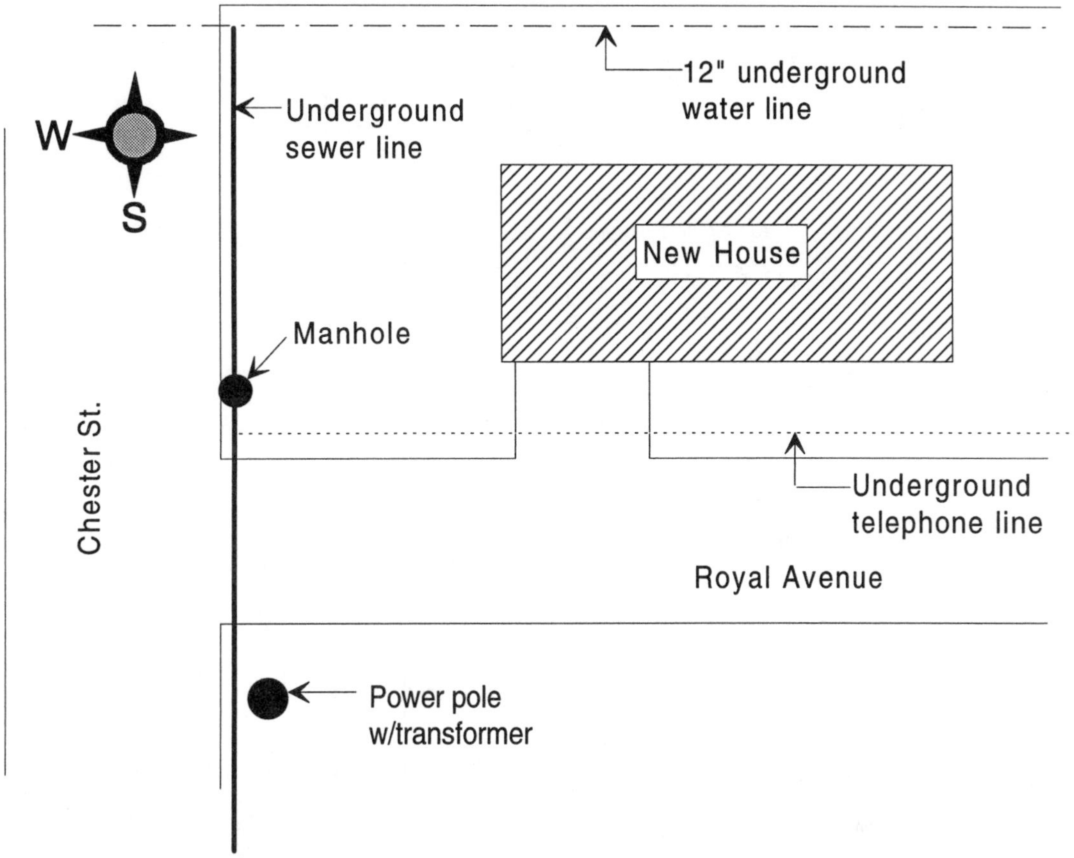
12" underground
water line
W
S
Underground
sewer line
New House
Manhole
Chester St.
Underground
telephone line
Royal Avenue
Power pole
w/transformer

Figure 5-12: Plot plan of sample residence

the local power company, let's assume that the service will be brought to the house from this pole by triplex which will "hit" the residence at a point on its left (west) end. The steel uninsulated conductor of triplex cable acts as both the grounded conductor (neutral) and as a support for the insulated (ungrounded) conductors, and is suitable for overhead use.

When type SE cable is used it will run directly from the point of attachment and service head to the meter base. However, since the carport is located on the west side of the building, a service mast (Figure 5-13) will have to be installed.

VERTICAL CLEARANCES OF SERVICE DROP

Besides the clearances over roofs, the *NE Code* also specifies the distance that service-drop conductors must clear the ground. These distances will vary with the surrounding conditions.

In general, the *NE Code* states that the vertical clearances of all service-drop conductors — 600 volts or under — are based on conductor temperature of 60°F (15°C), no wind, with final unloaded sag in the wire, conductor, or cable. Service-drop conductors must be at least 10 feet above the ground or other accessible surfaces at all times. More distance is required under most conditions. For example, if the service conductors pass over residential property and driveways or commercial property not subject to truck traffic, the conductors must be at least 15 feet above the ground. However, this distance may be reduced to 12 feet when the voltage is limited to 300 volts to ground, which covers all single- and two-family residences.

In other areas — public streets, alleys, roads, parking areas subject to truck traffic, driveways on

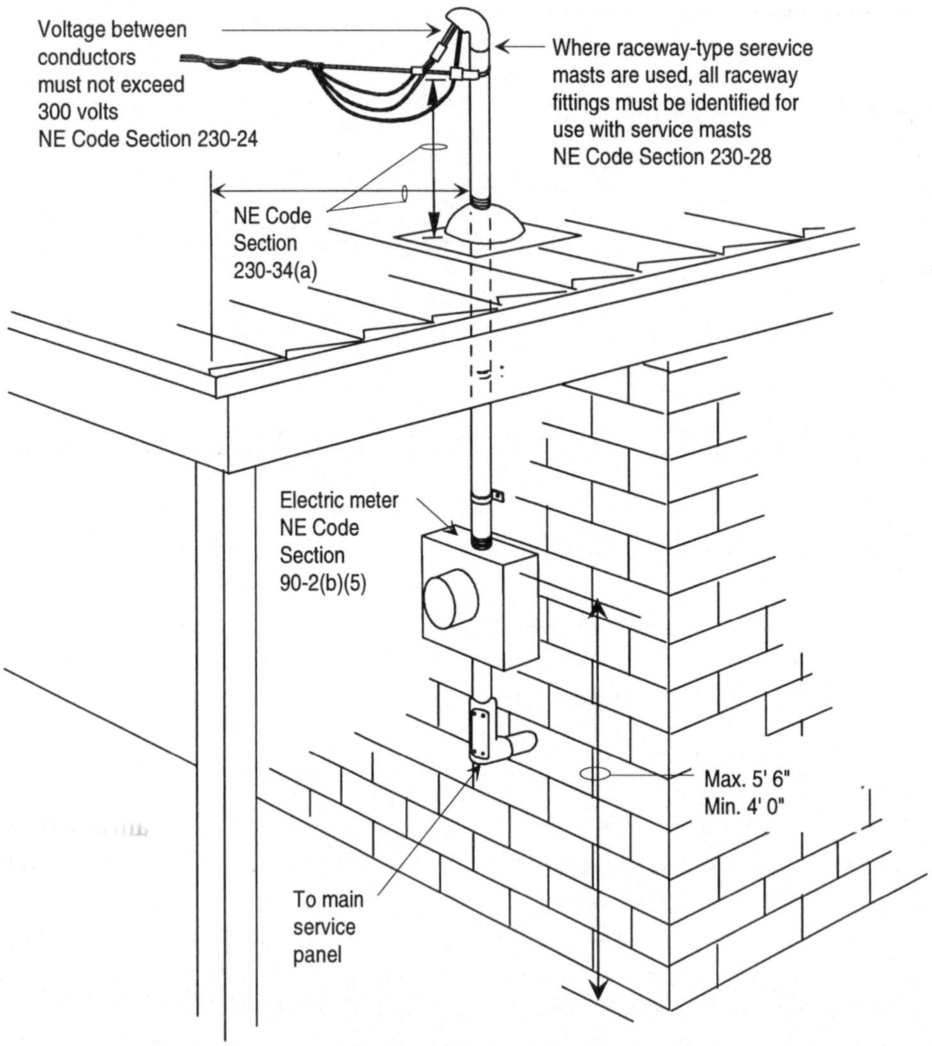
Voltage between conductors must not exceed 300 volts NE Code Section 230-24
Where raceway-type serevice masts are used, all raceway fittings must be identified for use with service masts NE Code Section 230-28
NE Code Section 230-34(a)
Electric meter NE Code Section 90-2(b)(5)
Max. 5' 6"
Min. 4' 0"
To main service panel

Figure 5-13: *NE Code* regulations governing service-mast installations

other than residential property, etc. — the minimum vertical distance is 18 feet. The conditions of our sample residence are shown in Figure 5-14.

PANELBOARD LOCATION

The main service disconnect or panelboard is normally located in a portion of an unfinished basement or utility room on an outside wall so that the service cable coming from the electric meter can terminate immediately into the switch or panelboard when the cable enters the building. In our case, however, there is no basement. The utility room is located in the center of the house with no outside walls. Consequently, a somewhat different arrangement will have to be used.

NE Code Section 230-70 requires that the service disconnecting means is installed in a readily accessible location — either outside or inside the building. If located inside the building, it must be located nearest the point of entrance of the service conductors. Consequently, to comply with this *NE Code* regulation, there are at least two methods of installing the panelboard in the utility room of our sample home.

The first utilizes a weatherproof 100-ampere disconnect (safety switch or circuit-breaker enclosure) mounted next to the meter base on the outside

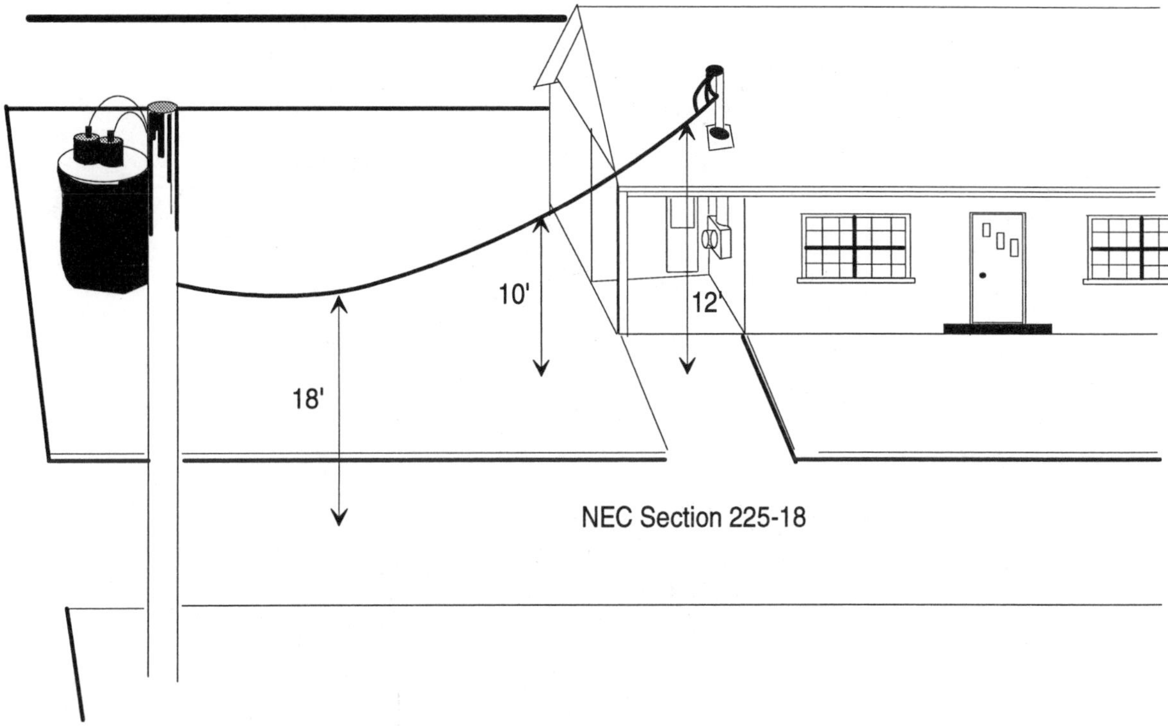
10'
12'
18'
NEC Section 225-18

Figure 5-14: Vertical clearances required

of the building. With this method, service conductors are provided with overcurrent protection; the neutral conductor is also grounded at this point, as this becomes the main disconnect switch. Three-wire cable with an additional grounding wire is then routed from this main disconnect to the panelboard in the utility room. All three current-carrying conductors (two ungrounded and one neutral) must be insulated with this arrangement; the equipment ground, however, may be bare. The panel-

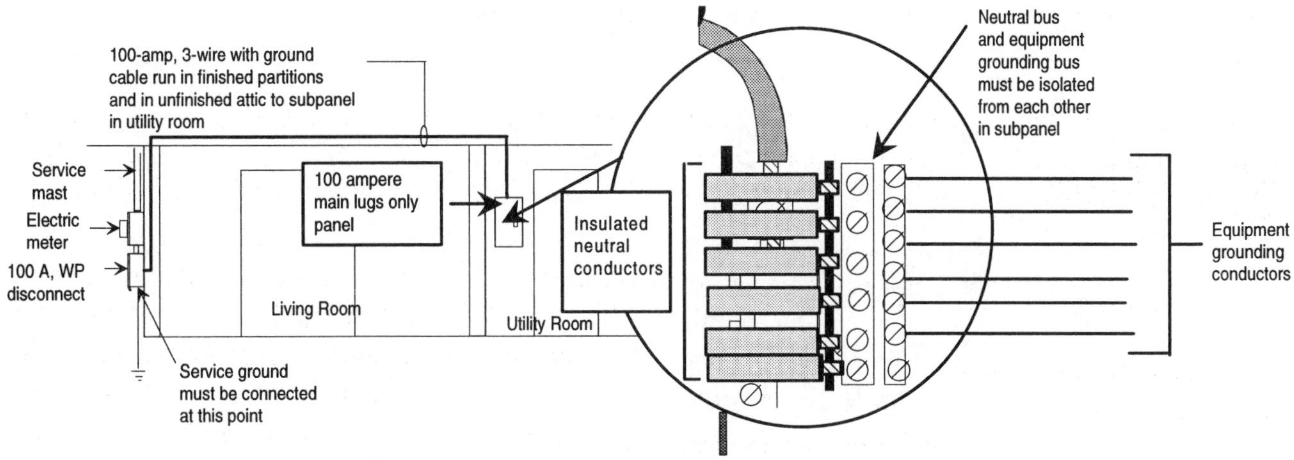
100-amp, 3-wire with ground cable run in finished partitions and in unfinished attic to subpanel in utility room
Service mast
Electric meter
100 A, WP disconnect
100 ampere main lugs only panel
Insulated neutral conductors
Living Room
Utility Room
Service ground must be connected at this point
Neutral bus and equipment grounding bus must be isolated from each other in subpanel
Equipment grounding conductors

Figure 5-15: One method of wiring panelboard for the sample residence

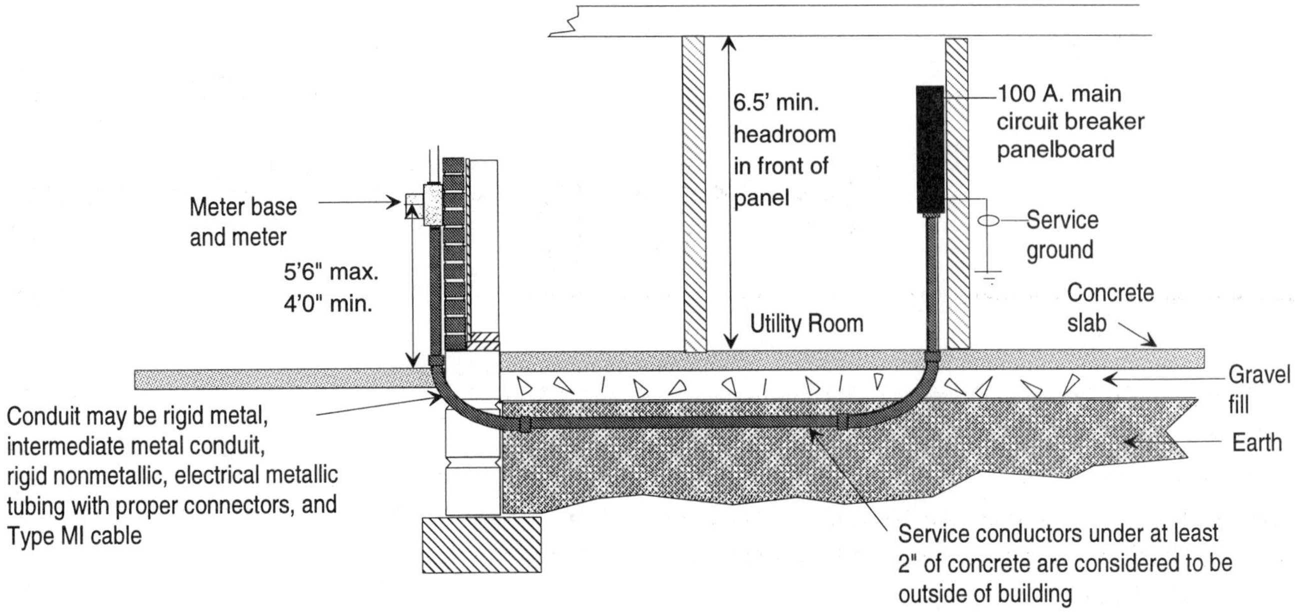
6.5' min.
headroom
in front of
panel
100 A. main
circuit breaker
panelboard
Meter base
and meter
5'6" max.
4'0" min.
Service
ground
Concrete
slab
Utility Room
Gravel
fill
Earth
Conduit may be rigid metal,
intermediate metal conduit,
rigid nonmetallic, electrical metallic
tubing with proper connectors, and
Type MI cable
Service conductors under at least
2" of concrete are considered to be
outside of building

Figure 5-16: Alternate method of service installation for sample residence

board containing overcurrent-protection devices for the branch-circuits, which is located in the utility room, now becomes a subpanel. Details of this installation are shown in Figure 5-15.

An alternate method utilizes conduit from the meter base, routed under the concrete slab to emerge up to a main panelboard located in the utility room. *NE Code* Section 230-6 considers conductors installed under at least 2 inches of concrete to be outside of the building. Our sample residence has a 4-inch thick reinforced concrete slab — well within the *NE Code* regulations. Therefore, the service conductors from the meter base which are installed under the concrete slab in conduit are considered to be outside of the house, and no disconnect is required at the meter base. When this conduit run emerges in the utility room, the conduit will run straight up into the bottom of the panelboard, again meeting the *NE Code* requirement, "the panel is located nearest the point of entrance of the service conductors." Details of this service arrangement are shown in Figure 5-16.

Note: Local ordinances in some areas may require a disconnect at the meter base, making the panel in the utility room a subpanel.

Chapter 6
Wiring Methods

Branch circuits and feeders are used in residential construction to provide power to operate electrically-operated components and equipment, and control wiring to regulate the equipment. Wiring may be further subdivided into either open or concealed wiring.

In open wiring systems, the cable and/or raceways are installed on the surface of the walls, ceilings, columns, and the like where they are in view and readily accessible. Open wiring is often used in areas where appearance is not important like in unfinished basements, attics, and garages.

Concealed wiring systems have all cable and raceway runs inside of walls, partitions, ceilings, columns, and behind baseboards or molding where they are out of view and not readily accessible. This type of wiring is generally used in all new construction with finished interior walls, ceilings, and floors and is the preferred type where good appearance is important.

In general, there are two basic wiring methods used in the majority of modern residential electrical systems. They are:

- Sheathed cables of two or more conductors
- Raceway (conduit) systems

The method used on a given job is determined by the requirements of the *NE Code*, the type of building construction, the location of the wiring in the building, and the relative costs of different wiring methods.

In most applications, either of the two methods may be used, and both methods are frequently used in combination.

CABLE SYSTEMS

Several types of cable systems are used in wiring systems for building construction to feed or supply power to equipment, and include the following:

Nonmetallic Sheathed Cable

Type NM cable is manufactured in two or three wires, and with varying sizes of conductors. In both two- and three-wire cables, conductors are color-coded: one conductor is black while the other is white in two-wire cable; in three-wire cable, the additional conductor is red. Both types will also have a grounding conductor which is usually bare, but it is sometimes covered with a thin green plastic insulation. The jacket or covering consists of rubber, plastic, or fiber. Most will also have markings on this jacket giving the manufacturer's name or trademark, the wire size, and the number of conductors. For example, "NM 12-2 W/GRD" indicates that the jacket contains two No. 12 AWG conductors along with a grounding wire; "NM 12-3 W/GRD" indicates three conductors

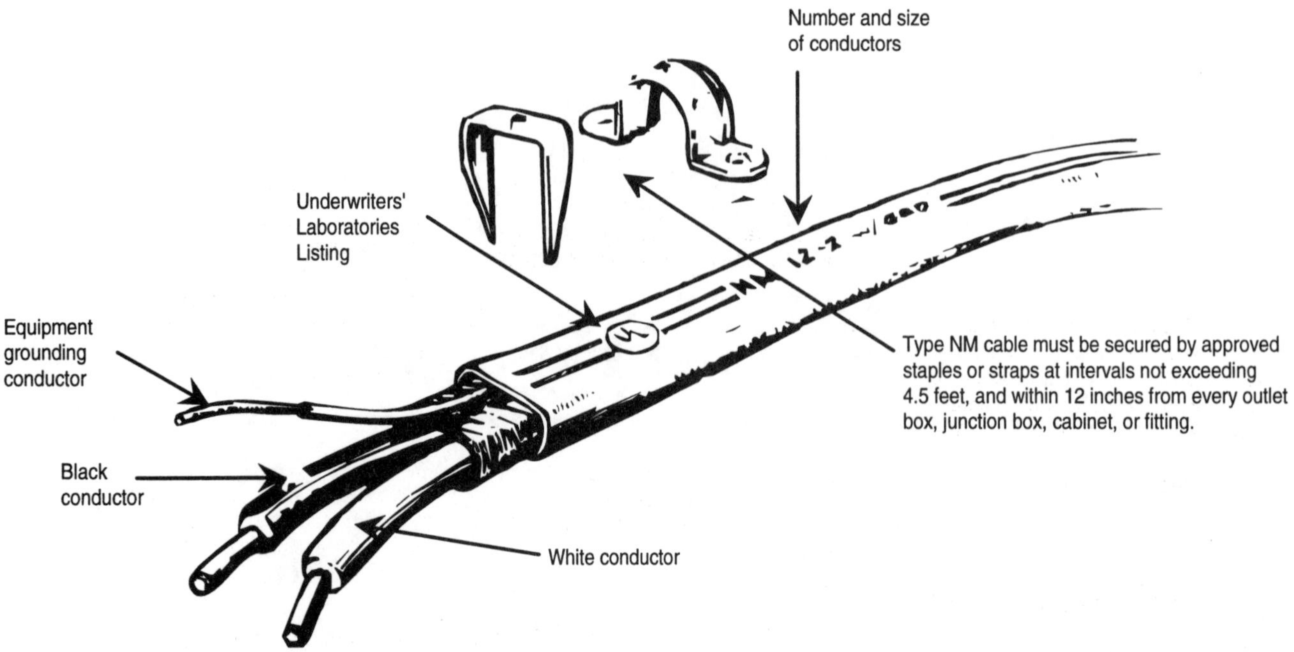
Number and size
of conductors
Underwriters'
Laboratories
Listing
Equipment
grounding
conductor
Type NM cable must be secured by approved staples or straps at intervals not exceeding 4.5 feet, and within 12 inches from every outlet box, junction box, cabinet, or fitting.
Black
conductor
White conductor

Figure 6-1: Characteristics of Type NM cable (Romex)

plus a grounding wire. This type of cable may be concealed in the framework of buildings, or in some instances, may be run exposed on the building surfaces. It may not be used in any building exceeding three floors above grade, as a service-entrance cable, in commercial garages having hazardous locations, in theaters and similar locations, in places of assembly, in motion picture studios, in storage battery rooms, in hoistways, embedded in poured concrete or aggregate, or in any hazardous location except as otherwise permitted by the *NE Code*. Nonmetallic sheathed cable is frequently referred to as *Romex* on the job. See Figure 6-1.

Since Type NM cable is the least expensive type for residential use, this wiring method will be the type most often encountered. Figure 6-2 shows additional *NE Code* installation requirements pertaining to NM cable.

Type AC (Armored) Cable

Type AC cable — commonly called "BX" — is manufactured in two-, three-, and four-wire assemblies, with varying sizes of conductors, and is used in locations similar to those where Type NM cable is allowed. The metallic spiral covering on BX cable offers a greater degree of mechanical protection than with NM cable, and the metal jacket also provides a continuous grounding bond without the need for additional grounding conductors.

BX cable may be used for under-plaster extensions, as provided in the *NE Code*, and embedded in plaster finish, brick, or other masonry, except in damp or wet locations. It may also be run or "fished" in the air voids of masonry block or tile walls, except where such walls are exposed or subject to excessive moisture or dampness or are below grade. See Figure 6-3.

Underground Feeder Cable

Type UF cable may be used underground, including direct burial in the earth, as a feeder or branch-circuit cable when provided with overcurrent protection at the rated ampacity as required by the *NE Code*. When Type UF cable is used above

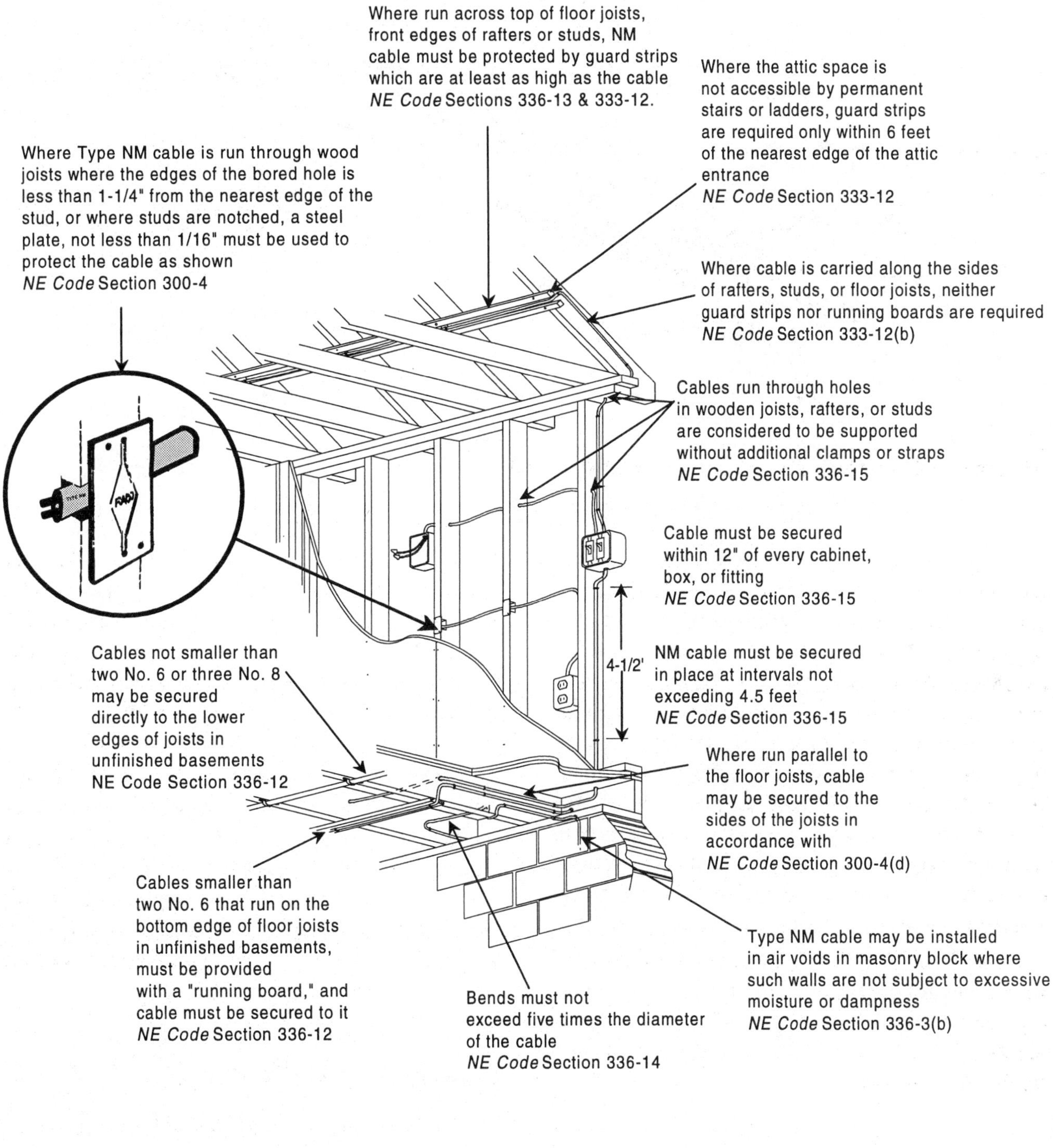
Where run across top of floor joists, front edges of rafters or studs, NM cable must be protected by guard strips which are at least as high as the cable *NE Code* Sections 336-13 & 333-12.
Where the attic space is not accessible by permanent stairs or ladders, guard strips are required only within 6 feet of the nearest edge of the attic entrance *NE Code* Section 333-12
Where Type NM cable is run through wood joists where the edges of the bored hole is less than 1-1/4" from the nearest edge of the stud, or where studs are notched, a steel plate, not less than 1/16" must be used to protect the cable as shown *NE Code* Section 300-4
Where cable is carried along the sides of rafters, studs, or floor joists, neither guard strips nor running boards are required *NE Code* Section 333-12(b)
Cables run through holes in wooden joists, rafters, or studs are considered to be supported without additional clamps or straps *NE Code* Section 336-15
Cable must be secured within 12" of every cabinet, box, or fitting *NE Code* Section 336-15
4-1/2'
NM cable must be secured in place at intervals not exceeding 4.5 feet *NE Code* Section 336-15
Cables not smaller than two No. 6 or three No. 8 may be secured directly to the lower edges of joists in unfinished basements NE Code Section 336-12
Where run parallel to the floor joists, cable may be secured to the sides of the joists in accordance with *NE Code* Section 300-4(d)
Cables smaller than two No. 6 that run on the bottom edge of floor joists in unfinished basements, must be provided with a "running board," and cable must be secured to it *NE Code* Section 336-12
Type NM cable may be installed in air voids in masonry block where such walls are not subject to excessive moisture or dampness *NE Code* Section 336-3(b)
Bends must not exceed five times the diameter of the cable *NE Code* Section 336-14

Figure 6-2: *NE Code* installation requirements for Type NM cable

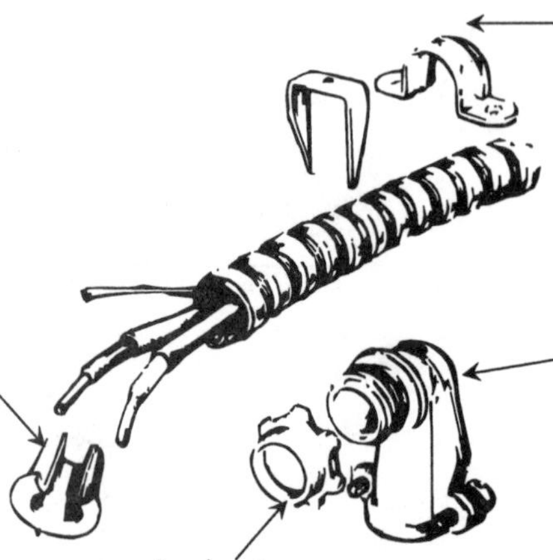
Type AC cable must be secured by approved staples or straps at intervals not exceeding 4.5 feet, and within 12 inches from every outlet box, junction box, cabinet, or fitting NE Code Section 333-7
An approved insulating bushing or its equivalent must be provided between the conductors and the armor at all points where the armor of the cable terminates NE Code Section 333-9
At all points where the armor of AC cable terminates, a fitting shall be provided to protect wires from abrasion, unless the design of the outlet boxes or fittings provides equivalent protection NE Code Section 333-9
Locknut
Connector

Figure 6-3: Characteristics of Type AC (BX) cable

grade where it will come in direct contact with the rays of the sun, its outer covering must be sun resistant. Furthermore, where Type UF cable emerges from the ground, some means of mechanical protection must be provided. This protection may be in the form of conduit or guard strips. Type UF cable resembles Type NM cable in appearance. The jacket, however, is constructed of weather resistant material to provide the required protection for direct-burial wiring installations.

Cable Color-Coding

The *NE Code* specifies certain methods of identifying conductors used in wiring systems of all types. For example, the high leg of a 120/240-volt grounded three-phase delta system must be marked with an orange color for identification; a grounded conductor must be identified either by the color of its insulation, by markings at the terminals, or by other suitable means. Unless allowed by *NE Code* exceptions, a grounded conductor must have a white or natural gray finish. When this is not practical for conductors larger than No. 6 AWG, marking the terminals with white color is an acceptable method of identifying the conductors.

Conductors contained in cables are color-coded so that identification may be easily made at each access point. The following table lists the color-coding for cables up through four-wire cable. Although some control-wiring and communication cables contain 60, 80, or more pairs of conductors — using a combination of colors — the ones listed are the most common and will be encountered the most on electrical installations.

Number of Conductors in Cable	Color of Conductors
Two-wire cable	One black (ungrounded phase conductor) One white (grounded conductor)
Two-wire cable with ground	One black (ungrounded phase conductor) One white (grounded conductor) One bare (equipment grounding conductor)
Three-wire cable	One black (ungrounded phase conductor) One white (grounded conductor) One red (ungrounded phase conductor)
Three-wire cable with ground	One black (ungrounded phase conductor) One white (grounded conductor) One red (ungrounded phase conductor) One bare (equipment grounding conductor)
Four-wire cable	One black (ungrounded phase conductor) One white (grounded conductor) One red (ungrounded phase conductor) One blue (ungrounded phase conductor)
Four-wire cable with ground	One black (ungrounded phase conductor) One white (grounded conductor) One red (ungrounded phase conductor) One blue (ungrounded phase conductor) One bare (equipment grounding conductor)

When conductors are installed in raceway systems, any color insulation is permitted for the ungrounded phase conductors except the following:

- White or gray: reserved for use as the grounded circuit conductor
- Green: reserved for use as a grounding conductor only

Changing Colors

Should it become necessary to change the actual color of a conductor to meet *NE Code* requirements or to facilitate maintenance on circuits and equipment, the conductors may be reidentified with colored tape or paint.

For example, assume that a two-wire cable containing a black and white conductor is used to feed a 240-volt, two-wire single-phase motor. Since the white colored conductor is supposed to be reserved for the grounded conductor, and none is required in this circuit, the white conductor may be marked with a piece of black tape at each end of the circuit so that everyone will know that this wire is not a grounded conductor.

Service-Entrance Cable

Type SE cable, when used for electrical services, must be installed as specified in *NE Code* Article 230. This cable is available with the grounded conductor bare for outside service conductors, and also with an insulated grounded conductor for interior wiring systems.

Type SE cable is permitted for use on branch circuits or feeders provided all current-carrying conductors are insulated; this includes the grounded or neutral conductor. When Type SE cable is used for interior wiring, all *NE Code* regulations governing the installation of Type NM cable also apply to Type SE cable. There are, however, some exceptions. Type SE cable with an uninsulated grounded conductor may be used on the appliances listed on the next page provided the voltage is not more than 150 volts to ground.

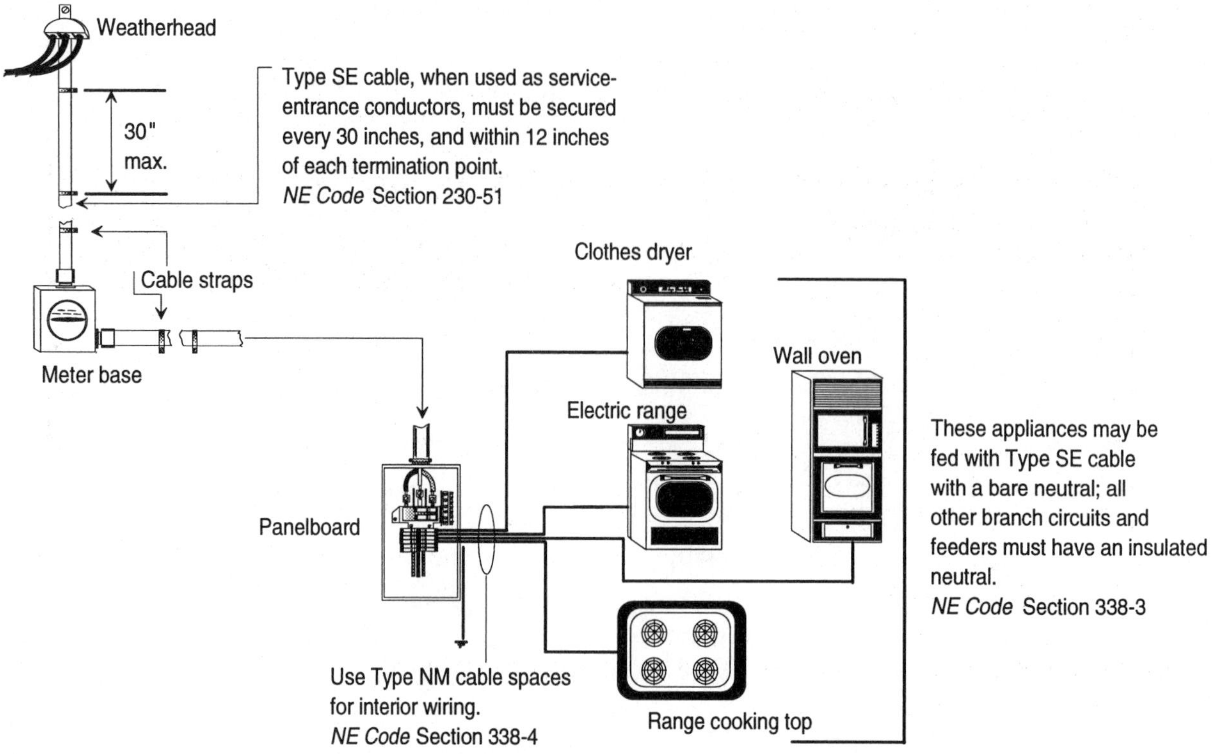
Weatherhead
Type SE cable, when used as service-entrance conductors, must be secured every 30 inches, and within 12 inches of each termination point.
NE Code Section 230-51
30" max.
Cable straps
Meter base
Clothes dryer
Wall oven
Electric range
Panelboard
These appliances may be fed with Type SE cable with a bare neutral; all other branch circuits and feeders must have an insulated neutral.
NE Code Section 338-3
Use Type NM cable spaces for interior wiring.
NE Code Section 338-4
Range cooking top

Figure 6-4: *NE Code* requirements for installing Type SE cable

- Electric range
- Wall-mounted oven
- Counter-mounted cooking unit
- Clothes dryer

Figure 6-4 summarizes the installation rules for Type SE cable — for both exterior and interior wiring.

RACEWAYS

A raceway is any channel used for holding wires, cables, or busbars, which is designed and used solely for this purpose. Types of raceways include rigid metal conduit, intermediate metal conduit (IMC), rigid nonmetallic conduit, flexible metallic conduit, electrical metallic tubing (EMT), and auxiliary gutters. Raceways are constructed of either metal or insulating material such as PVC (plastic). Metal raceways are joined by either threaded, compression, or set-screw couplings; nonmetallic raceways are joined with cement-coated couplings. Where a raceway terminates in an outlet box, junction box, or other enclosure, an approved connector must be used.

Raceways provide mechanical protection for the conductors that run in them and also prevent accidental damage to insulation and the conducting material. They also protect conductors from the harmful chemical attack of corrosive atmospheres and prevent fire hazards to life and property by confining arcs and flames due to faults in the wiring system. Conduit or raceways are used in residential applications for service masts, underground wiring embedded in concrete, and sometimes in unfinished basements, shops or garage areas. Figure 6-5 gives a practical application of conduit used in a residential application.

Another function of metal raceways is to provide a continuous equipment grounding system throughout the electrical system. To maintain this feature, it is extremely important that all raceway

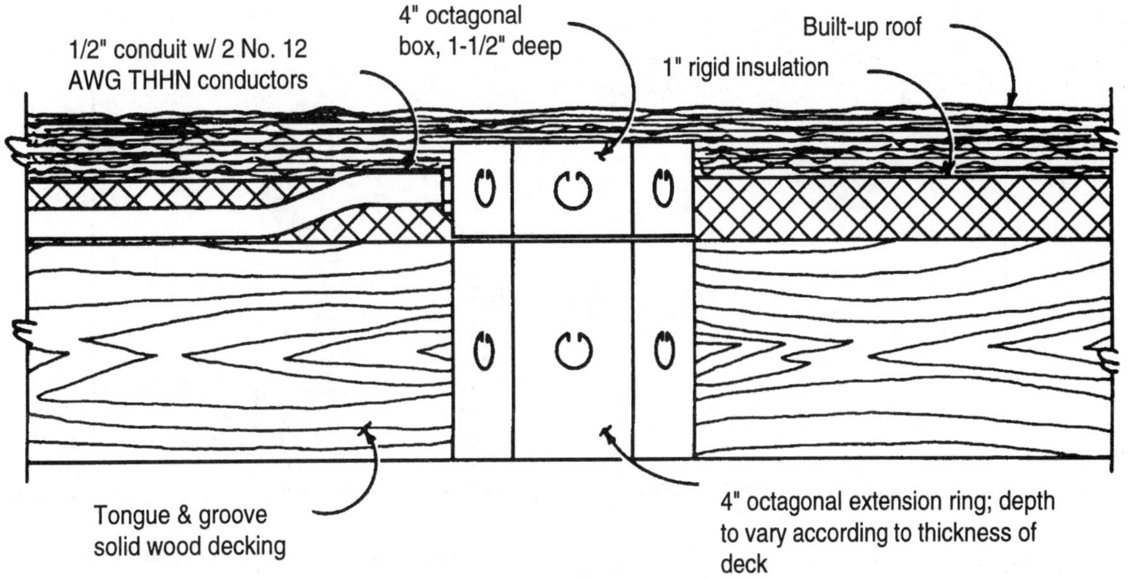
4" octagonal
box, 1-1/2" deep
1/2" conduit w/ 2 No. 12
AWG THHN conductors
Built-up roof
1" rigid insulation
Tongue & groove
solid wood decking
4" octagonal extension ring; depth
to vary according to thickness of
deck

Figure 6-5: Conduit used to feed lighting outlets on a solid wooden deck

systems be securely bonded together into a continuous conductive path and properly connected to the system ground.

Figure 6-6 summarizes the equipment grounding rules for most types of equipment that are required to be grounded. These general *NE Code* regulations apply to all installations except for specific equipment (special applications) as indicated in the *NE Code*. The *NE Code* also lists specific equipment that is to be grounded regardless of voltage.

In all occupancies, major appliances and many handheld appliances and tools are required to be grounded. The appliances include refrigerators, freezers, air conditioners, clothes dryers, washing machines, dishwashing machines, sump pumps, and electrical aquarium equipment. Other tools likely to be used outdoors and in wet or damp locations must be grounded or have a system of double insulation.

Although most appliance circuits require an equipment grounding conductor, the frames of electric ranges, clothes dryers and similar appliances that utilize both 120 and 240 volts may be grounded via the grounded circuit conductor (neutral) under most conditions. In addition, however, the grounding contacts of any receptacles on the equipment must be bonded to the equipment. If these specified conditions are met, it is not necessary to provide a separate equipment grounding conductor, either for the frames or any outlet or junction boxes which are part of the circuit for these applications.

A bonding jumper is sometimes used to assure electrical conductivity between metal parts. When the jumper is installed to connect two or more portions of the equipment grounding conductor, the jumper is referred to as an *equipment bonding jumper*. Some specific cases in which a bonding jumper is required are also listed below:

- Metal raceways, cable armor, and other metal noncurrent-carrying parts that serve as grounding conductors must be bonded whenever necessary in order to assure electrical continuity.
- When flexible metal conduit is used for equipment grounding, an equipment bonding jumper is required if the length

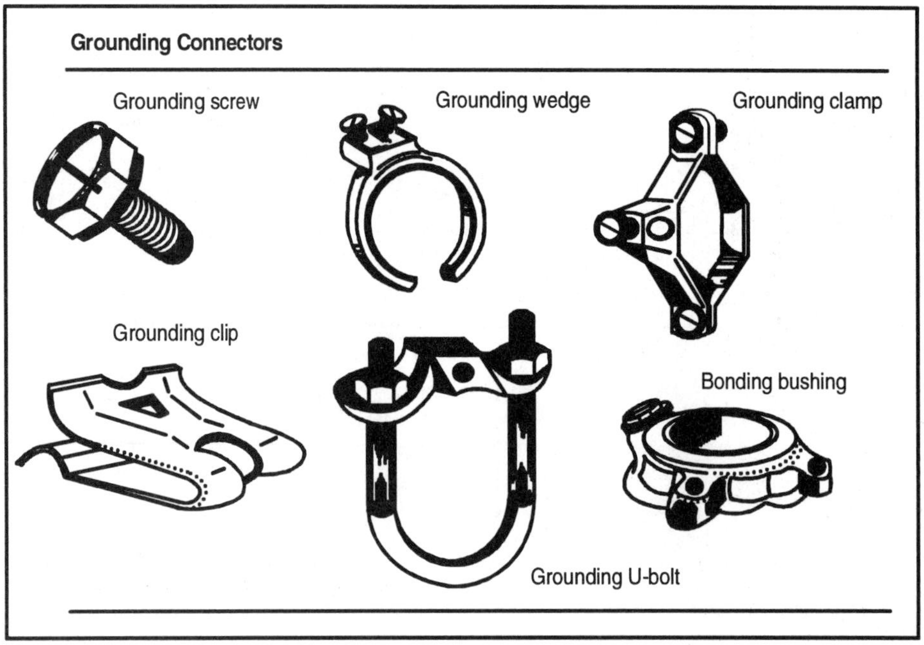
Grounding Connectors
Grounding screw
Grounding wedge
Grounding clamp
Grounding clip
Bonding bushing
Grounding U-bolt

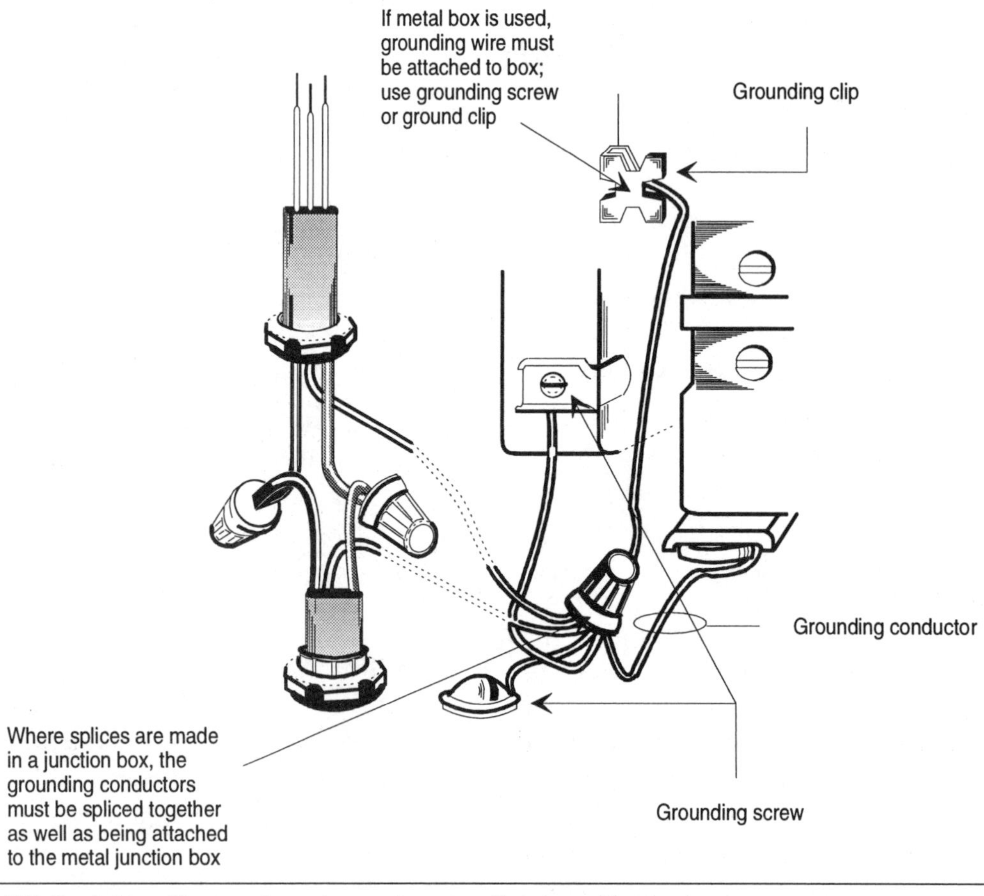
If metal box is used, grounding wire must be attached to box; use grounding screw or ground clip
Grounding clip
Grounding conductor
Where splices are made in a junction box, the grounding conductors must be spliced together as well as being attached to the metal junction box
Grounding screw

Figure 6-6: *NE Code* bonding and equipment-grounding requirements

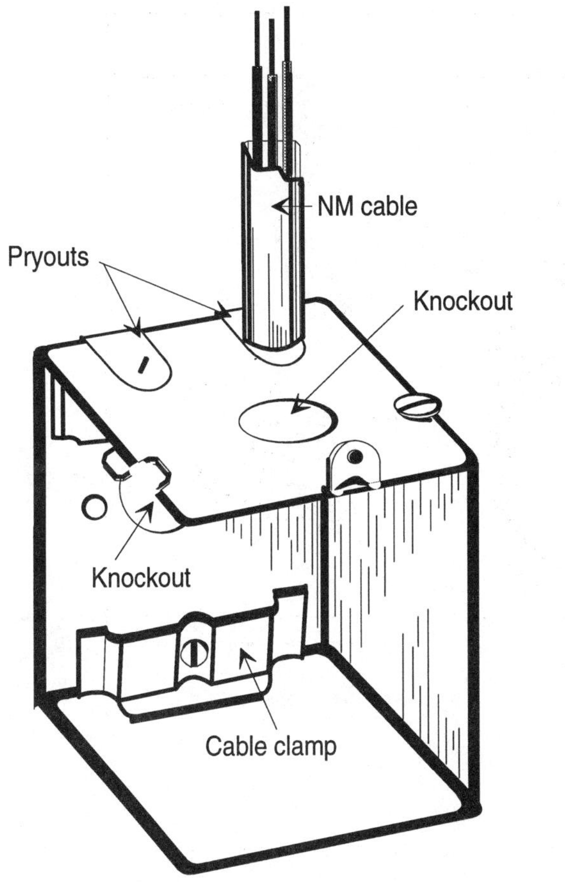
NM cable
Pryouts
Knockout
Knockout
Cable clamp

Figure 6-7: Location of knockouts and pryouts

of the ground return path exceeds 6 feet or if the circuit enclosed is rated over 20 amperes. When the path exceeds 6 feet, the circuit contains an equipment grounding conductor and the bonding may be accomplished by approved fittings.

OUTLET BOXES

Regardless of the wiring method used, an outlet box, cabinet, conduit body or other approved enclosure must be used every time a circuit is spliced or terminated; that is, at each outlet or junction point for all wiring installations. Outlet boxes have traditionally been made from galvanized sheet metal, but many nonmetallic boxes are now finding their way into residential construction. The most common nonmetallic outlet boxes are constructed from PVC and Bakelite (a fiber-reinforced plastic).

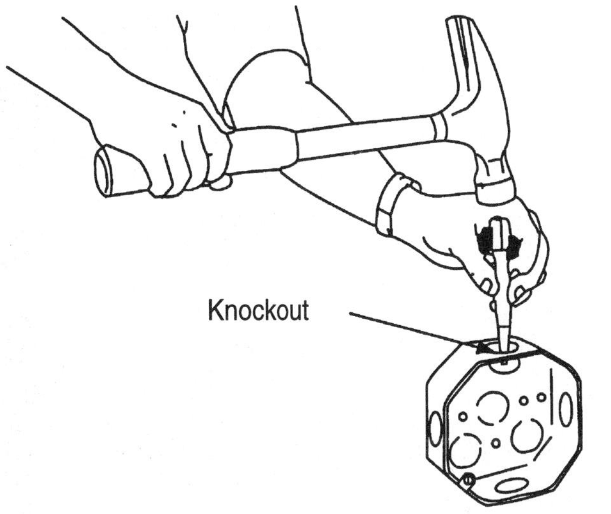
Knockout

Figure 6-8: Removing knockout from metallic box

Metal boxes are made with removable circular sections called knockouts or pryouts as shown in Figure 6-7. These sections are removed to allow conduit or cable to enter the box so that wire connections may be made within the box. When cable is used in metallic boxes, either cable connectors or cable clamps must be used to secure the cable to the box.

Knockouts are easily removed by hitting the knockout section with a screwdriver blade or center punch as shown in Figure 6-8; pryouts are removed by inserting a screwdriver blade into the slot and twisting it to break the solid tabs. See Figure 6-9.

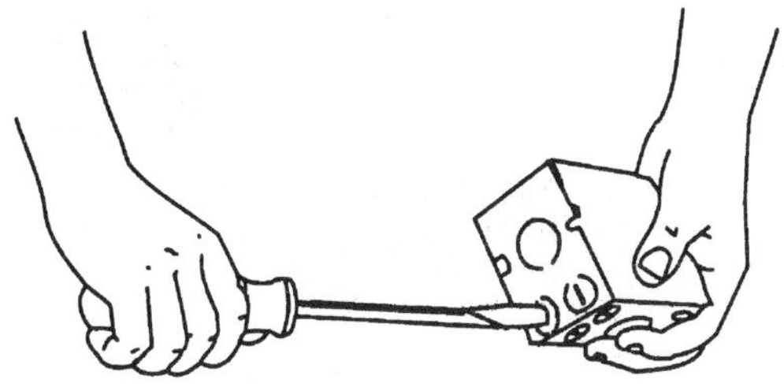

Figure 6-9: Removing pryout from metallic box

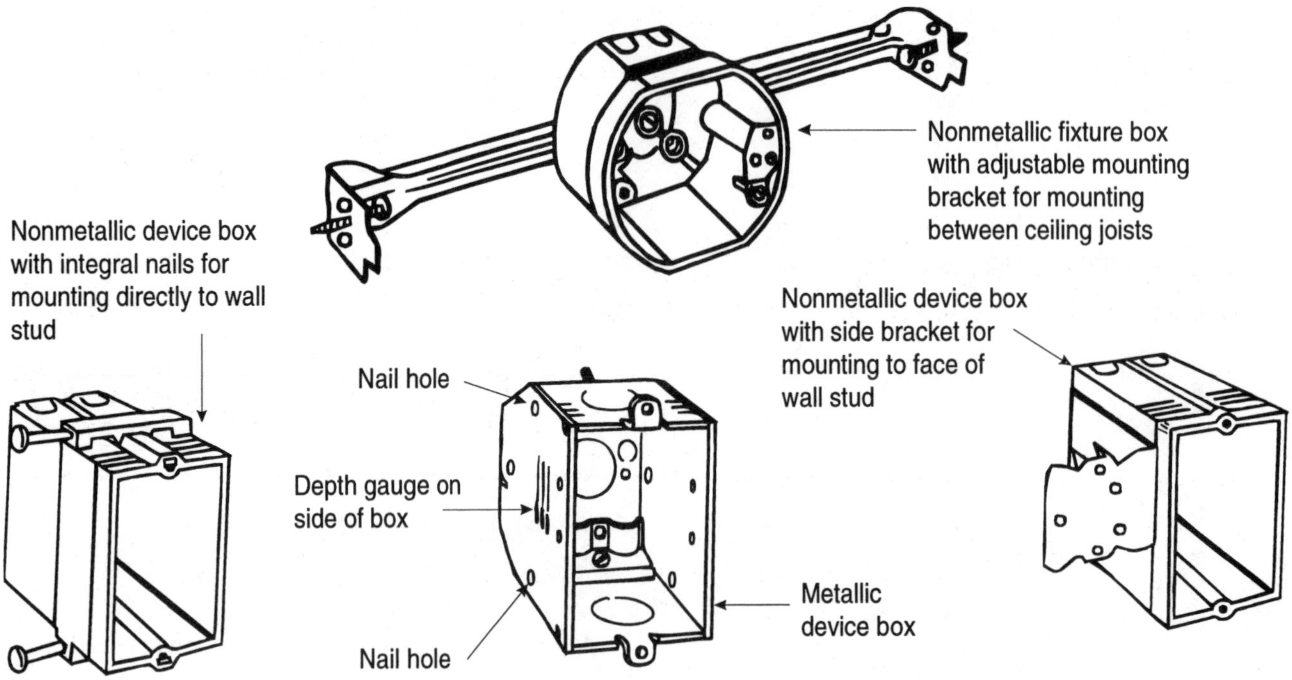
Nonmetallic fixture box with adjustable mounting bracket for mounting between ceiling joists
Nonmetallic device box with integral nails for mounting directly to wall stud
Nonmetallic device box with side bracket for mounting to face of wall stud
Nail hole
Depth gauge on side of box
Nail hole
Metallic device box

Figure 6-10: Several types of outlet boxes used in residential construction

Nonmetallic boxes are not required to have cable clamps provided the cables are secured within 8 inches of the box; the box is no larger than $2\frac{1}{4}$ inches by 4 inches and the cable extends into the box no less than $\frac{1}{4}$ inch. Several types of outlet boxes are shown in Figure 6-10, but these are by no means the extent of those available. Review manufacturers' catalogs.

Chapter 7

Branch-Circuit Layout for Power

The point at which electrical equipment is connected to the wiring system is commonly called an outlet. There are many classifications of outlets: lighting, receptacle, motor, appliance, and the like. This chapter, however, deals with the power outlets normally found in residential electrical wiring systems.

When viewing an electrical drawing, outlets are indicated by symbols—usually a small circle with appropriate markings to indicate the type of outlet. The most common symbols for receptacles are shown in Figure 7-1.

BRANCH CIRCUITS AND FEEDERS

The conductors that extend from the panelboard to the various outlets are called *branch circuits* and are defined by the *NE Code* as ". . . that point of a wiring system extending beyond the final overcurrent device protecting the circuit . . . ." See Figure 7-2.

A *feeder* consists of all conductors between the service equipment and the final overcurrent device. See Figure 7-3.

In general, the size of the branch-circuit conductors varies depending upon the load requirements

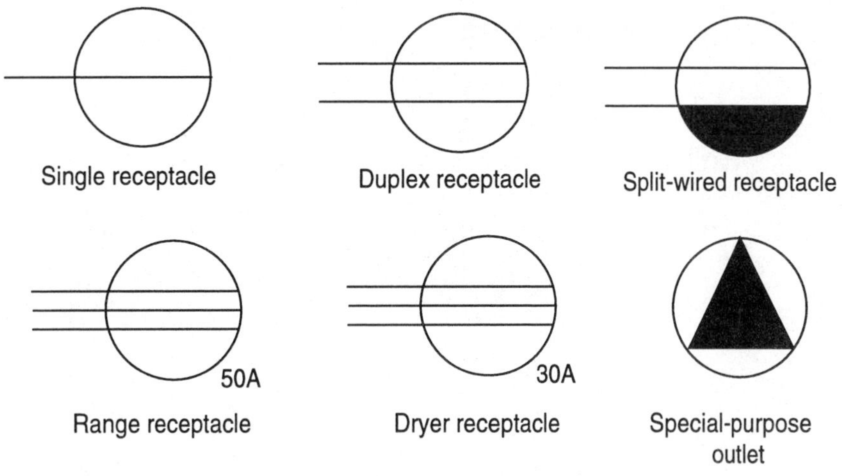

Single receptacle
Duplex receptacle
Split-wired receptacle
50A
Range receptacle
30A
Dryer receptacle
Special-purpose outlet

Figure 7-1: Typical outlet symbols appearing on electrical drawings

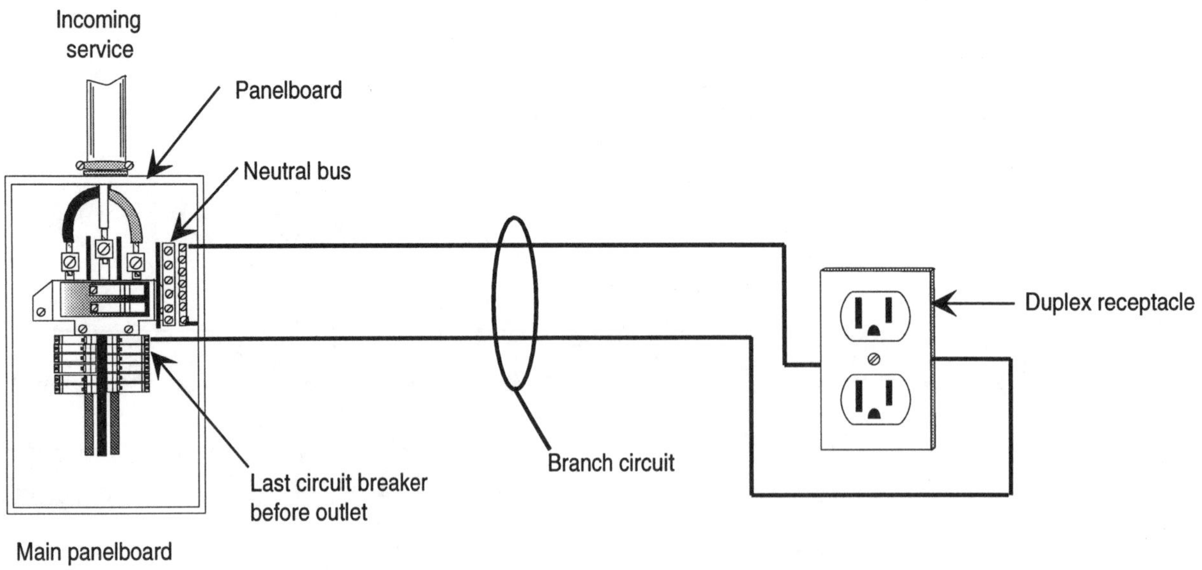
Incoming service
Panelboard
Neutral bus
Duplex receptacle
Branch circuit
Last circuit breaker before outlet
Main panelboard

Figure 7-2: Components of a duplex receptacle branch circuit

of the electrically-operated equipment connected to the outlet. But for residential use, most branch circuits consist of either No. 14 AWG, 12 AWG, 10 AWG, or 8 AWG.

The basic branch circuit requires two wires or conductors to provide a continuous path for the flow of electric current. The usual branch circuits feeding duplex receptacles (convenience outlets) operate at 120 volts utilizing a two-wire branch circuit with an equipment grounding conductor.

Fractional horsepower motors and small electric heaters usually operate at 120 volts also and are connected to 120-volt branch circuits either by means of a receptacle, a junction box, or a direct connection. Larger electric motors, air-conditioning and electric heating equipment, and major

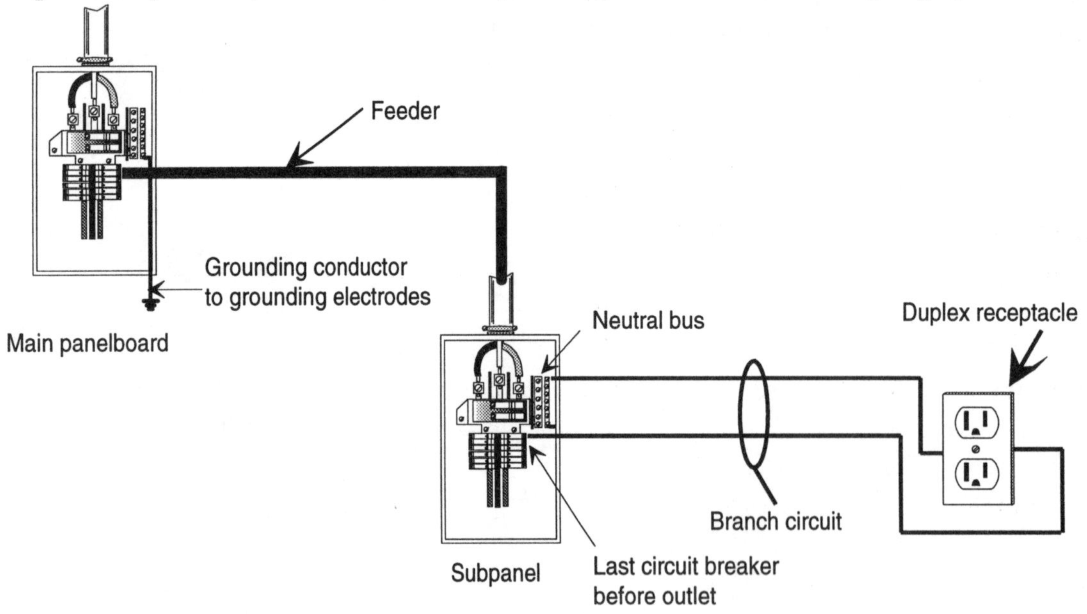
Feeder
Grounding conductor to grounding electrodes
Main panelboard
Neutral bus
Duplex receptacle
Branch circuit
Subpanel
Last circuit breaker before outlet

Figure 7-3: A "feeder" being used to feed a subpanel from the main panelboard

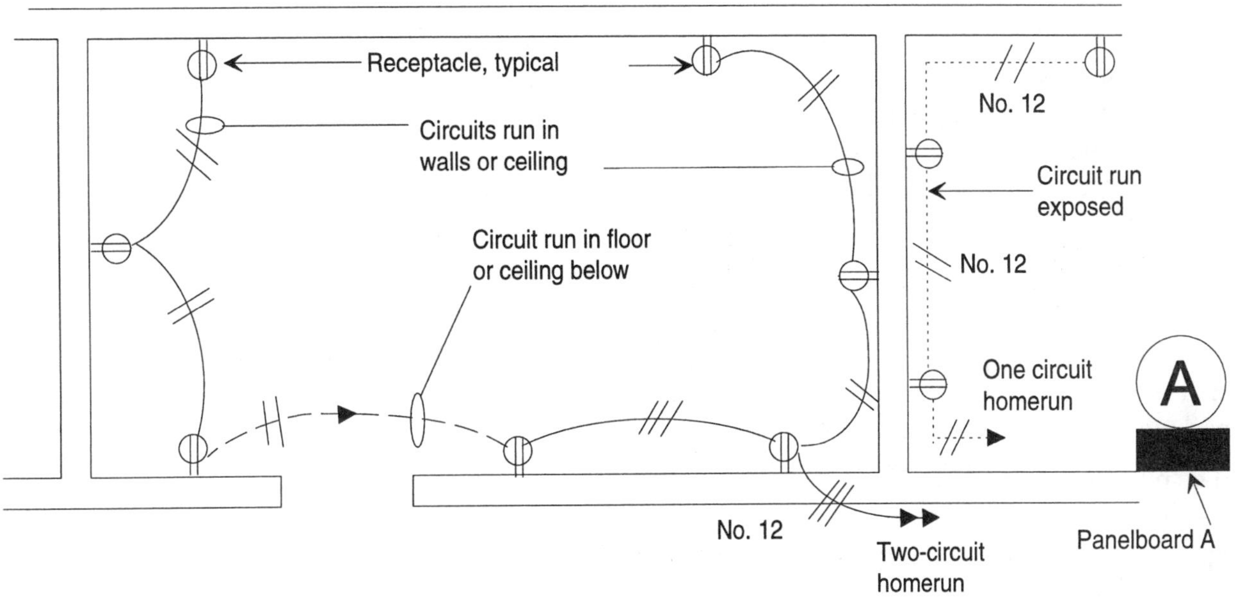

Receptacle, typical
Circuits run in walls or ceiling
Circuit run in floor or ceiling below
No. 12
Circuit run exposed
No. 12
One circuit homerun
A
No. 12
Two-circuit homerun
Panelboard A

Figure 7-4: Types of branch-circuit lines shown on electrical drawings

appliances operate on a two- or three-wire branch circuit — usually at 240 volts.

Branch circuits are shown on electrical drawings by means of a single line drawn from the panelboard (or by "homerun" arrowheads indicating that the circuit goes to the panelboard) to the outlet or from outlet to outlet where there is more than one outlet on the circuit.

The lines indicating branch circuits can be solid to show that the conductors are to be run concealed in the ceiling or wall, slightly broken to show that the conductors are to be run in the floor or ceiling below, or dashed to show that the wiring is to be run exposed. Figure 7-4 shows examples of these three types of branch-circuit lines.

The slash marks cutting the circuits in Figure 7-4 indicate the number of current-carrying conductors in the circuit. The number 12 indicates the wire size. Although two slash marks are shown in Figure 7-4, in actual practice, a branch circuit containing only two conductors usually contains no slash marks; that is, any circuit with no slash marks is assumed to have two conductors. However, three or more conductors are always indicated on electrical working drawings — either by slash marks for each conductor, or else by note.

Never take any instructional text as gospel when it comes to electrical symbols. Although great efforts have been made in recent years to standardize drawing symbols, architects, consulting engineers, and electrical drafters still modify existing symbols or devise new ones to meet their own needs. Always consult the "symbol list" or "legend" on electrical working drawings for an exact interpretation of the symbols used.

LOCATING RECEPTACLES

NE Code Section 210-52 specifically states the minimum requirement for the location of receptacles in residential buildings.

> . . . In every kitchen, family room, dining room . . ., receptacle outlets shall be installed so that no point along the floor line in any wall space is more than 6 feet, measured horizontally, from an outlet in that space, including any wall space 2 feet wide or greater and the wall space

> occupied by sliding panels in exterior walls . . . Receptacle outlets shall, insofar as practical, be spaced equal distance apart. Receptacle outlets in floors shall not be counted as part of the required number of receptacle outlets unless located close to the wall . . . .

The *NE Code* defines "wall space" as an unbroken wall along the floor line by doorways, fireplaces, and similar openings. Each wall space 2 feet or more wide must be treated individually and separately from other wall spaces within the room.

The purpose of the above *NE Code* requirement is to minimize the use of cords across doorways, fireplaces, and similar openings.

With this *NE Code* requirement in mind, outlets for our sample residence will be laid out. In laying out these receptacle outlets, the floor line of the wall is measured (also around corners) but not across doorways, fireplaces, passageways, or other spaces where a flexible cord extended across the space would be unsuitable. In general, duplex receptacle outlets must be no more than 12 feet apart. When spaced in this manner, a 6-foot extension cord will reach a receptacle from any point along the wall line — complying with the latest edition of the *NE Code*.

Figure 7-5 shows all of the duplex receptacles laid out in our sample residence. Note that at no point along the wall line is any receptacle more than 12 feet apart or more than 6 feet from any door or room opening. Where practical, no more than eight receptacles are connected to one circuit. However, *NE Code* Section 220-3(c) specifies a demand factor of 180 volt-amperes per receptacle or group of receptacles. Since a 15-ampere branch-circuit is rated at 1800 volt-amperes (15 amps × 120 volts), 10 general-purpose duplex receptacles may be connected to one circuit.

The utility room has two receptacles: one for the washer on a separate circuit in order to comply with *NE Code* Section 210-52(f), and another for ironing.

One duplex receptacle, connected to the living-room circuit, is located in the vestibule for cleaning purposes, such as for plugging in a portable vacuum cleaner or similar appliance.

The living-room outlets are split-wired — the lower half of each duplex receptacle is "hot" (energized) all the time, while the upper half can be switched on or off. The reason for this is that a great deal of the illumination for this area will be provided by portable table lamps and the split-wired receptacles provide a means of control for them. Note that these split-wired receptacles can be controlled from several locations — at each entry to the living room.

To comply with *NE Code* Section 220-4(b), the kitchen receptacles are laid out according to the following specifications:

> In addition to the number of branch circuits determined previously, two or more 20-ampere small appliance branch circuits must be provided to serve all receptacle outlets, including refrigeration equipment, in the kitchen, pantry, breakfast room, dining room, or similar area of the house. Such circuits, whether two or more are used, must have no other outlets connected to them.

To further meet *NE Code* requirements, one duplex receptacle is installed in the bathroom which is connected to the same circuit as the three outside receptacles. The bathroom receptacle and the three outside receptacles must be connected to a ground-fault circuit-interrupter.

240-VOLT CIRCUITS

The electric range, clothes dryer, and water heater in our sample residence all operate at 240 volts ac. Each will be fed with a separate circuit and connected to a 2-pole circuit breaker of the appropriate rating in the panelboard. To determine the conductor size and overcurrent protection for the range, proceed as described on the following page.

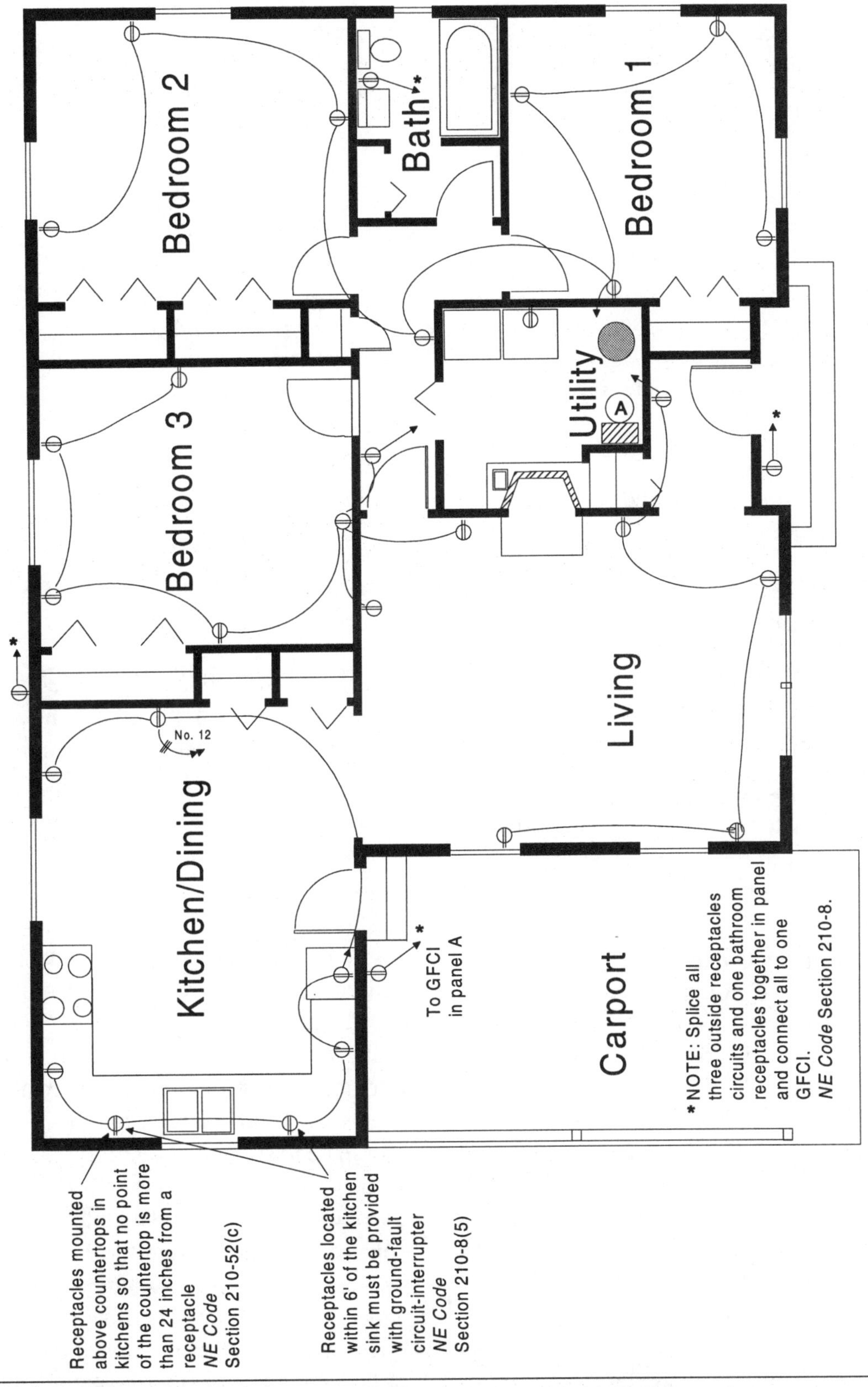
Bedroom 2
Bath
Bedroom 1
Utility
A
Bedroom 3
Living
Kitchen/Dining
No. 12
Carport
To GFCI in panel A
* NOTE: Splice all three outside receptacles circuits and one bathroom receptacles together in panel and connect all to one GFCI. NE Code Section 210-8.
Receptacles mounted above countertops in kitchens so that no point of the countertop is more than 24 inches from a receptacle NE Code Section 210-52(c)
Receptacles located within 6' of the kitchen sink must be provided with ground-fault circuit-interrupter NE Code Section 210-8(5)

Figure 7-5: Floor plan of sample residence with duplex receptacles laid out

- Find the nameplate rating of the electric range. This has previously been determined to be 12 kVA.
- Refer to *NE Code* Table 220-19. Since Column A of this table applies to ranges rated not over 12 kVA, this will be the column to use in this example.
- Under the "Number of Appliances" column, locate the appropriate number of appliances (1 in this case), and find the maximum demand given for it in Column A. Column A states that the circuit should be sized for 8 kVA (not the nameplate rating of 12 kVA).
- Calculate the required conductor ampacity as follows:

$$\frac{8000\ VA}{240\ V} = 33.33\ amperes$$

The minimum branch circuit, however, must be rated at 40 amperes since common residential circuit breakers are rated in steps of 15-, 20-, 30-, 40-, etc. amperes. A 30-ampere circuit breaker is too small, so a 2-P, 40-ampere circuit breaker is selected. The conductors must have a current-carrying capacity equal to or greater than the overcurrent protection. Therefore, No. 8 AWG conductors will be used.

If a cooktop and wall oven were used instead of the electric range, the circuit would be sized similarly. The *NE Code* specifies that a branch circuit for a counter-mounted cooking unit and not more than two wall-mounted ovens, all supplied from a single branch circuit and located in the same room, is computed by adding the nameplate ratings of the individual appliances and treating this total as equivalent to one range. Therefore, two appliances of 6 kVA each may be treated as a single range with a 12 kVA nameplate rating.

Figure 7-6 shows how the electric range circuit may appear on an electrical working drawing. The connection may be made directly to the range junction box, but more often a 50-ampere range receptacle is mounted at the range location whereas a range cord-and-plug set is used to make the connection. This facilitates moving the appliance later for maintenance or cleaning.

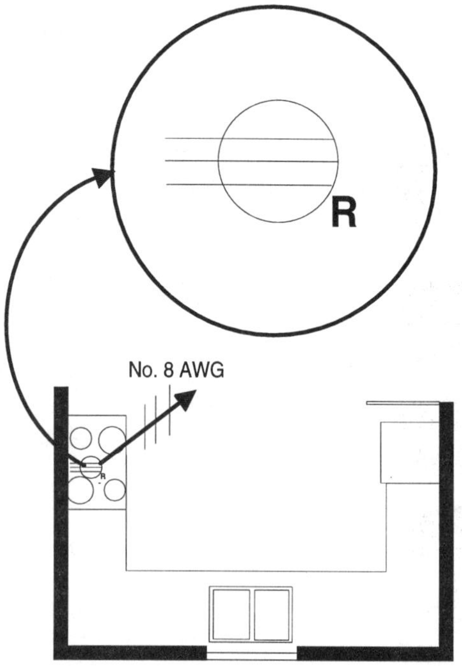
R
No. 8 AWG

Figure 7-6: Range circuit shown on blueprint

Figure 7-7 shows several types of receptacle configurations used in residential wiring applications. The electrical designer should eventually know these configurations at a glance.

The branch circuit for the water heater in the residence under consideration must be sized for its full capacity because there is no diversity or demand factor for this appliance. Since the nameplate rating on the water heater in the residence under consideration indicates two heating elements of 4500 watts each, the first inclination would be to size the circuit for a total load of 9000 watts (volt-amperes). However, only one of the two elements operates at a time. Look at Figure 7-8 for an explanation of how the water-heater controls operate. Note that each element is controlled by a separate thermostat. The lower element becomes energized when the thermostat calls for heat, and at the same time the thermostat opens a contact to prevent the upper element from operating. When the lower-element thermostat is satisfied, the

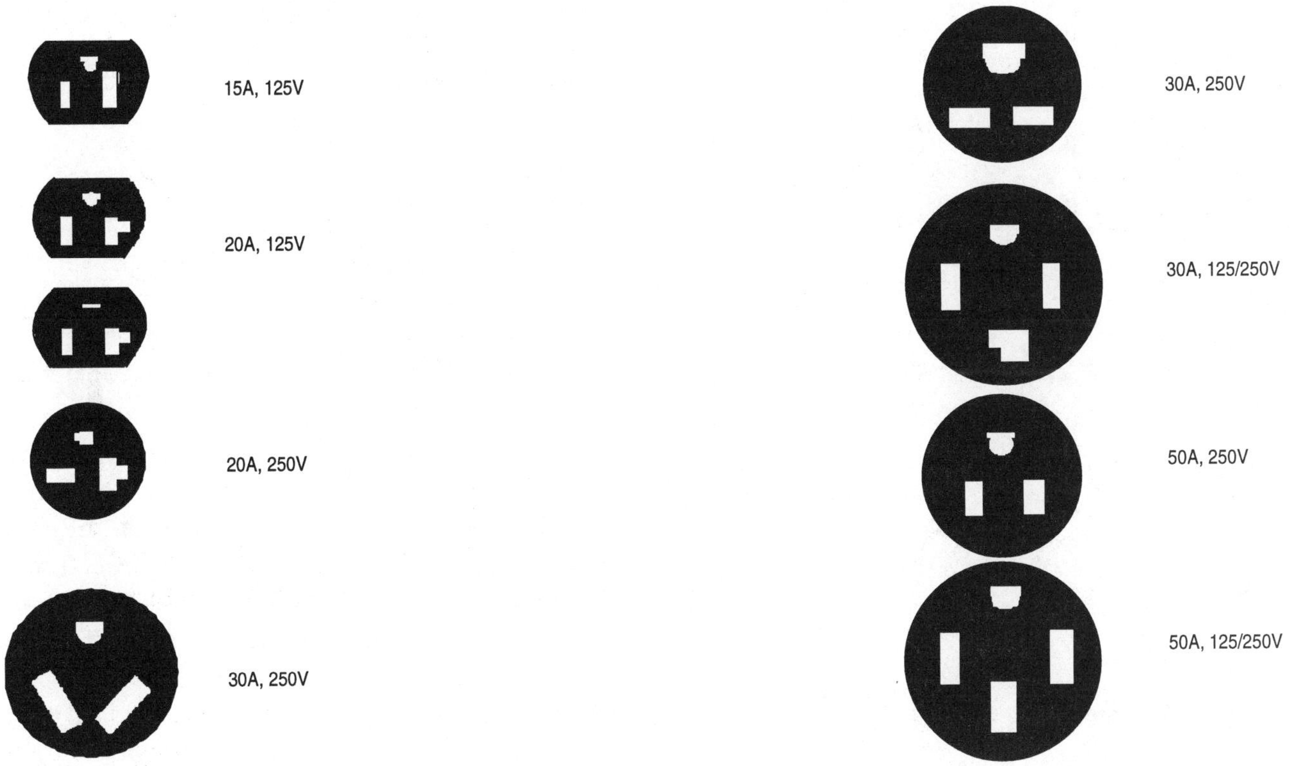
15A, 125V
20A, 125V
20A, 250V
30A, 250V
30A, 250V
30A, 125/250V
50A, 250V
50A, 125/250V

Figure 7-7: Residential receptacle configurations

lower contact opens, and at the same time the thermostat closes the contact for the upper element to become energized to maintain the water temperature.

With this information in hand, the circuit for the water heater may be sized by the equation:

$$\frac{4500\ VA}{240\ V} = 18.75\ A \times 1.25 = 23.43\ amperes$$

Since a water heater will more than likely fall under the "continuous load" category, the conductor and overcurrent protection should be rated at 125%. Therefore, No. 10 AWG wire should be used and protected with a 2-P, 30-ampere circuit breaker. A direct connection is made to water heaters at the integral junction box on top of the heater.

The *NE Code* specifies that electric clothes dryers must be rated at 5 kVA or the nameplate rating — whichever is greater. In our case, the dryer is rated at 5.5 kVA and the conductor current-carrying capacity is calculated as follows:

$$\frac{5500\ VA}{240\ V} = 22.91\ amperes$$

A 3-wire, 30-ampere circuit will be provided (No. 10 AWG wire) and also protected by a 2-P, 30-ampere circuit breaker. The dryer may be connected directly, but a 30-ampere dryer receptacle is normally provided for the same reasons as mentioned for the electric range.

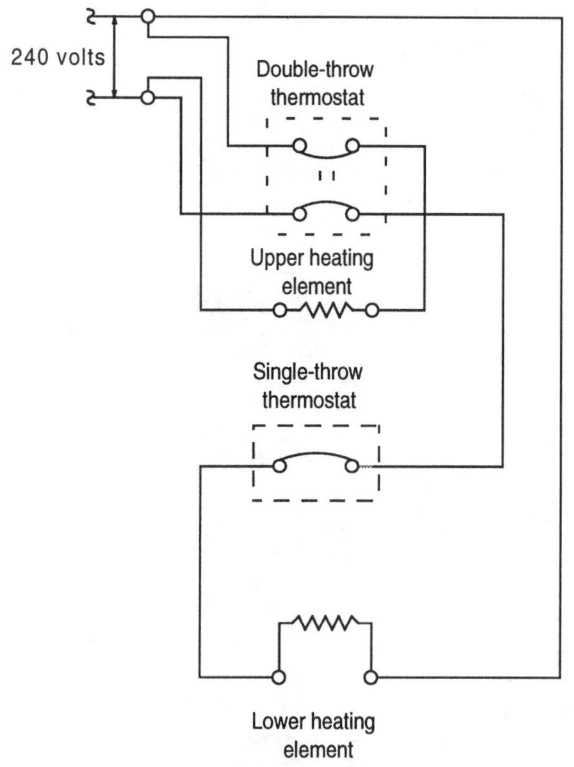
240 volts
Double-throw
thermostat
Upper heating
element
Single-throw
thermostat
Lower heating
element

Figure 7-8: Water-heater controls

Large-appliance outlets — rated at 240 volts — are frequently shown on the electrical floor plan, using lines and symbols to indicate the outlets and circuits. Sometimes, however, such circuits are shown in the form of riser diagrams as shown in Figure 7-9.

OUTLET BOXES

Electricians installing residential electrical systems must be familiar with outlet-box capacities, means of supporting outlet boxes, and other requirements of the *NE Code*. A general review of the rules and necessary calculations follows.

The maximum number of conductors permitted in standard outlet boxes is listed in *NE Code* Table 370-16(a). These figures apply where no fittings or devices such as fixture studs, cable clamps, switches, or receptacles are contained in the box and where no grounding conductors are part of the

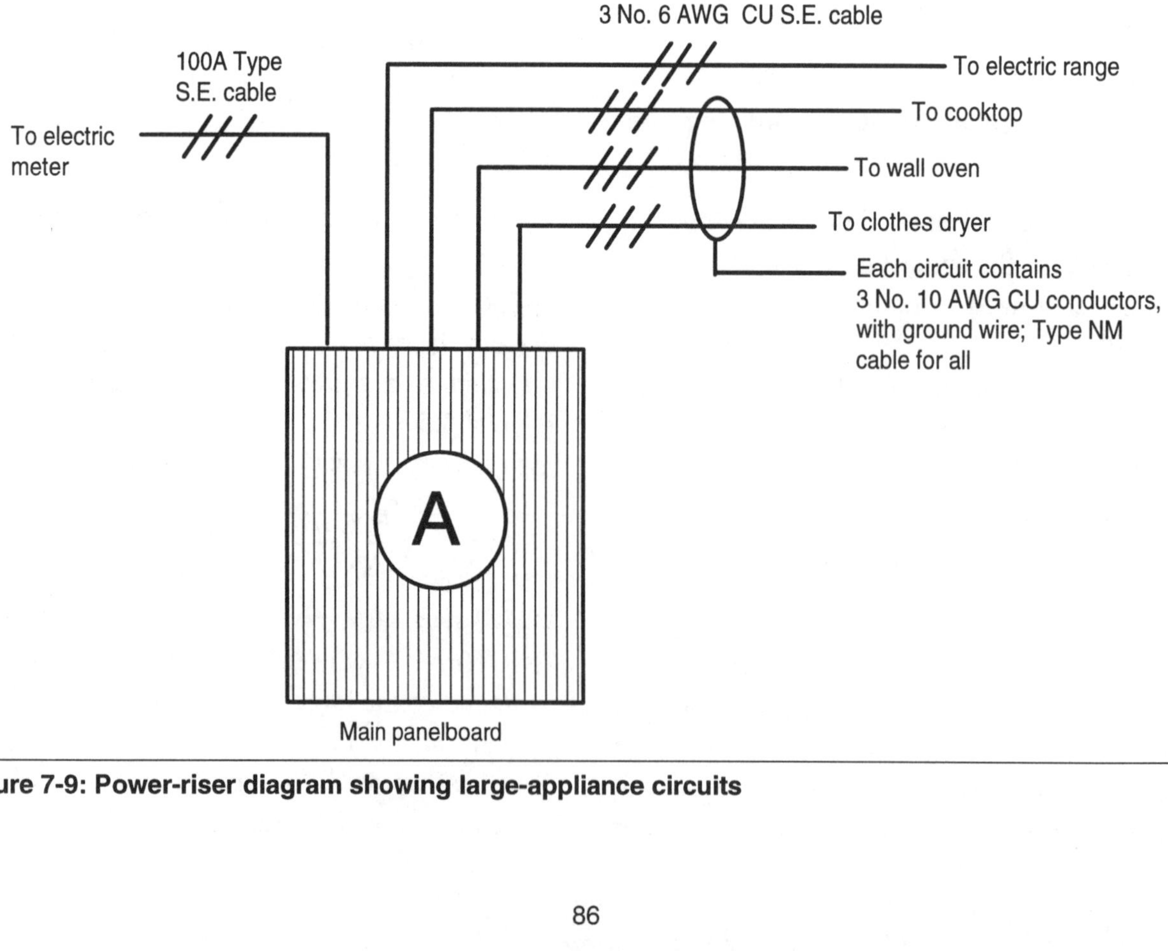
3 No. 6 AWG CU S.E. cable
100A Type
S.E. cable
To electric
meter
To electric range
To cooktop
To wall oven
To clothes dryer
Each circuit contains
3 No. 10 AWG CU conductors,
with ground wire; Type NM
cable for all
A
Main panelboard

Figure 7-9: Power-riser diagram showing large-appliance circuits

wiring within the box. Obviously, in all modern residential wiring systems there will be one or more of these items contained in every outlet box installed. Therefore, where one or more of the above-mentioned items are present, the number of conductors must be one less than shown in the table.

For example, a deduction of one conductor must be made for each strap containing a wiring device such as a switch or duplex receptacle; a further deduction of one conductor must be made for one or more equipment grounding conductors entering the box. A 3-inch × 2-inch × 2¾-inch box is listed in the table as containing a maximum number of six No. 12 wires. If the box contains cable clamps and a duplex receptacle, two wires will have to be deducted from the total of six—providing for only four No. 12 wires. If a ground wire is used—which is always the case in residential wiring—only three No. 12 wires may be used.

With the above rules in mind, let's see what sizes of outlet boxes are necessary for the receptacle circuits in our sample residence. Refer back to Figure 7-5 and note that the receptacle-box configurations consist of the following:

- Two No. 12 AWG conductors with ground wire
- Four No. 12 AWG conductors with ground wire
- Six No. 12 AWG conductors with ground wire
- Seven No. 12 AWG conductors with ground wire
- Two No. 14 AWG conductors with ground wire
- Four No. 14 AWG conductors with ground wire
- Six No. 14 AWG conductors with ground wire

To start with, let's take two No. 12 AWG conductors with ground wire and size a metallic outlet box that will meet *NE Code* requirements. In doing so, the first step is to count the total number of conductors and equivalents that will be used in the box — following the regulation specified in *NE Code* Section 370-16.

Note: In former editions of the NE Code, this section was 370-6. However, this has been changed in the 1993 NE Code to Section 370-16. Consequently, if an older edition of the NE Code is being used, refer to Section 370-6.

1. Calculate the total number of conductors and equivalents.

One ground wire	=	1
One cable clamp	=	1
One receptacle	=	2
Two #12 conductors	=	2
Total #12 conductors	=	6

2. Determine amount of space required for each conductor.

NE Code Table 370-16(b) gives the box volume required for each conductor:

No. 12 AWG = 2.25 cubic inches

3. Calculate the outlet box space required by multiplying the number of cubic inches required for each conductor by the number of conductors found in No. 1 above.

$$5 \times 2.25 = 11.25 \text{ cubic inches}$$

Once you have determined the required box capacity, again refer to *NE Code* Table 370-16(a) and note that a 3 × 2 × 2½-inch box comes closest to our requirements. This size box is rated for 12.5 cubic inches.

Where four No. 12 conductors enter the box with two ground wires, only the two additional No. 12 conductors must be added to our previous count for a total of (5 + 2 =) 7 conductors. Remember, any number of ground wires in a box counts as only one conductor; any number of cable clamps count as only one conductor. Therefore, the size box required for use with two No. 12 conductors may be calculated as follows:

$$7 \times 2.25 = 15.75 \text{ cubic inches}$$

Again, refer to *NE Code* Table 370-16(a) and note that a 3 × 2 × 3½-inch device box with a rated capacity of 18.0 cubic inches, is the closest device box that meets *NE Code* requirements. An alternate is to use a 4 × 1¼-inch square box with a single-gang plaster ring. This size box also has a capacity of 18.0 cubic inches.

The remaining box sizes are calculated in a similar fashion except when sizing boxes for No. 14 AWG conductors, the required free space within the box for each conductor drops to 2.0 cubic inches instead of 2.25 cubic inches for No. 12 conductors.

MOUNTING OUTLET BOXES

If you would look at any of the outlet box manufacturer's illustrated catalogs, you would be astonished! Outlet-box configurations are almost endless. If you research the various methods of mounting these boxes, you will be further astonished.

The conventional metallic device box, used for residential duplex receptacles and switches for lighting control, may be mounted to wall studs using 16d (penny) nails placed through the round mounting holes and passing through the interior of the box. The nails are then driven into the wall stud. When nails are used for mounting outlet boxes in this manner, the nails must be located within ¼ inch of the back or ends of the enclosure.

Nonmetallic boxes normally have mounting nails fitted to the box for mounting. Other boxes have mounting brackets of many different configurations for mounting the box. When mounting outlet boxes with brackets, use either wide-head roofing nails or box nails about 1¼ inches in length.

Before mounting any boxes during the rough wiring process, first find out what type and thickness of finish will be used on the walls. This will dictate the depth to which the boxes must be mounted to comply with *NE Code* regulations. For example, the finish on plastered walls or ceilings is normally ½-inch thick; gypsum board or drywall is either ½- or ⅝-inch thick, while wood paneling is only ¼ inch. Some tongue-and-groove wood paneling is ½- to ⅝-inch thick.

The *NE Code* specifies the amount of space permitted from the edge of the outlet box to the finished wall. When a noncombustible wall finish, such as plaster, masonry, or tile, is used, the box may be recessed ¼ inch. However, when combustible finishes are used, such as wood paneling, the box must be flush (even with) with the finished wall or ceiling.

When Type NM cable is used in either metallic or nonmetallic outlet boxes, the cable assembly, including the sheath, must extend into the box no less than ¼ inch (*NE Code* Section 370-17c). In all instances, all permitted wiring methods must be secured to the boxes by means of either cable clamps or approved connectors. There is, however, one exception to this rule:

> Where Type NM cable is used with nonmetallic boxes no larger than a nominal size 2¼ inch by 4 inch mounted in walls and where the cable is fastened within 8 inches of the box (this is very important), the cable does not have to be secured to the box. See *NE Code* Section 370-17(c), *Exception*.

WIRING DEVICES

Wiring devices include various types of receptacles and switches, the latter being used for lighting control. Switches are covered in *NE Code* Article 380, while regulations for receptacles may be found in *NE Code* Section 210-7 and Article 410-L.

Receptacles

Receptacles are rated by voltage and amperage capacity. *NE Code* Section 210-7 requires that receptacles connected to a 15- or 20-ampere circuit have the correct voltage and current rating for the application, and be of the grounding type.

Where there is only one outlet on a circuit, the receptacle's rating must be equal to or greater than the capacity of the conductors feeding it. Therefore, if one receptacle is connected to, say, a 20-ampere residential laundry circuit, the receptacle must be rated at 20 amperes or more. When more than one outlet is on a circuit, the total connected load must be equal to or less than the capacity of the branch-circuit conductors feeding the receptacles.

Receptacles have various symbols and information inscribed on them that help to determine their proper use and ratings. For example, Figure 7-10 shows a standard duplex receptacle and contains the following printed inscriptions:

- The testing laboratory label
- The CSA (Canadian Standards Association) label
- Type of conductor for which the terminals are designed
- Current and voltage ratings, listed by maximum amperage, maximum voltage, and current restrictions

The testing laboratory label is an indication that the device has undergone extensive testing by a nationally recognized testing lab and has met with the minimum safety requirements. The label does not indicate any type of quality rating. The receptacle in Figure 7-10 is marked with the "UL" label which indicates that the device type was tested by Underwriters' Laboratories, Inc. of Northbrook, IL. ETL Testing Laboratories, Inc. of Cortland, NY is another nationally recognized testing laboratory. They provide a labeling, listing and follow-up service for the safety testing of electrical products to nationally recognized safety standards or specifically designated requirements of jurisdictional authorities.

The CSA (Canadian Standards Association) label is an indication that the material or device has undergone a similar testing procedure by the Canadian Standards Association and is acceptable for use in Canada.

Current and voltage ratings are listed by maximum amperage, maximum voltage and current restriction. On the device shown in Figure 7-10, the maximum current rating is 15 amperes at 125 volts — the latter of which is the maximum voltage allowed on a device so marked.

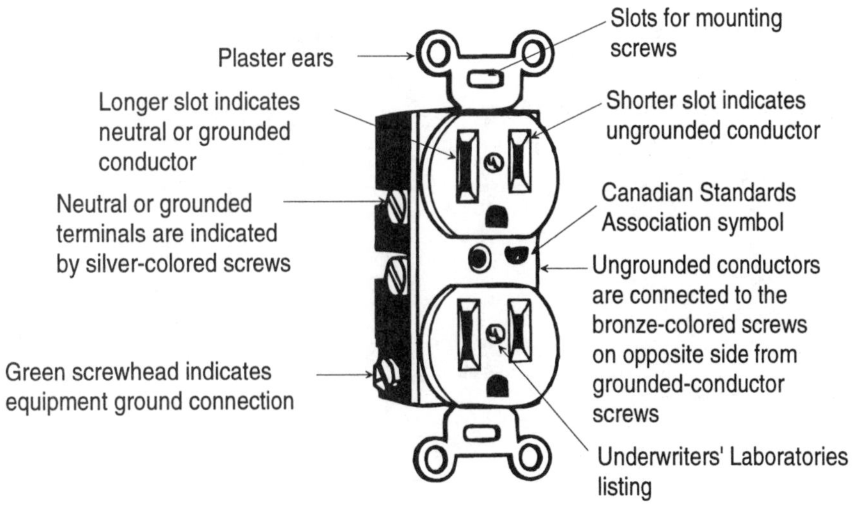
Plaster ears
Longer slot indicates neutral or grounded conductor
Neutral or grounded terminals are indicated by silver-colored screws
Green screwhead indicates equipment ground connection
Slots for mounting screws
Shorter slot indicates ungrounded conductor
Canadian Standards Association symbol
Ungrounded conductors are connected to the bronze-colored screws on opposite side from grounded-conductor screws
Underwriters' Laboratories listing

Figure 7-10: Characteristics of a typical duplex receptacle

Conductor markings are also usually found on duplex receptacles. Receptacles with quick-connect wire clips will be marked "Use #12 or #14 solid wire only." If the inscription "CO/ALR" is marked on the receptacle, either copper, aluminum, or copper-clad aluminum wire may be used. The letters "ALR" stand for "aluminum revised." Receptacles marked with the inscription "CU/AL" should be used for copper only, although they were originally intended for use with aluminum also. However, such devices frequently failed when connected to 15- or 20-ampere circuits. Consequently, devices marked with "CU/AL" are no longer acceptable for use with aluminum conductors.

The remaining markings on duplex receptacles may include the manufacturer's name or logo, "Wire Release" inscribed under the wire-release slots, and the letters "GR." beneath or beside the green grounding screw.

The screw terminals on receptacles are color-coded. For example, the terminal with the green screw head is the equipment ground connection and is connected to the U-shaped slots on the receptacle. The silver-colored terminal screws are for connecting the grounded or neutral conductors and are associated with the longer of the two vertical slots on the receptacle. The brass-colored terminal screws are for connecting the ungrounded or "hot" conductors and are associated with the shorter vertical slots on the receptacle.

Note: *The long vertical slot accepts the grounded or neutral conductor while the shorter vertical slot accepts the ungrounded or hot conductor.*

Mounting Receptacles

Although no actual *NE Code* requirements exist on mounting heights and positioning receptacles, there are certain installation methods that have become "standard" in the electrical industry. Figure 7-11 shows mounting heights of duplex receptacles used on conventional residential and small commercial installations. However, do not take these dimensions as gospel; they are frequently varied to suit the building structure. For example, ceramic tile might be placed above a kitchen or bathroom countertop. If the dimensions in Figure 7-11

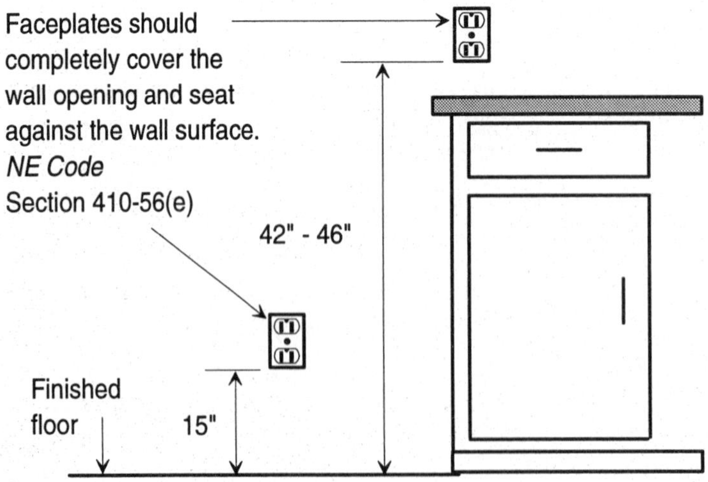
Faceplates should completely cover the wall opening and seat against the wall surface. *NE Code* Section 410-56(e)
42" - 46"
Finished floor
15"

Figure 7-11: Recommended mounting heights of duplex receptacles

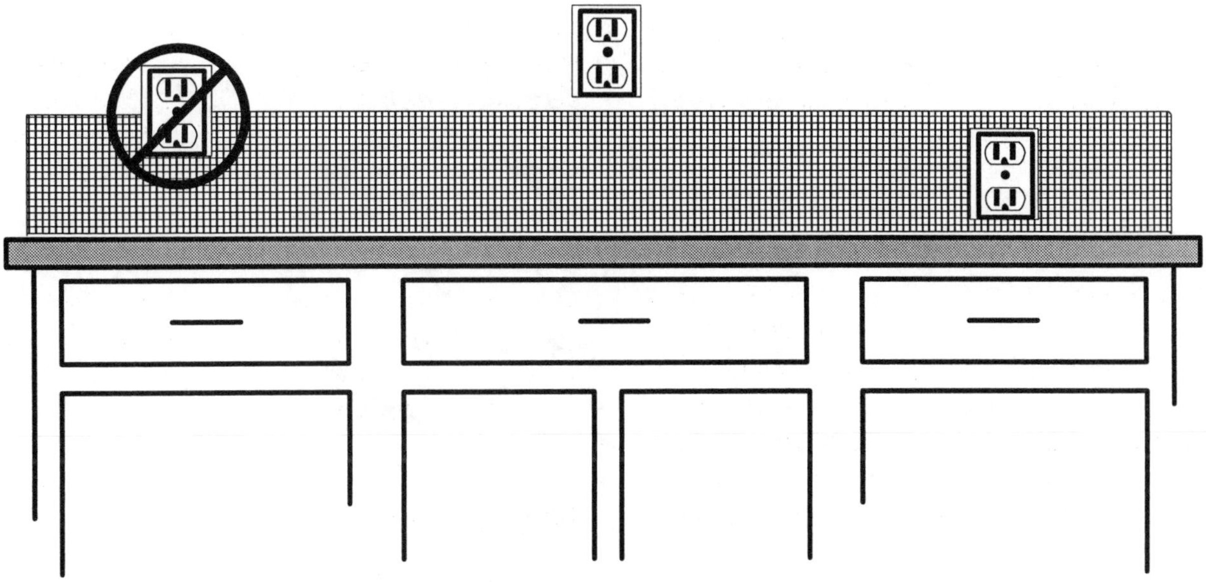

Figure 7-12: Adjust mounting heights so receptacles will either be completely in or completely out of the tile

puts the receptacle part of the way out of the tile, say, half in and half out, the mounting height should be adjusted to either place the receptacle completely in the tile or completely out of the tile as shown in Figure 7-12.

Refer again to Figure 7-11 and note that the mounting heights are given to the bottom of the outlet box. Many dimensions on electrical drawings are given to the center of the outlet box or receptacle. However, during the actual installation, workers installing the outlet boxes can mount them more accurately (and in less time) by lining up the bottom of the box with a chalk mark rather than trying to "eyeball" this mark to the center of the box.

A decade or so ago, most electricians mounted receptacle outlets 12 inches from the finished floor to the center of the outlet box. However, a recent survey taken of over 500 homeowners shows that they prefer a mounting height of 15 inches from the finished floor to the bottom of the outlet box. It is easier to plug and unplug the cord-and-plug assemblies at this height — especially among senior citizens and those homeowners who are confined to wheelchairs. However, always check the working drawings, written specifications, and details of construction for measurements that may affect the mounting height of a particular receptacle outlet.

There is always the possibility of a metal receptacle cover coming loose and falling downward onto the blades of an attachment plug cap that may be loosely plugged into the receptacle. By the same token, a hairpin, fingernail file, metal flyswatter handle, or any other metal object may be knocked off a table and fall downward onto the plug blades. Any of these objects could cause a short-circuit if the falling metal object fell on both the "hot" and grounded neutral blades of the plug at the same time. For these reasons, it is recommended that the equipment grounding slot in receptacles be placed at the top. In this position, any falling metal object would fall onto the grounding blade which would more than likely prevent a short-circuit. See Figure 7-13.

When duplex receptacles are mounted in a horizontal position, the grounded neutral slots should be on top for the same reasons as discussed previously. Again, see Figure 7-13.

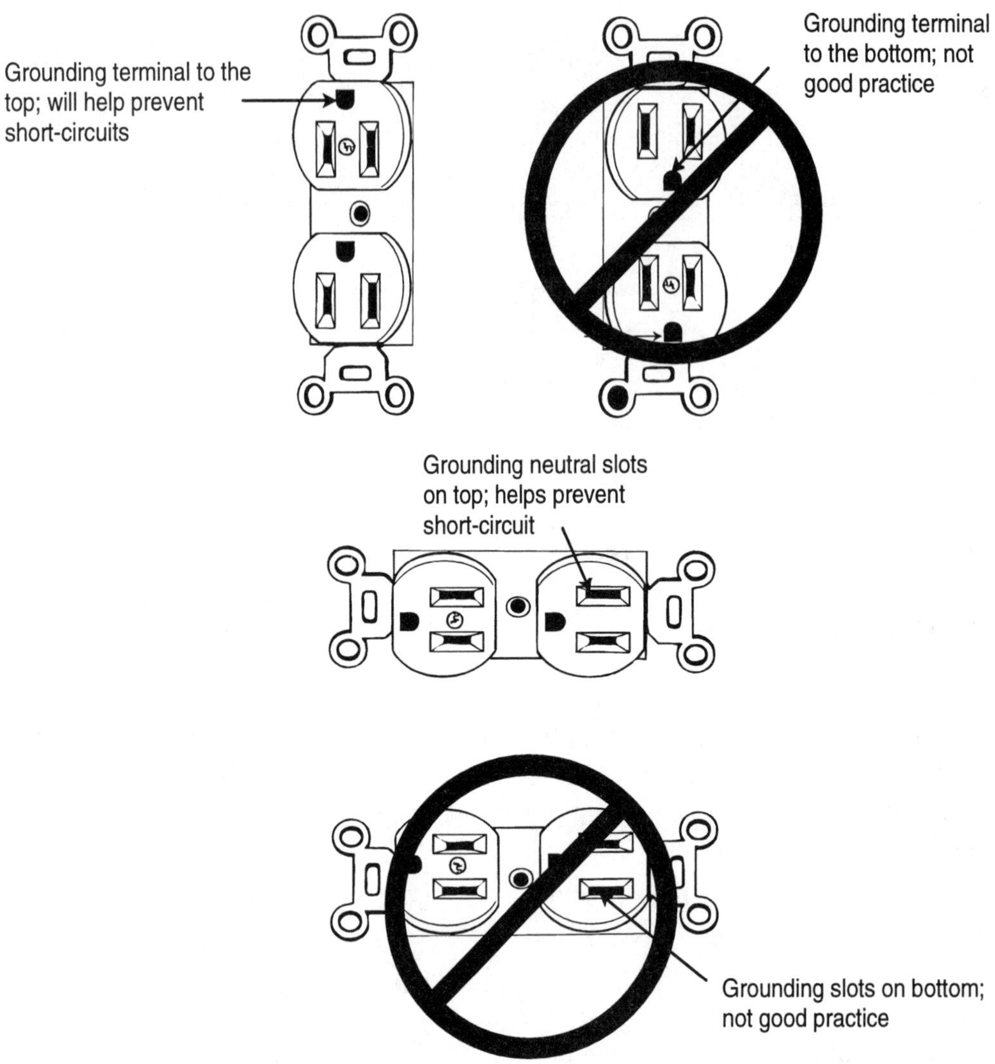
Grounding terminal to the top; will help prevent short-circuits
Grounding terminal to the bottom; not good practice
Grounding neutral slots on top; helps prevent short-circuit
Grounding slots on bottom; not good practice

Figure 7-13: Recommended mounting positions for duplex receptacles

NE Code Section 370-20 requires all outlet boxes installed in walls or ceiling of concrete, tile, or other noncombustible material such as plaster or drywall to be installed in such a matter that the front edge of the box or fitting will not set back of the finished surface more than $\frac{1}{4}$ inch. Where walls and ceilings are constructed of wood or other combustible material, outlet boxes and fittings must be flush with the finished surface of the wall. See Figure 7-14.

Wall surfaces such as drywall, plaster, etc. that contain wide gaps or are broken, jagged, or otherwise damaged must be repaired so there will be no gaps or open spaces greater than $\frac{1}{8}$ inch between the outlet box and wall material. These repairs should be made prior to installing the faceplate. Such repairs are best made with a noncombustible caulking or spackling compound. See Figure 7-15.

Types of Receptacles

There are many types of receptacles. For example, the duplex receptacles that have been discussed are the straight-blade type which accepts a

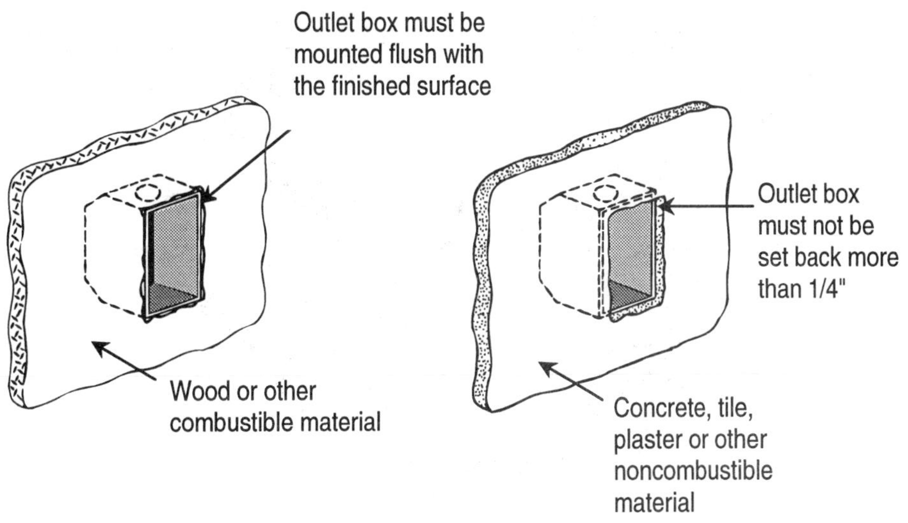
Outlet box must be mounted flush with the finished surface
Outlet box must not be set back more than 1/4"
Wood or other combustible material
Concrete, tile, plaster or other noncombustible material

Figure 7-14: *NE Code* requirements for mounting outlet boxes in walls or ceilings

straight blade connector or plug. This is the most common type of receptacle and such receptacles are found on virtually all electrical projects from residential to large industrial installations.

Twist lock receptacles: Twist lock receptacles are designed to accept a somewhat "curved blade" connector or plug. The plug/connector and the receptacle will lock together with a slight twist.

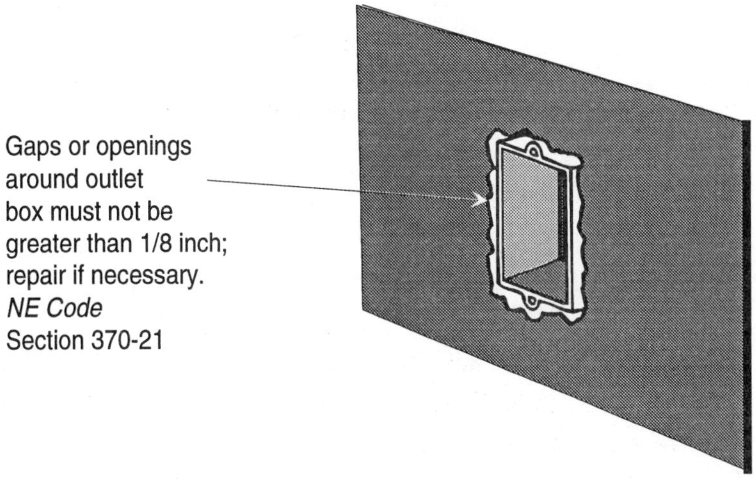
Gaps or openings around outlet box must not be greater than 1/8 inch; repair if necessary.
NE Code
Section 370-21

Figure 7-15: Gaps or openings around outlet boxes must be repaired

The locking prevents accidentally unplugging the equipment.

Pin-and-sleeve receptacles: Pin-and-sleeve devices have a unique locking feature. These receptacles are made with an extremely heavy-duty plastic housing that makes them highly indestructible. They are manufactured with long brass pins for long life and are color-coded according to voltage for easy identification.

Low-voltage receptacles: These receptacles are designed for both ac and dc systems where the maximum potential is 50 volts. Receptacles used for low-voltage systems must have a minimum current-carrying rating of 15 amperes.

Chapter 8
Residential Lighting

A simple lighting branch circuit requires two conductors to provide a continuous path for current flow. The usual lighting branch circuit operates at 120 volts; the white (grounded) circuit conductor is therefore connected to the neutral bus in the panelboard, while the black (ungrounded) circuit conductor is connected to an overcurrent protection device.

Lighting branch circuits and outlets are shown on electrical drawings by means of lines and symbols respectively; that is, a single line is drawn from outlet to outlet and then terminated with an arrowhead to indicate a homerun to the panelboard. Several methods are used to indicate the number and size of conductors, but the most common is to indicate the number of conductors in the circuit by slash marks through the circuit lines and then indicate the wire size by a notation adjacent to these slash marks. For example, two slash marks indicate two conductors; three slash marks indicate three conductors, etc. Some electrical designers omit slash marks for two-conductor circuits, stating in the symbol list or legend, "no slash marks indicate two No. 14 AWG conductors."

The circuits used to feed residential lighting must conform to standards established by the *NE Code* as well as by local and state ordinances. Most of the lighting circuits should be calculated to include the total load, although at times, this is not possible because the electrician cannot be certain of the exact wattage that might be used by the homeowner. For example, an electrician may install four porcelain lampholders for the unfinished basement area, each to contain one 100-watt incandescent lamp. However, in actual use the homeowners may eventually replace the original lamps with others rated at 150 watts or even 200 watts. Thus, if the electrician loaded the lighting circuit to full capacity initially, the circuit would probably become overloaded in the future.

From the above example, it is recommended that no residential branch circuit be loaded to more than 80% of its rated capacity. Since most circuits used for lighting are rated at 15 amperes, the total ampacity (in volt-amperes) for the circuit is 15 (amperes) × 120 (volts) = 1800 volt-amperes. Therefore, if the circuit is to be loaded to only 80% of its rated capacity, the maximum initial connected load should be no more than 1440 volt-amperes.

Electrical symbols are used to show the fixture types. Switches and lighting branch circuits are also shown by appropriate lines and symbols. The meaning of the symbols used on drawings in this chapter is explained in the symbol list in Figure 8-1. Figure 8-2 shows one possible lighting arrangement for our sample residence. All lighting fixtures are shown in their approximate physical location, as they should be installed in the residence. Other lighting possibilities follow.

Electric light in the home greatly improves the appearance of the home as well as people and

Symbol	Description
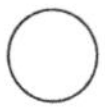	Surface-mounted ceiling lighting fixture with incandescent lamp.
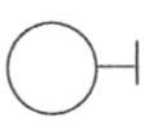	Surface-mounted wall lighting fixture with incandescent lamp.
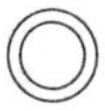	Recessed ceiling lighting fixture with incandescent lamp.
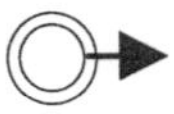	Directional recessed ceiling lighting fixture with incandescent lamp. Arrow indicates direction that lamp is pointed.
	Surface-mounting ceiling lighting fixture with fluorescent lamp.
S	Single-pole switch.
S_3	Three-way switch.
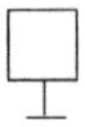 DS	Door actuated switch.

Figure 8-1: Electrical symbols used on drawings in this chapter

objects in the home. It also speeds up household chores, reduces eye strain, and makes it a pleasure for members of the family to work or play during evening hours. Electric lighting is not only cleaner, safer, and more convenient than any other form of artificial light, it is also low-cost enough to be within the means of almost every family.

In industry, electric lighting speeds up production, reduces errors, increases safety, and generally improves the morale of employees.

In stores, hotels, and office buildings, electric illumination is used on a large scale to improve the efficiency of employees, to aid in the selling of merchandise, and to reduce eye strain.

The exteriors of some buildings are beautifully floodlighted and streets are lighted brightly with electric lamps. The lighting of outdoor sport areas enables us to view football, baseball, and other sports at night. Television would not be possible without electricity and artificial light.

The cases are endless, and almost everyone today realizes the value of better lighting. This field also provides some of the most fascinating and enjoyable work for the electrician.

THE EYE AND VISION

Since the effect of light upon the eye gives us the sensation of sight, any study of lighting must begin with a consideration of the eye and the seeing process. An understanding of the eye's mechanism will help you to understand the primary function of illumination — to provide light for the performance of visual tasks with a maximum of comfort and a minimum of strain and fatigue.

The Seeing Mechanism

The human eye is a fine precision instrument that is often compared to a camera (see Figure 8-3). Both the eye and the camera have a *covering* or *housing*. Each has a *lens* that focuses an inverted image on a light-sensitive surface — the *retina* in the eye and the *film* in the camera. The camera

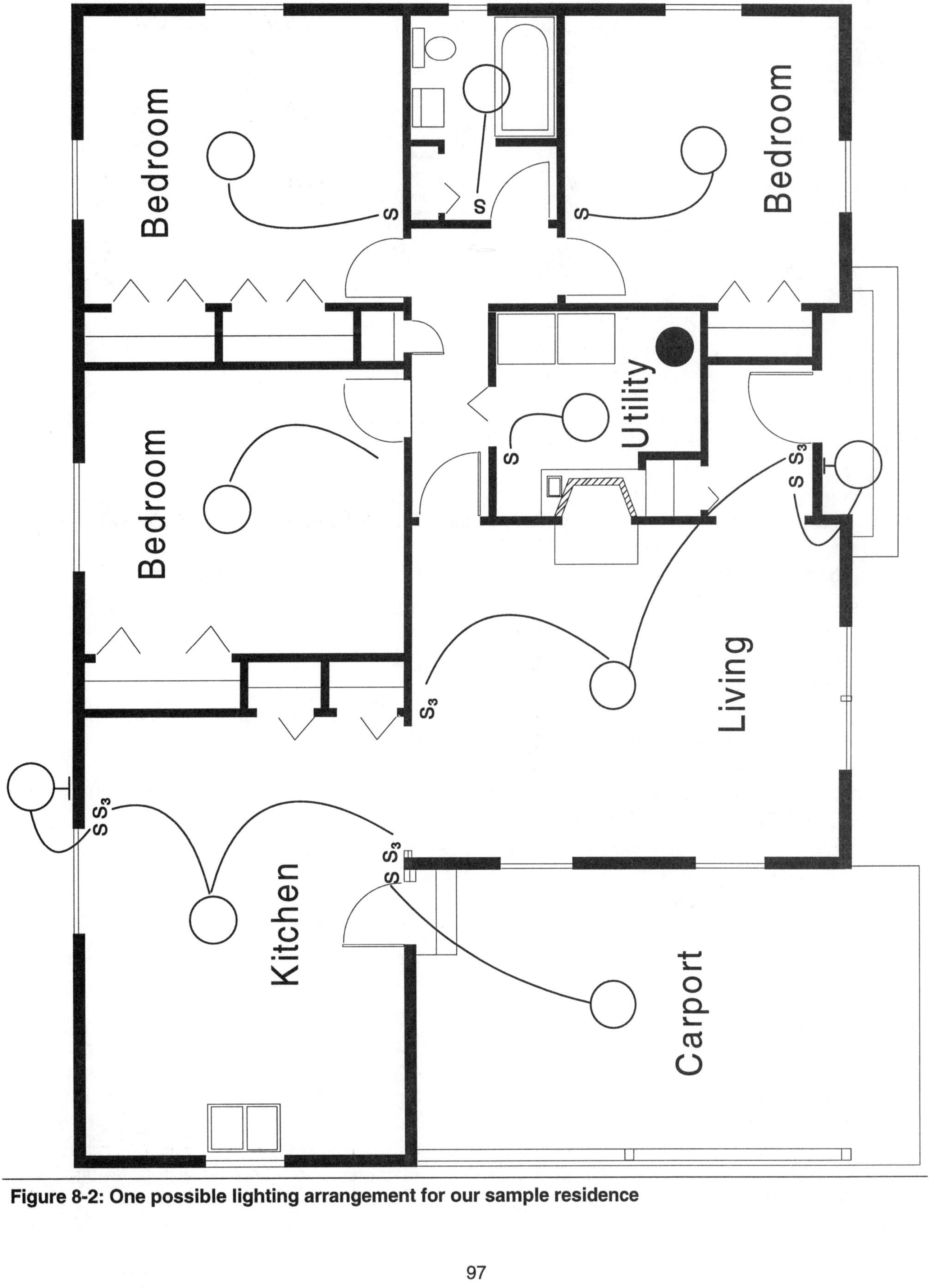
Bedroom
Bedroom
Bedroom
Utility
Living
Kitchen
Carport

Figure 8-2: One possible lighting arrangement for our sample residence

EYE	CAMERA
Sclera	Covering or housing
Retina	Film
Lid	Shutter
Lens	Lens
Iris	Diaphragm

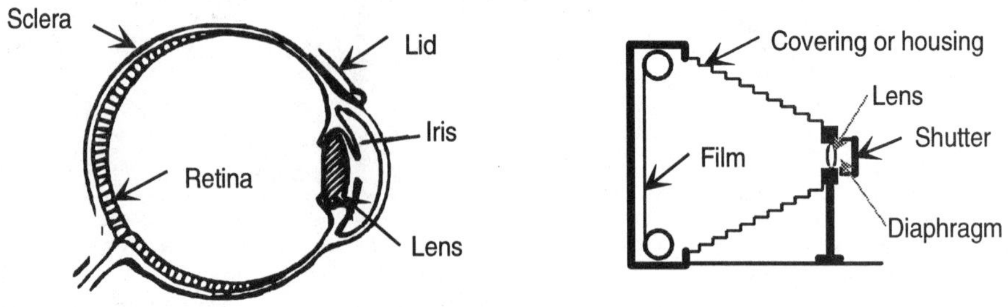
Sclera
Lid
Iris
Retina
Lens
Covering or housing
Lens
Shutter
Film
Diaphragm

Figure 8-3: Comparison of the eye to the camera

shutter corresponds to the *eyelid.* In front of the lens in the camera is a *diaphragm*, which may be opened or closed to regulate the amount of light entering the camera. The *iris* of the eye performs the same function.

There are, however, some important differences between the eye and the camera. The most important being the fact that the eye is a living organ. Taking pictures in poor-quality light will do no harm to the camera. But using the eyes under light of poor quality will result in unnecessary fatigue and may lead to headaches and inflammation of the eyes. Consistent misuse of the eyes can cause permanent damage to them and may also contribute to the development of disorders in other parts of the body. You should now begin to realize the importance of proper lighting design for visual tasks.

How We See

When a beam of light passes through the transparent protective outer layers of the eye, it is bent or refracted. The amount of light coming through the eye is controlled automatically by the contraction or expansion of the iris. The light continues on through the lens, which focuses the rays on to the retina. From this point on, the process is electrochemical. Pulsations are set up and are carried to the optic nerve, which, in turn, transmits them to the brain where they are interpreted as light, or where they cause the sensation of sight. Thus, the brain and the eye working together transform radiant energy (light) into the sensation of sight.

Objective Factors in the Process of Seeing

Investigation has shown that the quality of sight depends upon four primary conditions associated with the visual object in question:

- Size of object.
- Luminance (brightness) of object.
- The luminance contrast between the object and its immediate background.
- The time available for seeing the object.

Size: The size of the object is the most generally recognized and accepted factor in seeing. Everyone is familiar with the conventional eye test chart that is used by schools, optometrists, test for driving licenses, and others for testing visual defects. The larger the object in terms of *visual angle*, the better it can be seen. The following explains this

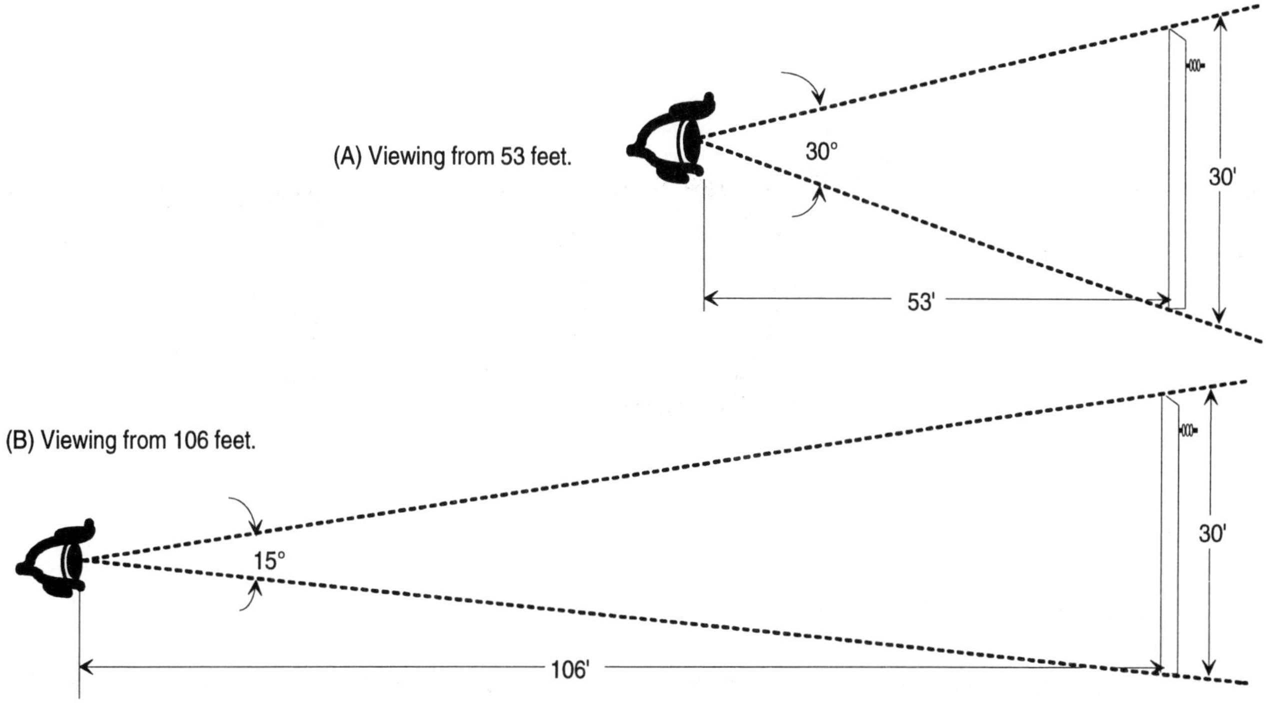
(A) Viewing from 53 feet.
30°
30'
53'
(B) Viewing from 106 feet.
15°
30'
106'

Figure 8-4: Viewing a 30-foot telephone pole from varying distances

principle while Figures 8-4 and 8-5 illustrate the explanation.

Figure 8-4A shows a 30-foot telephone pole approximately 53 feet away from the person viewing it. The angle of sight formed from the eye to the object is 30°. If the viewer then backs away from the telephone pole another 53 feet, (Figure 8-4B), the object has not become smaller; it is still a 30-foot telephone pole. However, since the angle from the eye to the object has become smaller, the object will appear smaller than it did in Figure 8-4A and cannot be seen as clearly.

To further illustrate, if we take two objects, one 12 inches high and the other 1 inch high (Figure 8-5A) and place both objects at the same distance from the viewer, the larger of the two objects can be seen more clearly since the angle from the eye to the object will be greater from the larger object.

However, if we move the smaller object closer to the viewer, so that the angle becomes the same as the larger object (Figure 8-5B), both objects can then be viewed with the same clarity.

Luminance (brightness): The brightness of an object depends upon the amount of light striking it, and the proportion of that light reflected in the direction of the eye. A light surface will have a much higher brightness than a dark surface. However, by adding enough light to the dark surface it is possible to make it as bright as the light object. Since the brighter object will be seen first, it will require a lesser amount of light than the darker object for good visibility. Thus, it would take a greater amount of general illumination to adequately light a room with walls painted a dark color, or a dark wood stain, than it would for one with the walls painted a light pastel color; the darker the wall finish, the more light that is required.

Contrast: The contrast in brightness or color between the visual object and its immediate background is as important for sight as the general brightness is. The difference in the visual effort required to read the two halves of the circle in Figure 8-6 demonstrates this fact; that is, black on white is easier to read than black on gray.

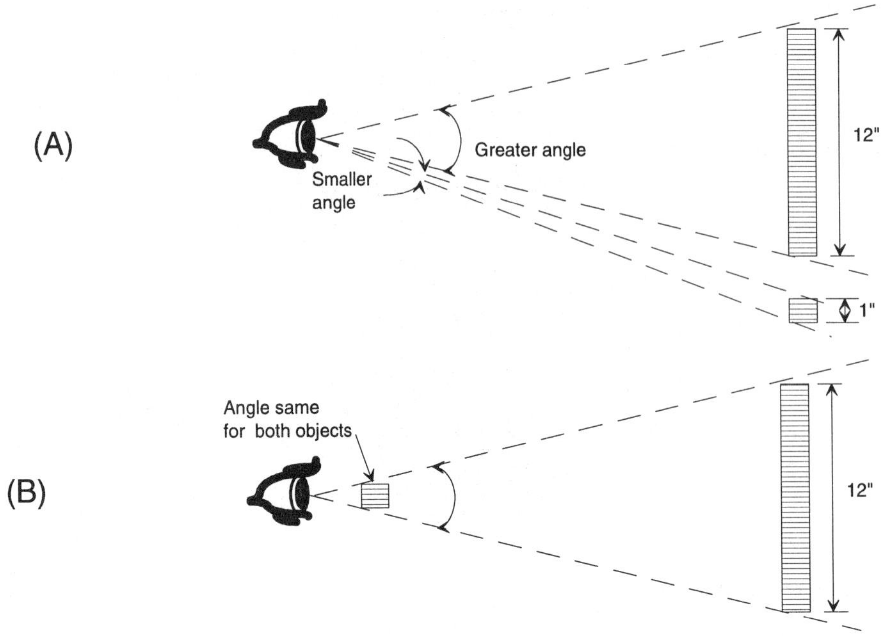
(A)
Greater angle
Smaller angle
12"
1"
(B)
Angle same for both objects
12"

Figure 8-5: Viewing two objects of different sizes

Again, the higher levels of illumination will partly compensate for brightness contrast where such conditions cannot be avoided.

Time: Seeing is not an instantaneous process; it requires *time*. The eye can see very well under low levels of brightness if sufficient time is allowed. If the object must be seen quickly, then more light is required. In fact, high lighting levels actually make moving objects appear to move slower, and this greatly increases their visibility. It then stands to reason that it would take a greater amount of illumination to properly light a baseball field than it would a football field, since the baseball usually will be traveling at a higher rate of speed than the football will.

Size, brightness, contrast, and time are mutually interrelated and interdependent. A deficiency in one can usually be corrected, within limits, by an adjustment in one or more of the others. Of these four conditions, brightness and contrast are usually under the direct control of the interior decorator. With proper control of brightness and contrast, unfavorable conditions, such as size of the object and time given for seeing this object, can be overcome.

Summary

- The purpose of lighting is to make vision possible.
- The mechanism of the human eye is similar to that of a camera.
- Proper illumination is necessary to protect the eyes.
- Good lighting can do much to relieve the eyestrain involved in the performance of difficult visual tasks.
- Research reveals that the advantages of high illumination are even more advantageous to older eyes than to young, normal eyes.
- Size, brightness, contrast, and time are the four basic conditions considered when evaluating the quality of sight.

CHARACTERISTICS OF LIGHT

Light may be defined as a radiant energy evaluated in terms of its capacity for producing the sensation of sight.

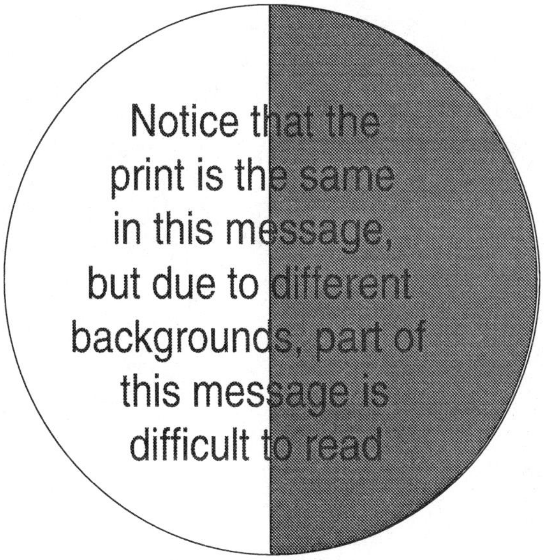
Notice that the
print is the same
in this message,
but due to different
backgrounds, part of
this message is
difficult to read

Figure 8-6: Contrast between the object and its immediate background

All light travels in a straight line unless it is modified or redirected by means of a reflecting, refracting, or diffusing condition.

Light Colors

The different colors of light are due to the different wave frequencies, which are considered to be of an electromagnetic nature, and are known to be of extremely high frequency and much shorter in length than the shortest television waves.

Ordinary sunlight, while it appears white, is actually made up of a number of colors. In 1666, Sir Isaac Newton passed a beam of light through a prism and discovered that it contained all colors of the rainbow. The three basic colors are red, blue, and green, but by continuously blending together, they also produce violet, yellow, and orange. See Figure 8-7.

Artificial white or daylight is generally the most desirable form of light for illuminating purposes, but it must contain a certain number of the colors which compose sunlight. It is the reflection to our eyes of these various colors from the object they strike that enables us to see objects and to get an impression of their color. Certain surfaces and

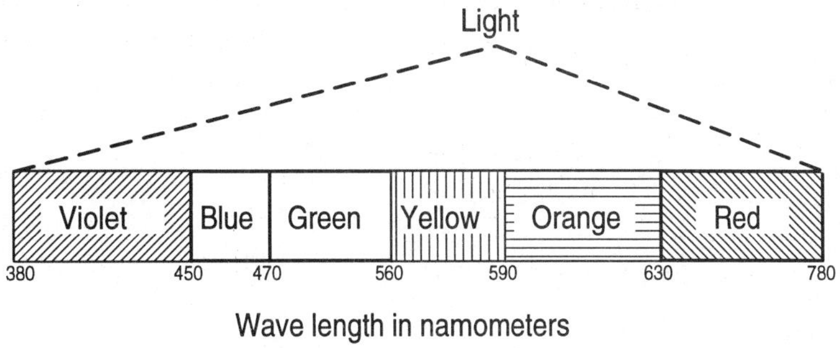
Light
Violet
Blue
Green
Yellow
Orange
Red
380
450
470
560
590
630
780
Wave length in namometers

Figure 8-7: The visible light spectrum

materials absorb light of one color and frequency and reflect that of another color; this gives us our color distinction in seeing different things.

White and light-colored surfaces reflect more light than dark surfaces do. (Remember *contrast* in the first part of this chapter.)

The ordinary tungsten-filament electric lamp (the ordinary light bulb) is a good example of nearly white artificial light that is excellent for most applications. The molecules of the tungsten wire are caused to vibrate rapidly and produce heat when an electric current is applied. When enough current is passed through the wire, it becomes incandescent (white light).

UNITS OF LIGHT MEASUREMENT

Before you undertake to perform actual lighting layouts or select lighting fixtures for certain applications, it is necessary to learn more about actual quantities of light, units of light, etc. An understanding of these interesting units and principles will enable you to better understand the nature of light.

When you purchase an incandescent lamp, you would normally refer to the rating of the lamp in terms of *watts* (a 60-watt lamp, 100-watt lamp, 150-watt lamp, etc.). While the rating in watts usually gives a general idea of the lamp size, it does not tell how much light a certain lamp can be expected to produce.

For example, one might expect a 100-watt incandescent lamp to produce more light than a 40-watt fluorescent lamp. However, the average inside-frosted 100-watt incandescent lamp emits light at the rate of about 1490 lumens, while the flow of light from a 40-watt fluorescent lamp is about 3200 lumens — over twice the amount of light given off by the incandescent lamp.

The preceding example implies that the total amount of light actually given off by a light source is measured in terms of the unit *lumen*.

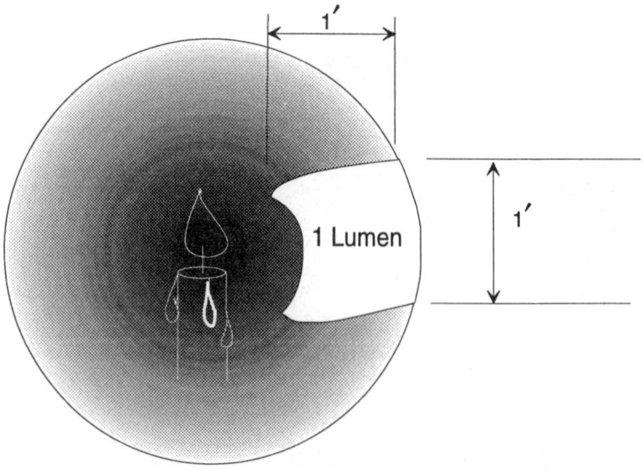
1′
1′
1 Lumen

Figure 8-8: Using a sphere to illustrate one lumen

Lumen

A lumen may be defined as the quantity of light that will strike a surface of one square foot, all points of which are one foot distance from a light source of one candlepower (one standard candle for our purposes).

A lumen of light may be visualized by placing a standard candle inside a hollow sphere that has a radius of one foot or a diameter of two feet, and the inside of which is completely black to prevent any reflection of light. If a one foot square is cut out of the sphere, as shown in Figure 8-8, the amount of light that will escape through this hole will be one lumen.

If the area of the opening was $\frac{1}{4}$ square foot, then the light emitted from the hole would be $\frac{1}{4}$ lumen; if the hole was $\frac{1}{2}$ square foot, the escaping light would be $\frac{1}{2}$ lumen; $\frac{3}{4}$ square would enable $\frac{3}{4}$ lumen to escape, and so on. Figure 8-9 illustrates this principle.

A sphere with a one-foot radius has a total area of 12.57 square feet, so if the entire sphere was removed from the candle, the total lumens emitted by the standard candle would be 12.57. From this we find that we can determine the approximate amount of lumens given off by any lamp by multiplying the average candlepower of the lamp by 12.57.

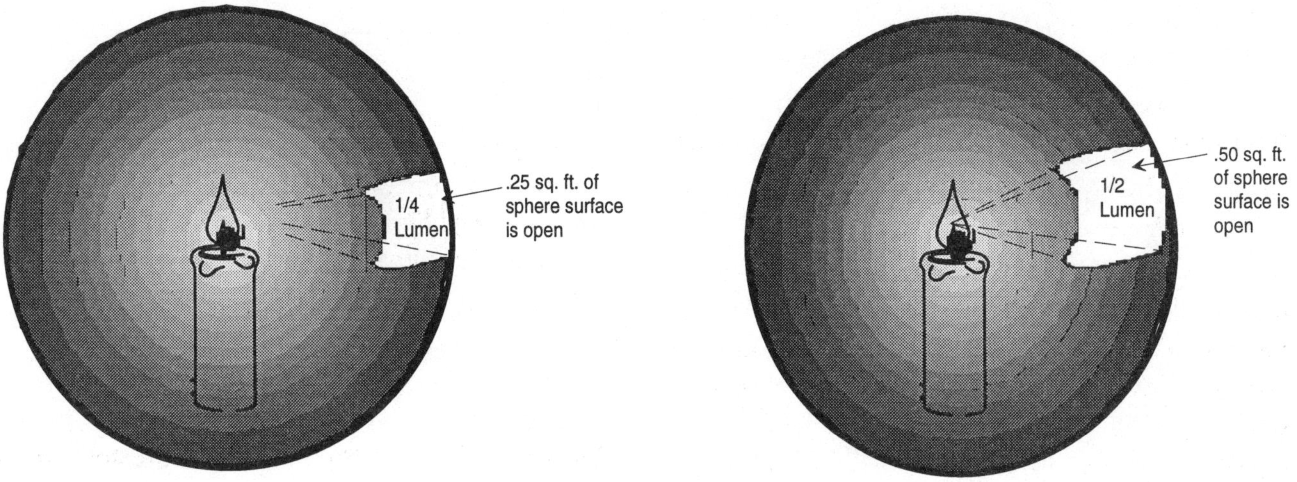
1/4
Lumen
.25 sq. ft. of
sphere surface
is open
1/2
Lumen
.50 sq. ft.
of sphere
surface is
open

Figure 8-9: Using a sphere to illustrate a portion of a lumen

Footcandles

Electric lamps are a source of light, and the result of this light striking various surfaces is called *illumination*.

While the lumen serves as a unit to measure the total light obtained from any light source, we must also have a unit to measure the intensity of this light on a given surface. (Desk tops and work benches would be good examples of such surfaces.) The unit used for this purpose is called the *footcandle*.

A footcandle is a unit of measurement which represents the intensity of illumination that will be produced on a surface that is one foot away from a source of one candlepower, and at right angles to the light rays from the source. This means that if you should hold a sheet of $8\frac{1}{2} \times 11$ paper one foot from a candle, the edges of the paper must be bent towards the candle so that all surfaces are one foot away (Figure 8-10); otherwise, if the center is one foot away, the edges will be a slight greater distance from the candle.

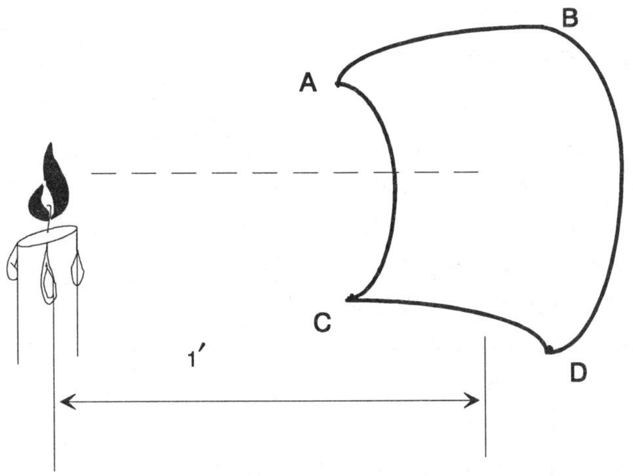
B
A
C
1′
D

Figure 8-10: One footcandle

The footcandle is the unit used in illumination calculations to determine the proper level of illumination on any working plane or surface.

Referring again to Figure 8-10, the surface A, B, C, D is illuminated at every point with an intensity of 1 footcandle. The total amount of light striking this same surface is also 1 lumen. Thus, when 1 lumen of light is evenly distributed over a surface of 1 square foot, that same area is illuminated to an intensity of 1 footcandle.

The preceding paragraph shows that if we know the area of a surface that is to be lighted and the intensity in footcandles of the desired illumination level, we can then calculate the number of lumens that will be required to light the area. For example, if we desire to illuminate a surface of 100 square feet to an average intensity of 10 footcandles by a light source 1 foot from the surface, we multiply the desired footcandles by the area of the surface to be illuminated — 10 × 100. Therefore, the source of light must produce 1000 lumens in order to light the surface to an intensity of 10 footcandles. In actual practice, more lumens would be required to produce 10 footcandles of illumination on this surface, since the efficiency of any light source is rarely 100%. More light will also be required as the distance between the light source and surface increases.

Inverse Square Law for Light

Footcandle units are used to indicate the illumination level at a specific point, or the average illumination on a surface or working plane. The inverse square law is the basis of calculation in the point-by-point method of lighting design. The interior designer will have use for this method in many commercial applications. Therefore, a brief description of the inverse square law is warranted.

The inverse square law states that the illumination on a surface varies directly with the candlepower of the source of light, and inversely with the square of the distance from the source. Whew! This really means that a small change in distance from a light source will make a great change in the illumination level on a surface. Figure 8-11 illustrates the reasons for this.

In Figure 8-11 we have a light source of 1 candlepower, and since the surface "A" is 1 foot from the candle, its illumination intensity will be 1 footcandle. If we move the surface or plane to "B," which is 2 feet from the source, the same number of light rays will have to spread over four times the area, as that area increases in all directions. Then the illumination intensity at twice the distance is only ¼ the amount it was before, as the distance of 2 squared is 4, and this is the number of times the illumination is reduced.

If we continue to move the surface to "C," which is 3 feet away from the light source, the rays

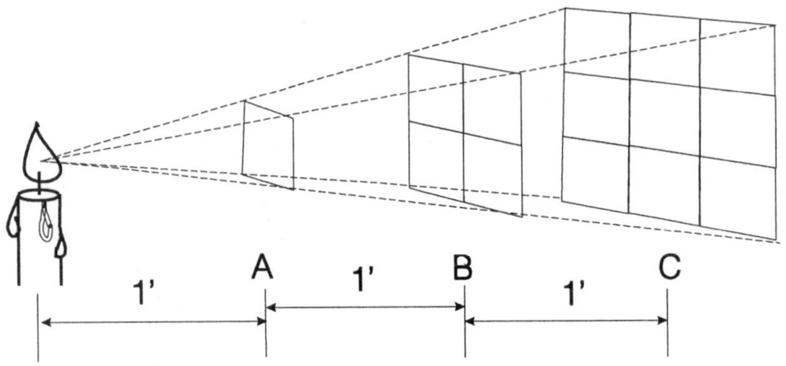
A
B
C
1'
1'
1'

Figure 8-11: The inverse square law illustrated

are now spread over 9 times the original area ($3^2 = 9$), and the intensity of illumination on the surface will now be only 1/9 of its former value.

The farther away any surface is from a source of light, the less light it receives from that source, since the same light rays must be distributed over a greater area.

Various models of convenient portable footcandle meters are available to measure the footcandle level on surfaces or planes.

Light Reflection

Light can be reflected from certain light-colored or highly polished surfaces. This fact must also be considered when performing lighting calculations.

Some surfaces and materials are much better reflectors than others. Usually, the lighter colors reflect more light and absorb less light than darker colors.

The percentage of light that will be reflected from many common materials are as follows:

White plaster	90% to 92%
Mirrored glass	80% to 90%
White paint	75% to 90%
Metallized plastic	75% to 85%
Polished aluminum	75% to 85%
Stainless steel	55% to 65%
Limestone	35% to 65%
Marble (white)	45%
Concrete	40%
Dark red glazed bricks	30%

The better classes of reflectors are used in directing light. The colors of walls, ceilings, and floors and their reflecting ability are also considered in interior lighting design — the darker these areas, the more light required.

Field Measurements

There are a number of large and elaborate devices used in laboratories for making exact tests and measurements on light and lighting fixtures. But for practical use in the field, a portable light or footcandle meter is quite satisfactory and relatively inexpensive.

To use a footcandle meter, first remove the cover. Hold the meter in a position so the cell is facing toward the light source and at the level of the work plane where the illumination is required. The shadow of your body should not be allowed to fall on the meter cell during tests. A number of such tests at various points in a room will give the average illumination level in footcandles—indicating if additional lighting fixtures are required. Such tests, however, are normally done by lighting engineers, but the practice is a good way for the electricians to observe the type of lighting system, and then find out the amount of light produced. Much can be learned from instructional material, but not nearly as much (or as quickly) as by actually putting your knowledge to practical use.

Summary

- While ordinary sunlight appears white, it is actually made up of a number of colors: violet, blue, green, yellow, orange, and red.
- Artificial white or daylight is generally the most desirable form of light for illuminating purposes.
- White and light-colored surfaces reflect more light than dark surfaces do.
- The total amount of light actually given off by a light source is measured in terms of the unit *lumen.*
- The level of illumination on any surface or work plane is measured in terms of the unit *footcandle.*
- The instrument used to measure the illumination level is called the *footcandle meter.*
- The farther any surface is from a light source, the less light it receives from that source.

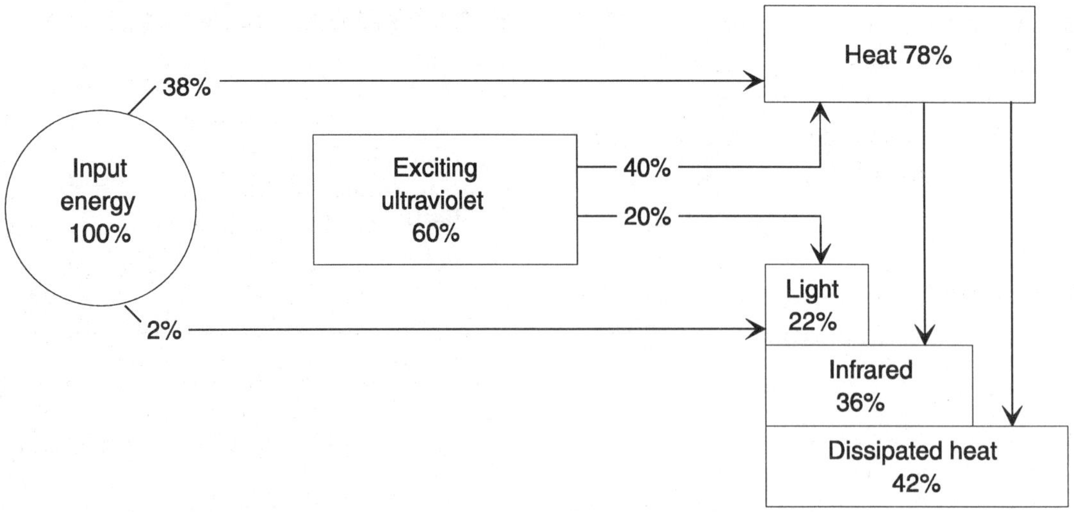
38%
Input
energy
100%
Exciting
ultraviolet
60%
40%
20%
Heat 78%
2%
Light
22%
Infrared
36%
Dissipated heat
42%

Figure 8-12: Energy distribution of a typical cool-white fluorescent lamp

LIGHT SOURCES

Now that we know something of the nature of light and the fundamentals of good illumination, we can discuss the three common sources of electric light:

- Incandescent lamps
- Electric discharge lamps
- Electroluminescent lamps

Despite continuous improvement, none of these light sources have a high overall efficiency. The very best light source converts only approximately ¼ of its input energy into visible light. The remaining input energy is converted to heat or invisible light. Figure 8-12 illustrates the energy distribution of a typical cool white fluorescent lamp.

INCANDESCENT LAMPS

Incandescent lamps are made in thousands of different types and colors, from a fraction of a watt to over 10,000 watts each, and for practically any conceivable lighting application.

Extremely small lamps are made for instrument panels, flashlights, etc., while large incandescent lamps, over 20 inches in diameter, have been used for spotlights and street lighting.

Regardless of the type or size, all incandescent filament lamps consist of a sealed glass envelope containing a filament (Figure 8-13). The incandescent filament lamp produces light by means of a filament heated to incandescence (white light) by its resistance to a flow of electric current. Most of these elements are capable of producing 11 to 22 lumens per watt, and some produce as high as 33 lumens per watt.

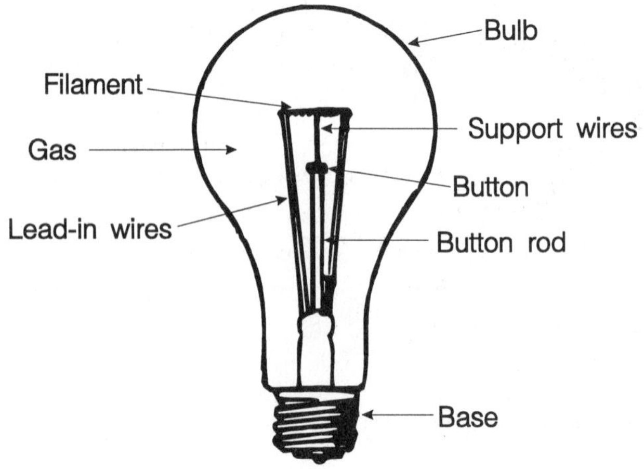
Bulb
Filament
Support wires
Gas
Button
Lead-in wires
Button rod
Base

Figure 8-13: Parts of an incandescent lamp

The filaments of incandescent lamps were originally made of carbon. Now, tungsten is used for virtually all lamp filaments because of its higher melting point, better spectral characteristics, and strength—both hot and cold.

Tungsten-halogen Lamps

The quartz-iodine tungsten-filament lamp is basically an incandescent lamp, since light is produced from the incandescence of its coiled tungsten filament. However, the lamp envelope, made of quartz, is filled with an iodine vapor which prevents the evaporation of the tungsten filament. This evaporation is what normally occurs in conventional incandescent lamps. When the bulb begins to blacken, light output deteriorates, and eventually the filament burns out. While the quartz-iodine lamp has approximately the same efficiency as an equivalent conventional incandescent lamp, it has the advantages of double the normal life, low lumen depreciation, and a smaller bulb for a given wattage. Figure 8-14 shows the basic components of a quartz-iodine lamp. *Caution: Never touch a quartz lamp with the bare hands; oil from the hands can adversely affect the lamp envelope.*

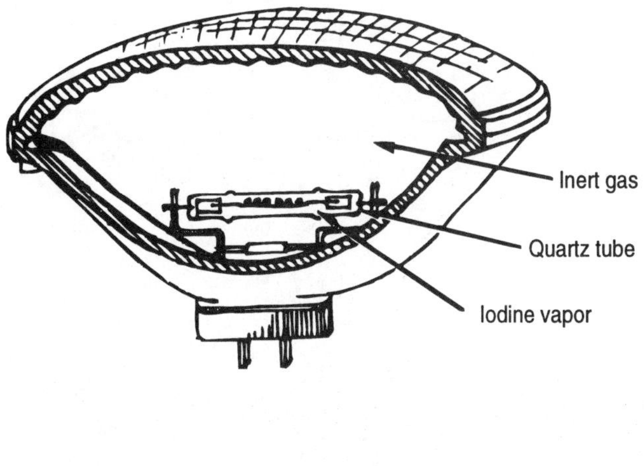
Inert gas
Quartz tube
Iodine vapor

Figure 8-14: Parts of a tungsten-halogen lamp

ELECTRIC DISCHARGE LAMPS

The electric discharge lamp category includes the well-known fluorescent, neon, and mercury-vapor lamps as well as the newer metal-halide and sodium lamps. In this group of lamps, light is produced by the passage of an electric current through a tungsten filament. The application of an electrical voltage ionizes the gas and permits current to flow between two electrodes located at opposite ends of the lamp. This arc discharge accelerates to tremendous speeds, and when the current collides with the atoms of the gas it temporarily alters the atomic structure. Light results from the energy given off by these altered atoms as they return to their normal state — resulting in a most efficient lamp output.

The incandescent lamp has certain characteristics that make it inherently inefficient as a source of light; maximum possible values for this type of lamp have probably already been approached. However, the electric discharge lamp produces light by an entirely different process and is capable of achieving a much higher efficiency of light output.

Fluorescent Lamps

Of all the electric discharge light sources, the fluorescent lamp is the best known and most widely used (Figure 8-15). Since fluorescent lighting was introduced to the general public, during the 1933 Chicago Centennial Exposition, it has almost completely taken the place of the incandescent lamp in all branches of construction, except for specialty lighting and possibly for residential use.

One of the reasons for the popularity of fluorescent lighting is its high efficiency as compared to incandescent lamps. The average 40-watt incandescent lamp delivers approximately 470 initial lumens, while a 40-watt fluorescent lamp delivers about 3150 lumens. This power efficiency not only saves on the cost of power consumed, but it lessens the heat and reduces air conditioning loads (another saving). Fluorescent lighting allows more

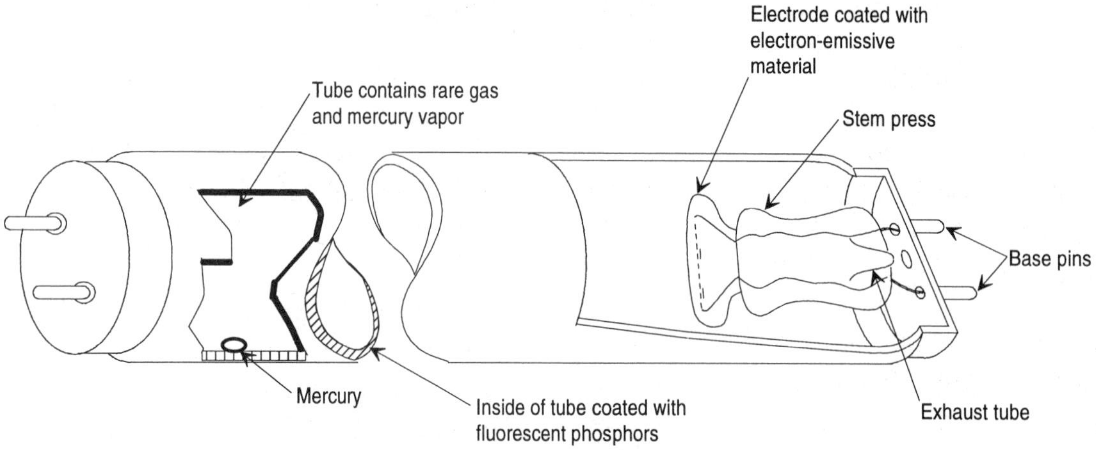
Electrode coated with electron-emissive material
Tube contains rare gas and mercury vapor
Stem press
Base pins
Mercury
Inside of tube coated with fluorescent phosphors
Exhaust tube

Figure 8-15: Main components of a fluorescent lamp

comfort for those working under bright lights during warm weather.

Fluorescent lamps are made with long glass tubes sealed at both ends and containing a rare inert gas, generally argon, and low pressure mercury vapor. Built into each end is a cathode which supplies the electrons to start and maintain the gaseous discharge.

The inside of the lamp tubing is coated with a thin layer of materials called *phosphors.* A phosphor is a substance that becomes luminous or which glows with visible light when struck by streams of electrons, that are caused to pass through the space between filaments inside the lamp. When the phosphors are thus made luminous, the action is called fluorescent — giving the lamp its name.

Figure 8-16 shows that fluorescent lamps come in many different configurations: straight, U-shaped, and circular.

Fluorescent Colors

Color is achieved in fluorescent lamps by mixing various phosphors and a wide range of visible light colors is possible.

Cool white: This lamp is often selected for offices, factories, and commercial areas where a psychologically cool working atmosphere is desirable. This is the most popular of all fluorescent lamp colors since it gives a natural outdoor lighting effect and is one of the most efficient fluorescent lamps manufactured today.

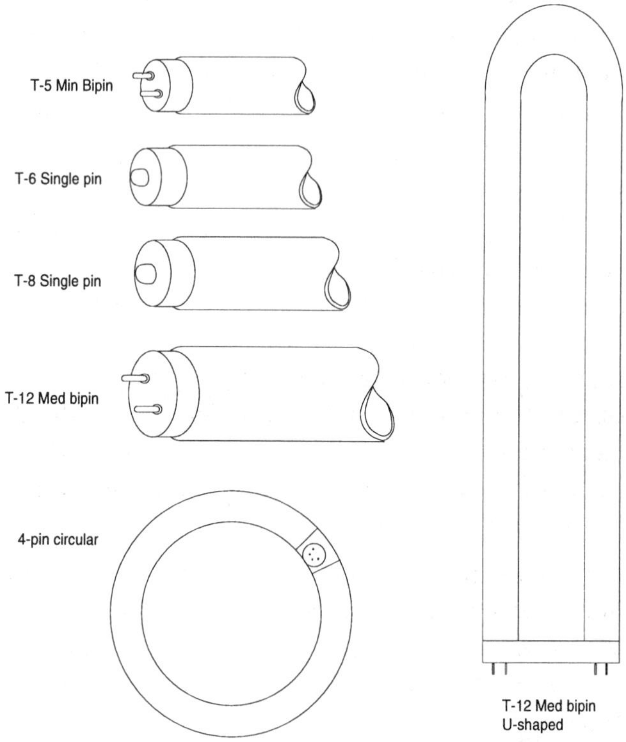
T-5 Min Bipin
T-6 Single pin
T-8 Single pin
T-12 Med bipin
4-pin circular
T-12 Med bipin
U-shaped

Figure 8-16: Sizes and shapes of fluorescent lamps

Deluxe cool white: This lamp is used for the same general applications as the cool white, but contains more red which emphasizes pink skin tones and is therefore more flattering to the appearance of people. Deluxe cool white is also used in food display because it gives a good appearance to lean meat; keeps fats looking white; and emphasizes fresh, crisp appearance of green vegetables. This type of lamp is generally chosen wherever very uniform color rendition is desired, although it is less efficient than cool white.

Warm white: Warm white lamps are used whenever a warm social atmosphere is desirable in areas that are not color critical. It approaches incandescent in color and is suggested whenever a mixture of fluorescent and incandescent lamps is used. While it gives an acceptable appearance to people, it has some tendency to emphasize shallowness. Yellow, orange, and tan interior finishes are emphasized by this lamp, and its beige tint gives a bright warm appearance to reds; brings out the yellow in green; and adds a warm tone to blue. It imparts a yellowish white or yellowish gray appearance to neutral surfaces.

Deluxe warm white: Deluxe warm white lamps are more flattering to complexions than warm white and are very similar to incandescent lamps in that they impart a ruddy or tanned hue to the skin. It is generally recommended for home or social environment applications and for commercial use where flattering effects on people and merchandise are considered important. This type of lamp enhances the appearance of poultry, cheese, and baked goods. These lamps are approximately 25% less efficient than warm white lamps.

White: White lamps are used for general lighting applications in offices, schools, stores, and homes where either a cool working atmosphere or warm social atmosphere is not critical. They emphasize yellow, yellow green, and orange interior finishes.

Daylight: Daylight lamps are used in industry and work areas where the blue color associated with the "north light" of actual daylight is preferred. While it makes blue and green bright and clear, it tends to tone down red, orange, and yellow.

In general, the designations "warm" and "cool" represent the differences between artificial light and natural daylight in appearance they give to an area. Their deluxe counterparts have a greater amount of red light, supplied by a second phosphor within the tube. The red light shows colors more naturally, but at a sacrifice in efficiency.

Other colors of fluorescent lamps are available in sizes that are interchangeable with white lamps. These colored lamps are best used for flooding large areas with colored light; where a colored light of small area must be projected at a distant object, incandescent lamps using colored filters are best.

Ballast

Every fluorescent lamp needs a ballast in order to operate. The ballast is simply a coil of insulated wire wound on a frame, or core, made up of thin layers of iron stampings, and it performs any or all of the following functions:

- Limits the current flow through the lamp to the value for which the lamp is designed.
- Causes a drop in line voltage and provides the desired lamp voltage, which in turn determines the rated current in the lamp.
- Provides power factor correction.
- Provides radio interference suppression.

Dimming and Flashing

Rapid-start fluorescent lamps can be dimmed from full brightness to approximately zero output by a number of special electrical and electronic circuits. This has made it possible to greatly increase the flexibility of fluorescent-lighting systems. However, a special rapid-start ballast in conjunction with dimming control devices is required. See Figure 8-17.

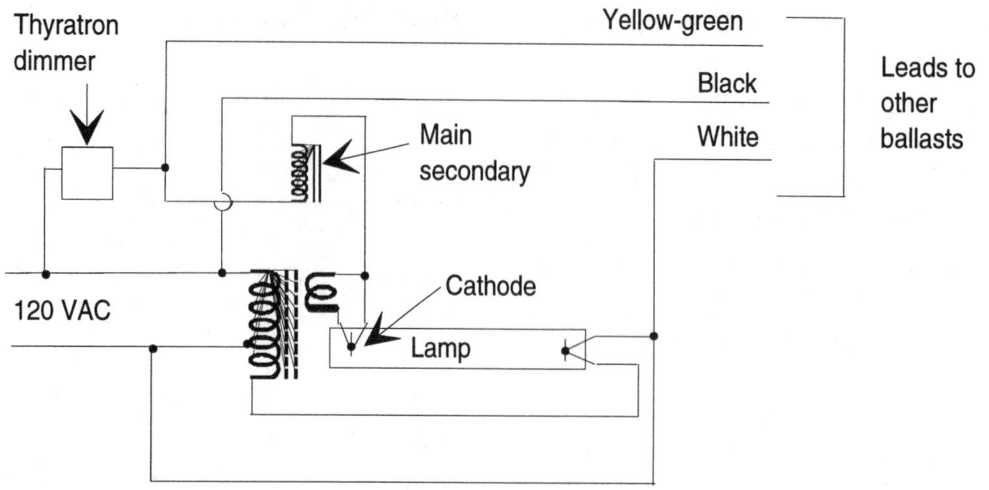
Thyratron
dimmer
Yellow-green
Black
White
Leads to
other
ballasts
Main
secondary
Cathode
120 VAC
Lamp

Figure 8-17: A dimmer circuit using a special ballast transformer

Rapid-start and cold-cathode fluorescent lamps can be flashed (as for sign use) without any appreciable loss in lamp life, since a special ballast can be used to provide continuous electrode heating while the current to the arc is interrupted. Additional information on outdoor lighting and other specialty lighting is presented in other chapters. In the meantime, manufacturer's data are very helpful learning sources, and it is recommended that each electrician accumulate as much of this information as possible for current and future reference.

Advantages and Disadvantages of Fluorescent Lamps

Advantages

- High efficiency.
- Long life.
- A linear source of light.
- Variety of colors available.
- Relatively low surface brightness.
- Economy in operation.

Disadvantages

- Very sensitive to temperature and humidity.
- Radio interference.
- Difficult to control.
- Higher initial installation cost.

HIGH-INTENSITY DISCHARGE LAMPS

High-intensity discharge lamps are used mostly for commercial and industrial establishments, and used only for outdoor lighting in residential applications. Therefore, only a brief coverage of these lamps is included in this book.

Mercury-vapor lamps: The traditional mercury-vapor lamp produces light with a predominance of yellow and green rays and a small percentage of violet and blue.

Mercury-vapor lamps operate by passing an arc through a high-pressure mercury vapor that is contained in an arc tube made of quartz or glass. This action produces visible light. As with all arc discharge lamps, ballasts are required to start the lamp, and thereafter to control the arc.

Metal-halide lamps: The metal-halide lamp is basically a mercury lamp that has been altered by adding iodine compounds to the mercury and argon gas in the arc tube. These iodine compounds are of metals such as indium, sodium, thallium, or dysprosium. The addition of these salts causes the emission of light which is of a better color than the basic mercury colors, although life and lumen maintenance may be decreased in the process.

Metal-halide lamps furnish approximately 75 to 90 lumens per watt of white light, which is much warmer than the mercury light and is suitable for many commercial indoor applications.

High-pressure sodium-vapor lamps: This type of lamp uses an arc tube of ceramic material such as polycrystalline alumina. The lamp operates in a similar manner to the other discharge lamps, producing a warm yellow-orange tone at the rate of over 100 lumens per watt. This type of lamp is mostly used for outdoor applications.

INTERIOR LIGHTING DESIGN

The basic requirement for any lighting design is to determine the amount of light that should be provided and the best means of providing it. However, since individual tastes and opinions vary greatly, there can be many suitable solutions to the same lighting problem. Some of these solutions will be dull and commonplace, while others will show imagination and resourcefulness. The lighting designer should always strive to select lighting equipment that will provide the highest visual comfort and performance that is consistent with the type of area to be lighted and the budget provided.

Lighting design for commercial, institutional, and industrial applications is a highly specialized field, best left to consulting engineers who are trained in the science of lighting and lighting design. For such installations, the electrician must only know how to interpret the working drawings and lighting-fixture schedules, plus have a good knowledge of wiring practices in general. However, electricians are frequently called upon to design lighting layouts for residential applications, and every electrician should have a knowledge of basic residential lighting design. Knowing these basic principles of illumination will also help the electrician to better understand why and how more sophisticated lighting applications are designed.

RESIDENTIAL LIGHTING

Properly designed lighting is one of the greatest comforts and conveniences that any homeowner can enjoy. In building new homes or remodeling old ones, the lighting should be considered equally as important as the heating/air-conditioning system, the furniture placement, and as one of the most important features of both interior and exterior decorations.

As a rule, residential lighting does not require a large quantity of elaborate calculations, as does a school or office building. However, electricians must apply their talent and ingenuity in selecting the best types of lighting fixtures for various locations in order to obtain a desirable effect, as well as the proper amount of illumination at the desired quality. Light has certain characteristics that can be used to change the apparent shape of a room, to create a feeling of separate areas within one room, or to alter architectural lines, form, color, pattern, or texture. Light also affects the mood and atmosphere within the area where used.

While calculations of any quantity are unnecessary, the electrician must use some guide until he or she has gained the necessary experience to improvise. The methods described in this section are suitable for selecting the proper amount of light as well as the proper types of lighting fixtures.

Let's review two definitions before continuing.

- *Lumen:* A lumen is the quantity of light that will strike a surface of 1 square foot, all points of which are a distance of 1 foot from a light source of 1 candlepower.
- *Footcandle:* A footcandle is a unit of measurement which represents the intensity of illumination that will be produced on a surface that is 1 foot distant from a source of 1 candlepower, and at right angles to the light rays from the source.

From the preceding statements, we can say that 1 lumen per square foot equals 1 footcandle.

Thus, the following method will be called the *lumens-per-square-foot method* of residential lighting design.

In using this method, it is important to remember that lighter room colors reflect light and darker colors absorb light. This method is based on rooms with light colors; therefore, if the room surfaces are dark — like one that has its walls covered with dark wood-grain paneling — the total lumens should be multiplied by a factor of at least 1.25.

The table in Figure 8-18 gives the required lumens per square foot for various areas in the home and also the required illumination in footcandles for those who desire to use a different method of calculating required illumination. The recommended footcandle level is "fixed" and will apply regardless of the type of lighting fixtures used. However, the recommended lumens given in this table are based on the assumption that portable table lamps, surface-mounted fixtures, or efficient structural lighting techniques will be used. If the majority of the lighting fixtures in an area will be recessed, the lumen figures in the table should be multiplied by 1.8.

Note that this method produces only approximate results; yet, quite adequate for most residential lighting applications. One of the most important considerations is to avoid shadows on work planes.

Living Room

As the living room is the social heart of most homes, lighting should emphasize special architectural features such as fireplace, bookcases, paintings, etc. The same is true of draperied walls, planters, or any other special room accents.

Dramatizing fireplaces with accent lights brings out texture of bricks, adds to overall room light level, and eliminates bright spots that cause subconscious irritation over a period of time. Use 75- to 150-watt lamps in wallwash-type fixtures — either recessed or surface mounted — for this application.

While recessed downlights, cornice, or valance lighting all add life to draperies, they also supplement the general living room lighting level. Position downlights 2.5 to 3 feet apart and 8 to 10 inches from the wall. Valances are always used at windows, usually with draperies (Figure 8-19). They provide up-light, which reflects off the ceiling for general room lighting, and down-light for drapery accent. Cornices direct all their light downward to give dramatic interest to wall coverings, draperies, etc., and are good for low-ceiling rooms.

Pulldown fixtures or table lamps are used for reading areas. While the pulldown fixtures are more dramatic, the electrician must know the furniture arrangement prior to fixture placement.

Area	Lumens Required per ft^2	Average Footcandles Required
Living room	80	70
Dining room	45	30
Kitchen	80	70
Bathroom	65	50
Bedroom	70	30
Hallway	45	30
Laundry	70	50
Work bench	70	70

Figure 8-18: Required lumens and footcandles per square foot for various areas in the home

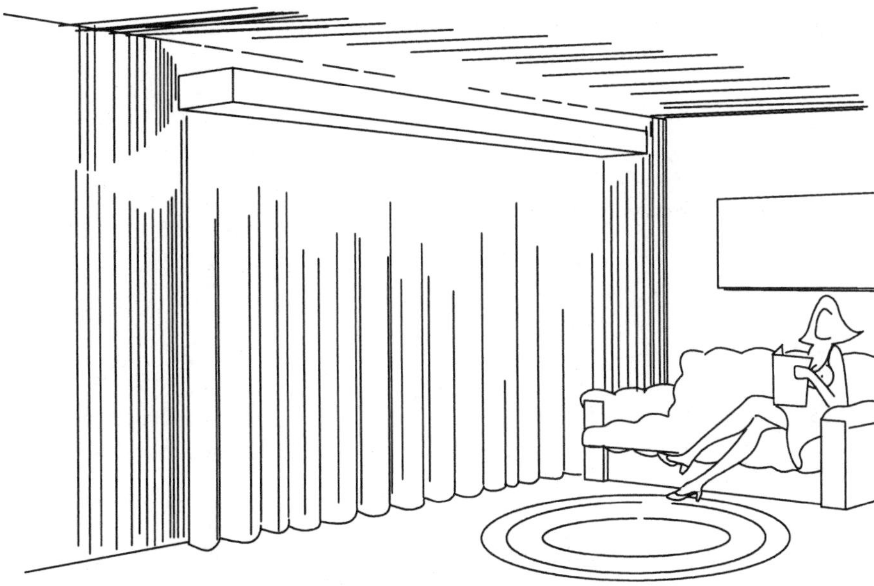

Figure 8-19: Valance lighting provides both uplight and downlight

As a final touch, add dimmers to vary the lighting levels exactly to the living room activities — low for a relaxed mood, bright for a merry, party mood.

The floor plan in Figure 8-20 shows a living room for a small residence. Let's see how one electrician went about designing a simple, yet highly attractive and functional, lighting layout for this area. Remember that more than one solution to a lighting design is usually available.

The first step is to scale the drawings to find the dimensions of the area in question. In doing so, we find that the area is 13.75 feet wide and 19 feet long. Thus, 13.75 × 19 = 261.25 which is rounded off to 261 square feet. The table in Figure 8-18 recommends 80 lumens per square foot for the living area. Therefore, 80 × 261 = 20,880 lumens required in this area.

The next step is to refer to a lamp catalog in order to select lamps that will give the required lumens. A sample sheet of lamp data is shown in Figure 8-21. At the same time, the designer should have a good idea of the type of lighting fixtures that will be used as well as their location.

Referring again to the floor plan of the living room, note that two recessed spotlights, designated by the symbol are mounted in the ceiling above the fireplace. Each fixture contains two 75-watt R-30 lamps, rated at 860 lumens each, for a total of 1720 lumens. However, since these are recessed into the ceiling (and not surface mounted), the total lumens will have to be reduced. This is accomplished by multiplying the total lumens by a factor of 0.555 to obtain the efficient lumens.

$$1720 \times 0.555 = 955 \text{ effective lumens}$$

This means that we now need 19,925 more lumens in this area to meet the recommended level of illumination.

The next section of this area will be the window area on the front side of the house. It was decided to use a drapery cornice from wall to wall which would contain four 40-watt warm-white fluorescent lamps, rated at 2080 lumens each, for a total of 8320 lumens. Combining this figure with the 955 effective lumens from the recessed lamps give a total of 9275 lumens, or 11,605 more lumens to account for.

Two 3-way (100-, 200-, 300-watt) bulbs in table lamps will be used on end tables, one on each side

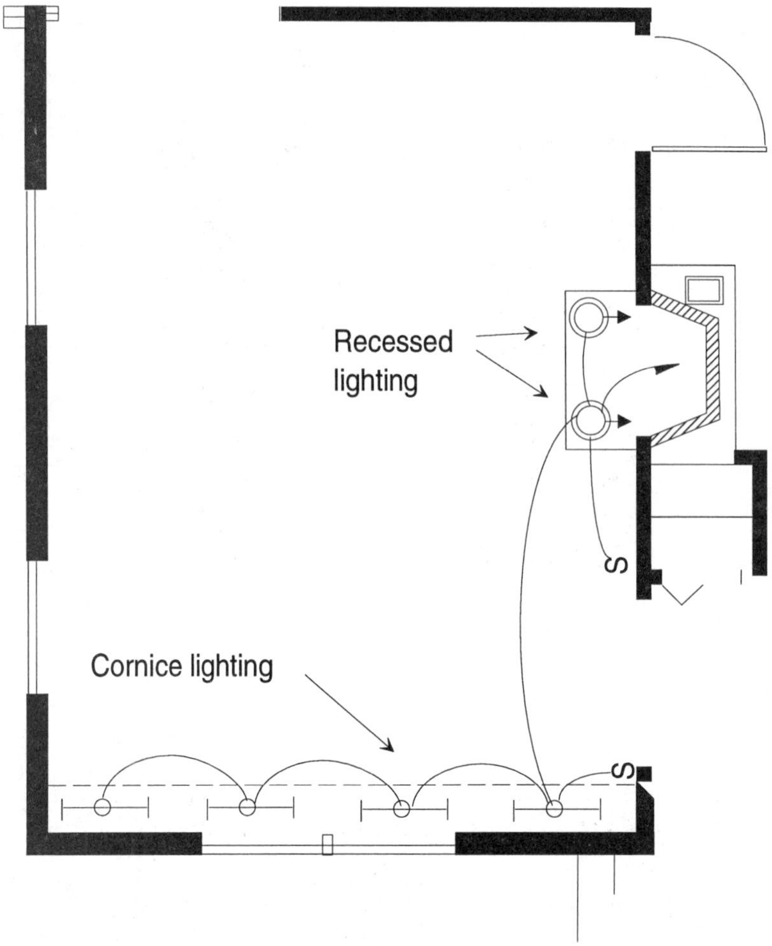
Recessed
lighting
Cornice lighting
S
S

Figure 8-20: Floor plan of living-room lighting layout

of a sofa for a total of 9460 lumens; this means that only 2145 lumens are unaccounted for. However, one 3-way (50-, 100-, 150-watt) bulb will be used in a lamp on a chairside table. Since this lamp is rated at 2190 lumens, we now have the total lumens required for the living-room area.

It can be seen that this method of residential lighting calculation makes it possible to quickly determine the light sources needed to achieve the recommended illumination level in any area of the home.

Kitchen

The lighting layout for the kitchen must always receive careful attention since the kitchen is the area where many homemakers will spend a great amount of time.

Good kitchen lighting begins with general illumination — usually one or more ceiling-mounted fixtures of a type that is close to the ceiling. The type of fixture for this general illumination should be a glare-free source that will direct light to every corner of the kitchen.

If a fluorescent lamp is selected for the source of illumination, it is recommended that Deluxe cool-white lamps be used. This type of fluorescent lamp contains more red than the standard cool-white lamp. It therefore emphasizes pink skin tones and is more flattering to the appearance of people. This type of lamp also gives a good appearance to lean meat; keeps fat white; and emphasizes

INCANDESCENT
General Service Lamps

Watts	Bulb/Base	Lumens (initial)	Life (Hours)
60W	A-19/Med.	870	1000
75W	A-19/Med.	1190	750
100W	A-19/Med.	1750	750
100W	A-21/Med.	1710	750
150W	A-21/Med.	2880	750
150W	A-23/Med.	2780	750
150W	PS-25/Med.	2680	750
200W	A-23/Med.	4010	750
200W	PS-30/Med.	3710	750
300W	PS-25/Med.	6360	750
300W	PS-30/Med.	6110	750

FLUORESCENT
Tubular Lamps

Watts	Bulb/Base	Description	Lumens (initial)	Life (Hours)
20W	T-12	CW	1300	9000
20W	T-12	CWX	850	9000
20W	T-12	WW	1300	9000
20W	T-12	WWX	820	9000
30W	T-12	CW	2300	15000
30W	T-12	CWX	1530	12000
30W	T-12	WW	2360	15000
30W	T-12	WWX	1480	12000
40W	T-12	CW	3150	18000
40W	T-12	CWX	2200	18000
40W	T-12	WW	3200	18000
40W	T-12	WWX	2150	18000
40W	T-12	Chroma 55	2020	18000
40W	T-12	Chroma 75	1990	18000
40W	T-12	Inc./Fluor.	1700	17000
40W U	T-12	CW(3⅝)	2800	12000
40W U	T-12	WW(3⅝)	2800	12000
40W U	T-12	CW(6")	2950	12000
40W U	T-12	WW(6")	3025	12000

Figure 8-21: Lamp data

Circline Lamps

Watts	Bulb/Base	Description	Lumens (initial)	Life (Hours)
22W	T-9	CW	950	7500
22W	T-9	CWX	755	7500
22W	T-9	WW	980	7500
22W	T-9	WWX	745	7500
32W	T-10	CW	1750	7500
32W	T-10	CWX	1250	7500
32W	T-10	WW	1800	7500
32W	T-10	WWX	1250	7500
40W	T-10	CW	2800	7500
40W	T-10	CWX	1780	7500
40W	T-10	WW	2350	7500
40W	T-10	WWX	1760	7500

Figure 8-21: Lamp data (Cont.)

the fresh, crisp appearance of green vegetables. Standard cool-white fluorescent lamps are seldom recommended.

The ideal general lighting system for a residential kitchen would be a luminous ceiling, such as that shown in Figure 8-22. This kitchen floor plan shows bare fluorescent strips, with dimming control, above ceiling panels with attractive diffuser patterns. This arrangement, while the most expensive, provides a "skylight" effect which makes seeing easier.

Another kitchen floor-plan layout appears in Figure 8-23. First, single-tube fluorescent light fixtures were located under the wall cabinets and behind a shielding board as shown in Figure 8-24. Then warm-white fluorescent lamps were selected as the best color for lighting countertops. This shadow-free light not only accents the colorful countertops but also makes working at the counter much more pleasant.

Two 75-watt R-30 floodlights installed in two recessed housings over the kitchen sink and spaced about 15 inches on center offer excellent light for work at the sink. However, a two-lamp fluorescent fixture using warm-white fluorescent lamps and concealed by a faceboard, as shown in Figure 8-25, will work equally well.

The light for the electric range is taken care of by a ventilating hood with self-contained lights. The hood contains two 60-watt incandescent lamps for proper illumination. If no hood is used, a 30-watt fluorescent light fixture mounted on the wall over the cooking area, as shown Figure 8-26, would be a good choice.

One surface-mounted ceiling fixture with an opal glass diffuser is used for general illumination; this fixture contains two 60-watt incandescent lamps. This light source accents and enriches the wood tones of the wall cabinets. In small kitchens, concealed fluorescent strip lighting mounted in a continuous cover around the perimeter will give the effect of a larger kitchen as well as provide an excellent source of general lighting.

In this particular example, the focal point of the kitchen is the dining area. Here, a versatile pull-down light completes the lighting layout. Using 150 watts of incandescent lamps, this fixture provides ample light on the table and also directs some

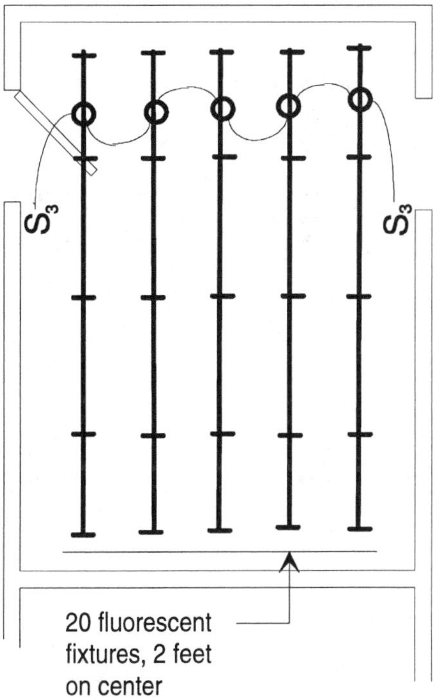
S3
S3
20 fluorescent fixtures, 2 feet on center

Figure 8-22: Kitchen floor-plan layout of luminous ceiling lighting

of the light upward for a pleasing effect. This pulldown lamp is controlled by a wall-mounted dimmer for added versatility.

In homes with separate dining rooms, a chandelier mounted directly above the dining table and controlled by a dimmer switch becomes the centerpiece of the room while providing general illumination. The dimmer, of course, adds versatility since it can set the mood of the activity — low brilliance (candlelight effect) for formal dining or bright for an evening of cards. When chandeliers with exposed lamps are used, the dimmer is essential to avoid a garish and uncomfortable atmosphere. The size of the chandelier is also very important; it should be sized in proportion to the size of the dining area.

Good planning calls for supplementary lighting at the buffet and sideboard areas. For a contemporary design, use recessed accent lights in these areas. For a traditional setting, use wall brackets to match the chandelier. Additional supplementary lighting may be achieved with a wide assortment

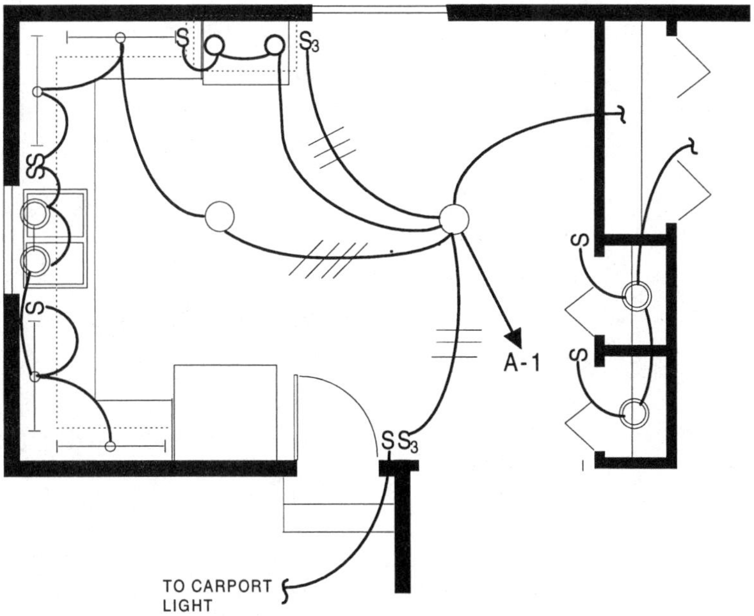
S
S3
SS
S
A-1
S
S
SS3
TO CARPORT LIGHT

Figure 8-23: Lighting floor-plan layout for a typical kitchen

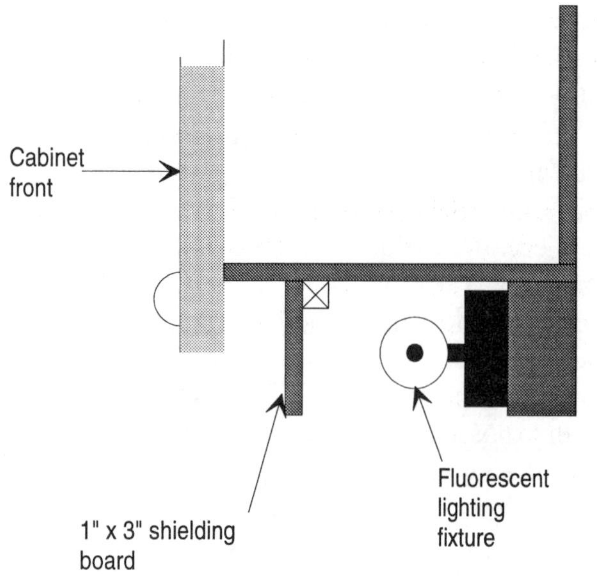
Cabinet front
1" x 3" shielding board
Fluorescent lighting fixture

Figure 8-24: Section view of a fluorescent lighting fixture located behind a shielding board under kitchen cabinets

of available fixtures, such as concealed fluorescent lighting in valances or cornices as discussed previously. Of course, there are several other possibilities. In fact, these possibilities are almost endless, limited only by the designer's knowledge, ability, and imagination.

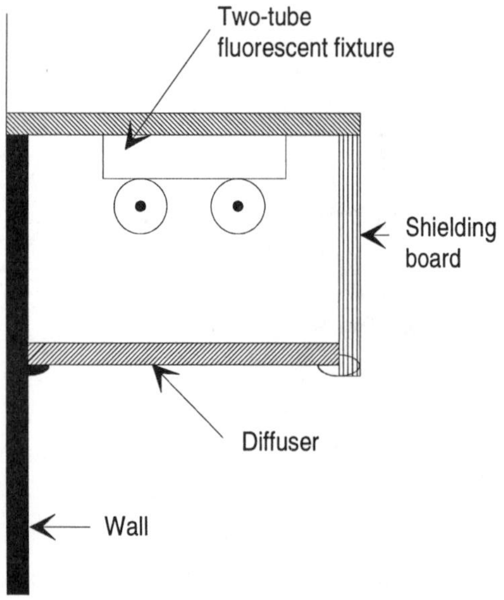
Two-tube fluorescent fixture
Shielding board
Diffuser
Wall

Figure 8-25: Example of a two-tube lighting fixture located above the kitchen sink

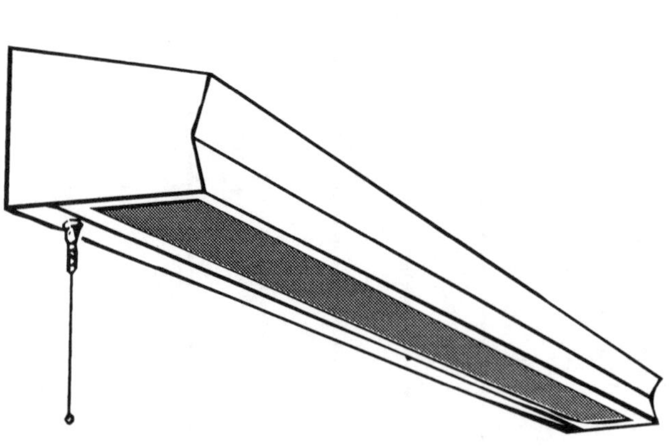

Figure 8-26: Fluorescent lighting fixture mounted on wall above range

Bathrooms

Lighting performs a wide variety of tasks in the bathroom of the modern residence. Good light is needed for good grooming and hygiene practices.

The bathroom needs as much general lighting as any other room. If the bath is small, usually the mirror and tub/shower lights will suffice for general illumination. However, in the larger baths, a bright central source is needed to transform it from the dim bath of the past to a smart, bright part of the house today.

For good grooming, the lavatory-vanity should be lighted to remove all shadows from faces and from under chins for shaving. Two wall-mounted fixtures on each side of the mirror or lighted soffits or downlights above the mirror will give the best results. Over a vanity table, pendants or downlights for concentrated light with a decorative touch may be used with equal results. For safety and health, a moistureproof, ceiling-mounted recessed fixture over the bath or shower should be included. Linen-closet lighting should also be considered.

For best results, bathroom lighting should be from three sources, like the three lighting fixtures illuminating the mirror in Figure 8-27. A theatrical effect may be obtained by using exposed-lamp fixtures across the top and sides of the mirror, as shown in Figure 8-28.

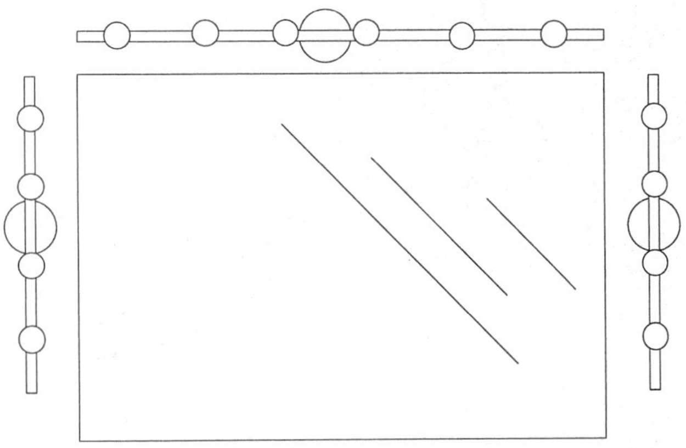

Figure 8-27: Bath mirror lighted from three different light sources

A small 7- to 15-watt night light is also recommended for the bath to permit occupants to see their way at night without turning on overhead lights that might disturb others.

A ceiling fixture with a sunlamp is another convenience that will be appreciated by sun worshippers. Such a lamp requires approximately two minutes to reach full ultraviolet output after it has been started and approximately three minutes to cool before it is restarted.

The sunlamp can be conveniently screwed into an ordinary household socket without the necessity of any other equipment. A good location for it in a residential bathroom is about 2 feet from the face, either over the shaving mirror or in a position where one would normally dry off after bathing.

A typical lighting layout for a residential bathroom appears in Figure 8-29. Note that recessed lighting fixtures are used in the linen closet and over the bath tub/shower — the latter equipped with a waterproof lens. A sunlamp is located in the center of the bathroom and is equipped with an automatic timer. Incandescent lamps are located on each side of the mirror. Also note that each lighting fixture, or group of lighting fixtures is individually controlled with a wall switch.

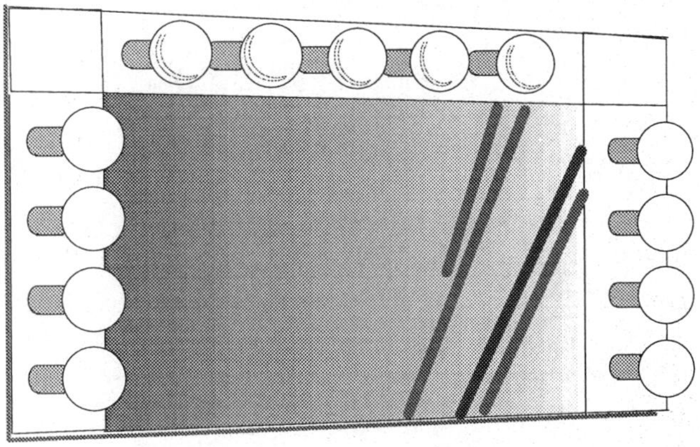

Figure 8-28: Theatrical lighting for a bathroom mirror

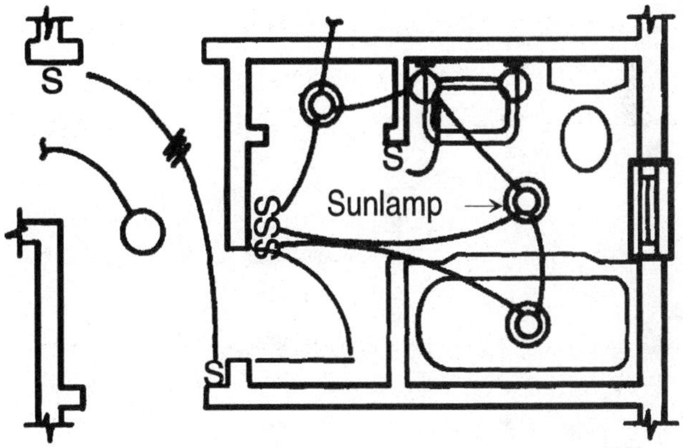
S
S
Sunlamp
SSS
S

Figure 8-29: Lighting layout for a typical bath and hallway

Bedroom Lighting

The majority of people spend at least a third of their lives in their bedroom. Still, the bedroom is often overlooked in terms of decoration and lighting as most homeowners would rather concentrate their efforts and money on areas that will be seen

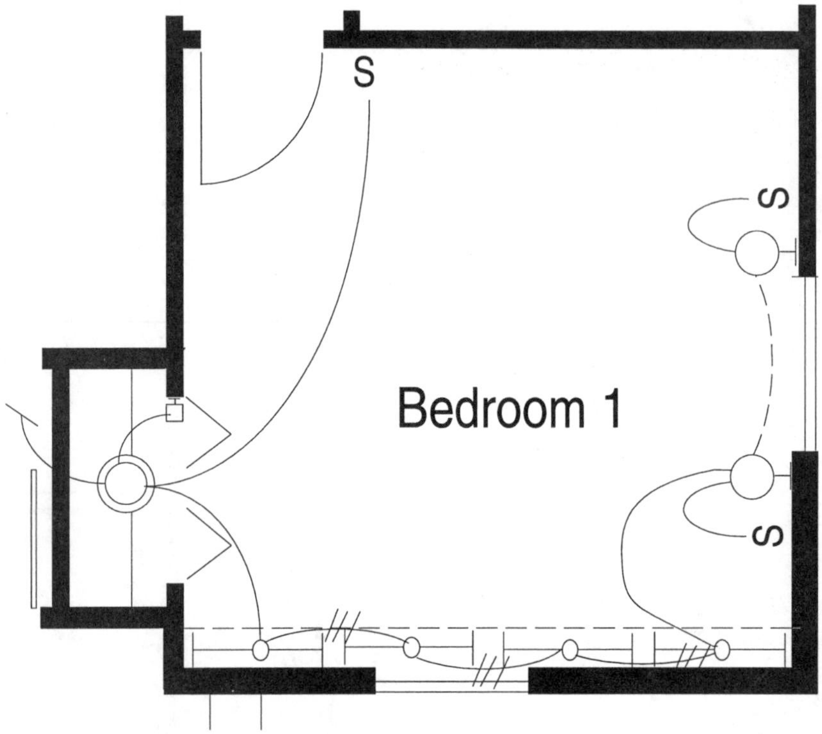
S
S
S
Bedroom 1

Figure 8-30: Lighting layout for master bedroom

more by visitors. However, proper lighting is equally important in the bedroom for such activities as dressing, grooming, studying, reading, and for a relaxing environment in general.

Basically, bedroom lighting should be both decorative and functional with flexibility of control in order to create the desired lighting environment. For example, both reading and sewing (two common activities occurring in the bedroom) require good general illumination combined with supplemental light directed onto the page or fabric. Other activities, however, like casual conversation or watching television, require only general nonglaring room illumination, preferably controlled by a dimmer switch/control.

Proper lighting in and around the closet area can do much to help in the selection and appearance of clothing, and supplementary lighting around the vanity will aid in personal grooming.

One master bedroom lighting layout is shown in Figure 8-30. Cornice lighting is used to highlight a colorfully draped wall and also to create an illusion of greater depth in this small bedroom. The wall-to-wall cornice board also lowers the apparent ceiling height in the room, which makes the room seem wider.

Two wall-mounted "swing-away" lamps on each side of the bed furnish reading light, while valance-type fluorescent fixtures furnish general illumination. The owners indicated that they preferred matched vanity lamps (table lamps) for grooming. This was handled with duplex receptacles located near the vanity.

A single recessed lighting fixture in the closet provides adequate illumination for selecting clothes and identifying articles on the shelves. The light is controlled by a door switch which turns the light on when the door is opened and turns it out

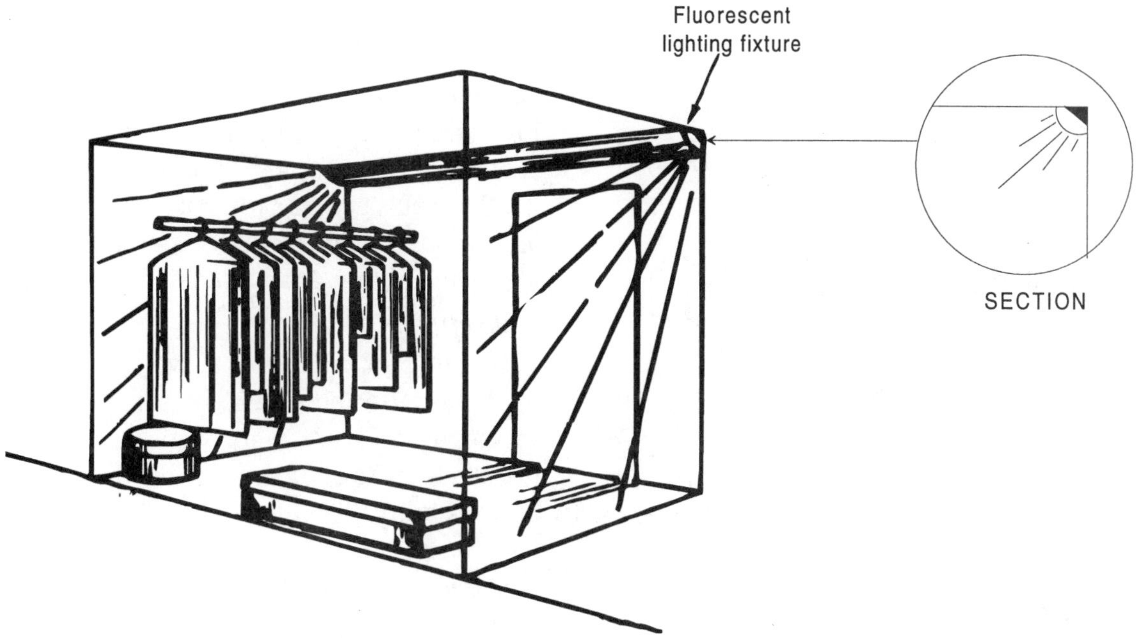
Fluorescent
lighting fixture
SECTION

Figure 8-31: Fluorescent fixture used for closet lighting

when the door is shut. This recessed lighting fixture, when combined with the general illumination of the bedroom, also illuminates a full-length mirror on the inside of the closet door.

If a closet is unusually long, two equally spaced recessed lighting fixtures may be required to provide adequate light distribution on the closet shelf; or, a fluorescent fixture mounted as shown in Figure 8-31 is an excellent choice.

To prevent closet lamps from coming in contact with clothing hung in the closet, which would be a potential fire hazard, certain requirements have been specified by the *National Electrical Code*.

A (lighting) fixture in a clothes closet shall be installed:

- On the wall above the closet door, provided the clearance between the fixture and a storage area where combustible material may be stored within the closet is not less than 18 inches, or
- On the ceiling over an area which is unobstructed to the floor, maintaining an 18-inch clearance horizontally between the fixture and a storage area where combustible material may be stored within the closet . . . .
- Pendants shall not be installed in clothes closets.

The recessed fixture used in the closet of the residence in question has a solid fresnel lens and is therefore considered to be located outside of the closet area. For this reason, they can be mounted anywhere on the ceiling or upper walls of the closet area.

To get a better understanding of why certain recessed lighting fixtures are considered to be located outside of the closet area, look at the drawing in Figure 8-32. This is a view of the fixture housing above the closet ceiling. Note that the make-up junction box, along with the housing for the lamp and reflector, are above the ceiling area. Consequently, if the proper lens and trim is installed in the closet area, all heat, electrical connections, and similar potential hazards (if installed within the closet) are actually outside of the closet area. Thus, the reason for the *NE Code's* allowance of such fixtures almost anywhere in a clothes closet.

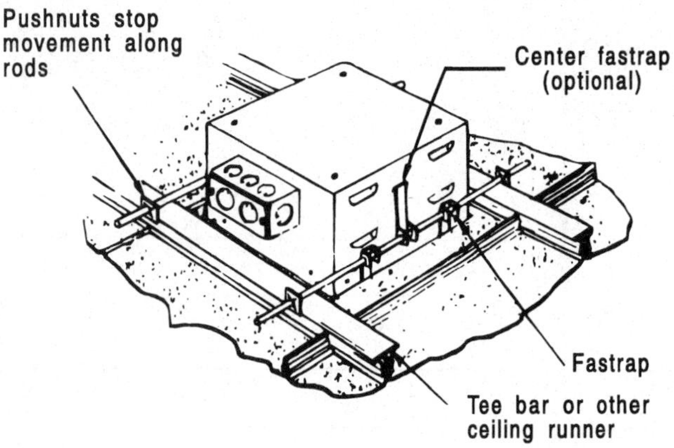
Pushnuts stop movement along rods
Center fastrap (optional)
Fastrap
Tee bar or other ceiling runner

Figure 8-32: Installation details of an incandescent recessed fixture housing

Other *NE Code* installation requirements for clothes closet lighting are depicted in Figure 8-33. Also carefully review the requirements in the actual code book; that is, *NE Code* Section 410-8.

Utility Rooms

The utility room in the floor plan in Figure 8-34 contains an electric water heater, clothes dryer and washer, and a pull-down ironing board. Therefore, this area will be treated as a laundry area requiring 70 lumens of illumination per square foot. Since the dimensions of the room are 6 feet 5 inches by 9 feet 2 inches, the area of the room is approximately 58 square feet (6.41×9.16). Thus, 58×70 (lumens) = 4060 required lumens in this area.

A two-lamp fluorescent lighting fixture was the interior decorator's choice for general illumination. This type of fixture was chosen for its high efficiency, long lamp life, linear source of light, and economical operation. However, since fluorescent lamps and ballasts are very sensitive to temperature and humidity, a vaporproof lens was specified on this fixture.

Warm-white lamps were used for the light source since this type of lamp approaches incandescent in color and is more flattering to complexions and clothes colors than cool-white fluorescent. Each of these lamps produces approximately 2150 lumens, giving a total of 4300 lumens for the area — just about right for our calculation. An alternate lighting arrangement for the utility room is shown in Figure 8-35.

Lighting for ironing is best when a shadow-casting light is used above and in front of the operator, as shown in Figure 8-36. With this arrangement, the shadows fall toward the operator, giving the best visibility. Therefore, these supplemental fixtures were specified. Use two louvered, swivel-type bullet housing or semirecessed "eyeball" fixtures for 150-watt R-40 reflector floodlight mounted on or in the ceiling 20 to 24 inches apart and 24 inches ahead of the front edge of the board — aimed at its surface.

If a portable ironing board will be used in various locations throughout the house, a "pole" lamp with three louvered bullets works very well. The middle and lower housings can direct sharp light at an angle, revealing wrinkles in the clothes that are being ironed while the top lamp can be aimed at the ceiling for indirect general illumination. The middle and lower housings should contain 50-watt lamps and the top housing (directed upward) should contain a 150-watt reflector lamp.

BASEMENTS

If the basement area is to be unfinished and used for utility purposes only, inexpensive lighting outlets should be located to illuminate designated work areas or equipment locations. All mechanical equipment, such as the furnace, pumps, etc. should be properly illuminated for maintenance. Laundry or work areas should have general illumination as well as areas where specific tasks are performed. At least one light near the stairs should be controlled by two 3-way switches, one at the top of the stairs and one at the bottom. Other outlets may be pull-chain porcelain lampholders.

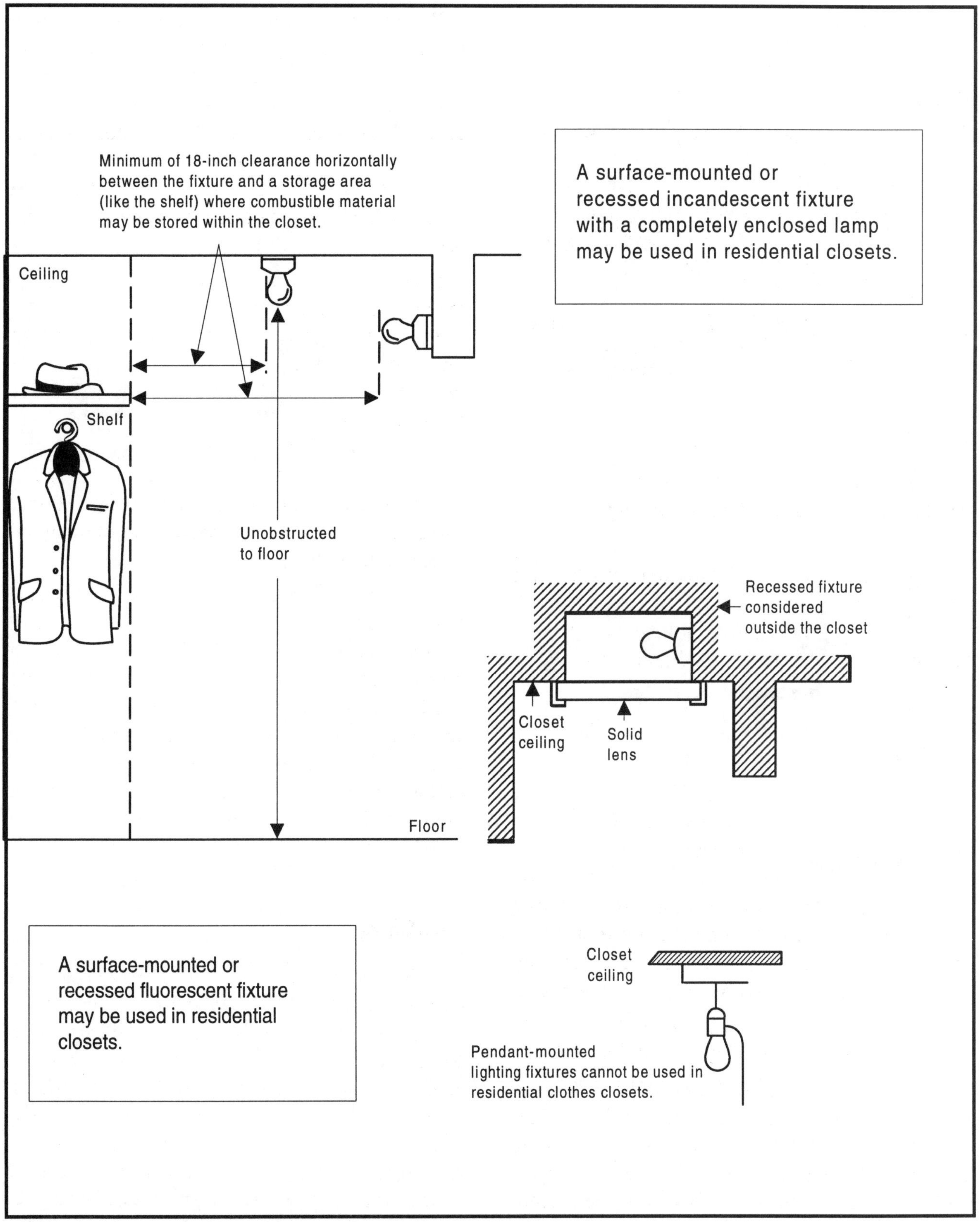
Minimum of 18-inch clearance horizontally
between the fixture and a storage area
(like the shelf) where combustible material
may be stored within the closet.
A surface-mounted or
recessed incandescent fixture
with a completely enclosed lamp
may be used in residential closets.
Ceiling
Shelf
Unobstructed
to floor
Floor
Recessed fixture
considered
outside the closet
Closet
ceiling
Solid
lens
A surface-mounted or
recessed fluorescent fixture
may be used in residential
closets.
Closet
ceiling
Pendant-mounted
lighting fixtures cannot be used in
residential clothes closets.

Figure 8-33: *NE Code* requirements for clothes closets

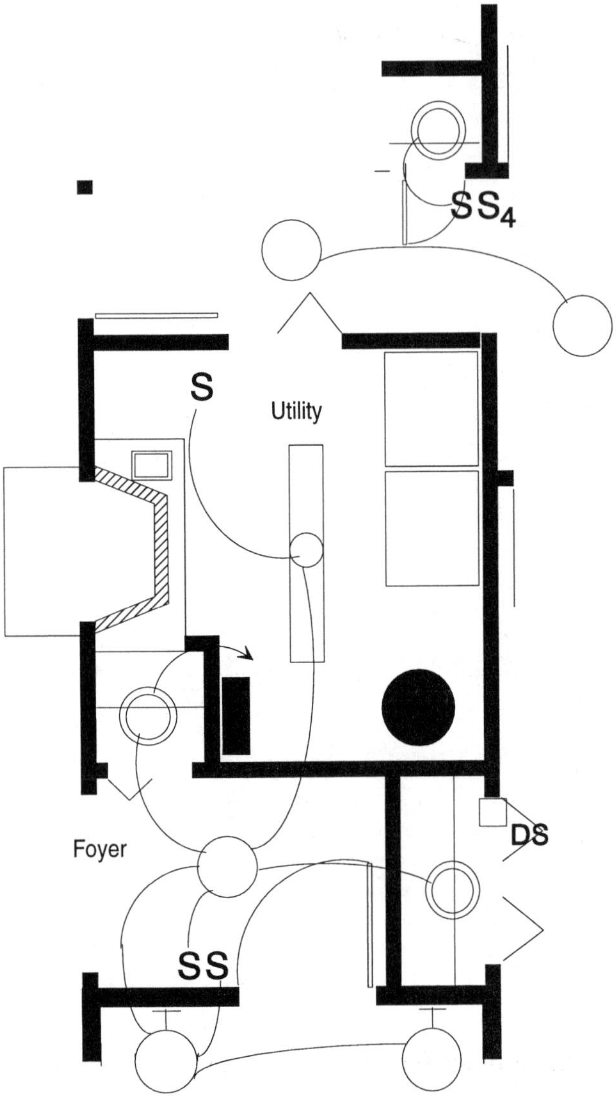
SS4
S
Utility
Foyer
DS
SS

Figure 8-34: Lighting layout for utility room

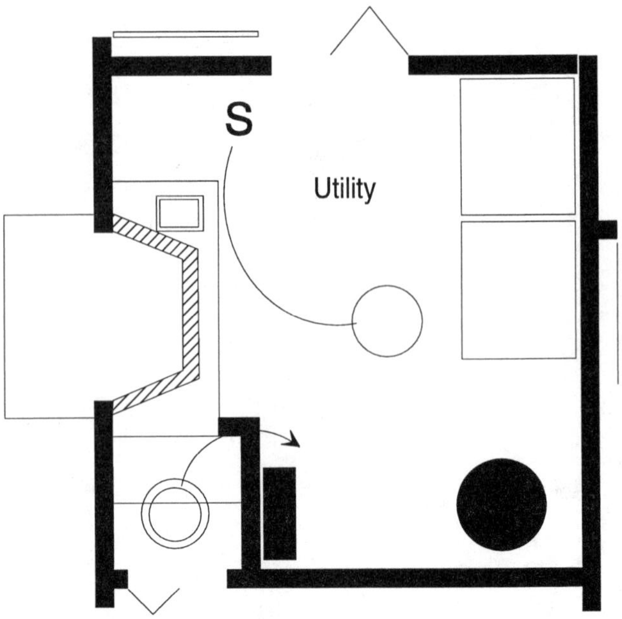
S
Utility

Figure 8-35: An alternate means of lighting the utility room in Figure 8-34

Today, many families have a portion of the basement converted into a family or recreation room. Lighting in this area should be designed for a relaxed, comfortable living atmosphere with the family's interests and activities as a starting point in design.

A typical well-designed lighting layout for a family room should include graceful blending of general lighting with supplemental lighting. For example, diffused incandescent lighting fixtures recessed in the ceiling furnish even, glare-free light throughout the room. The number of fixtures should be increased around game tables for added visual comfort.

Lamps concealed behind cornices near the ceiling enrich the natural beauty of paneled walls. This technique is particularly effective where the light shines down over books with colorful bindings. Fluorescent lamps installed end to end in a cove lighting system will not only furnish excellent general illumination for a family room, but will also give the impression of a higher ceiling; this is a very desirable effect in low-ceiling family rooms.

Light for reading can be accomplished by either table or floor lamps. Post lamps with two or three bullet fixtures are also helpful. One of the bullet housings (containing a reflector lamp) may be aimed at the proper angle for reading while the

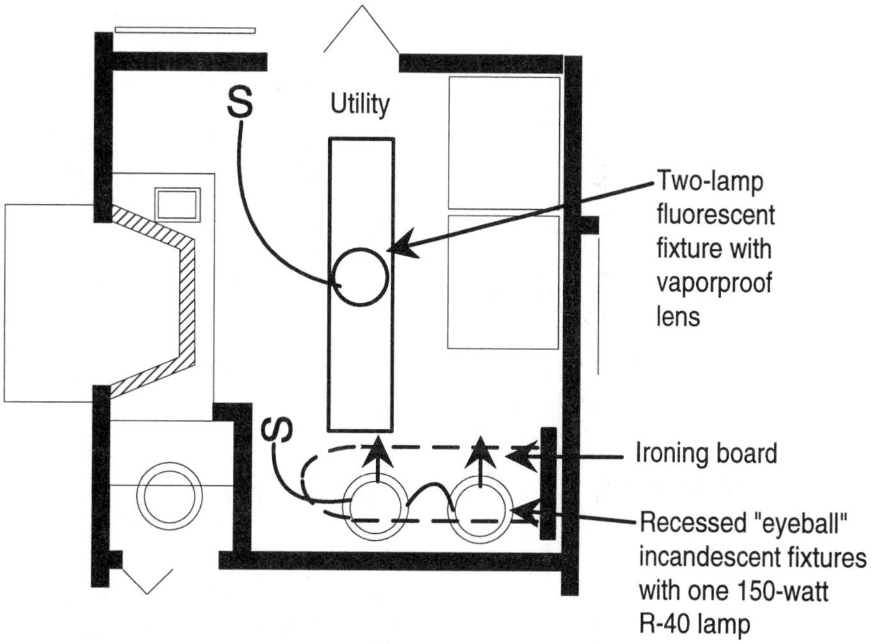
S
Utility
Two-lamp fluorescent fixture with vaporproof lens
S
Ironing board
Recessed "eyeball" incandescent fixtures with one 150-watt R-40 lamp

Figure 8-36: Plan view of lighting layout for ironing

other(s) may be aimed at the ceiling for indirect general lighting.

Directional light fixtures (Figure 8-37) mounted on the ceiling can be used to display mantel decorations or to illuminate a painting.

Fluorescent fixtures such as cabinet fixtures mounted on the underside of a bookcase or shelf create an excellent lighting source for displaying family portraits, collectables, hobby items, and the like.

Any family-room lighting scheme must be very flexible because most family rooms are in the scene of a variety of daily activities, and these activities require different atmospheres, which can be created by light. For instance, TV viewing requires softly lighted surroundings, while reading calls for a somewhat brighter setting with a light directed on the printed pages. Game participants feel more comfortable in a uniformly lighted room with additional glare-free light directed onto the playing area. Casual conversation flourishes amid subdued, complexion-flattering light such as incandescent or warm-white fluorescent.

Low-level lights over the bar area should be just bright enough for mixing a drink or having a late night snack.

A typical family room may be illuminated as follows. The general illumination is accomplished with recessed incandescent fixtures with fresnel lens. The electric circuit controlling these recessed fixtures is provided with a rheostat dimmer control to change the lighting level of the room as well as the atmosphere. A fluorescent lamp is installed in the center of built-in bookshelves to provide light on the counter below for writing, studying, etc. Wall-wash fixtures may also be used near the bookcase to highlight the colorful bindings of the various books.

Small recessed incandescent lamps with star-shaped lenses may be installed above a bar, while slimline fluorescent fixtures may be mounted on the inside and under the bar top to provide additional light on the work counter. If glass shelves are present behind the bar, these may be highlighted with fluorescent fixtures. With all of these lighting fixtures controlled by dimmers, many exciting effects can be achieved with this one lighting scheme.

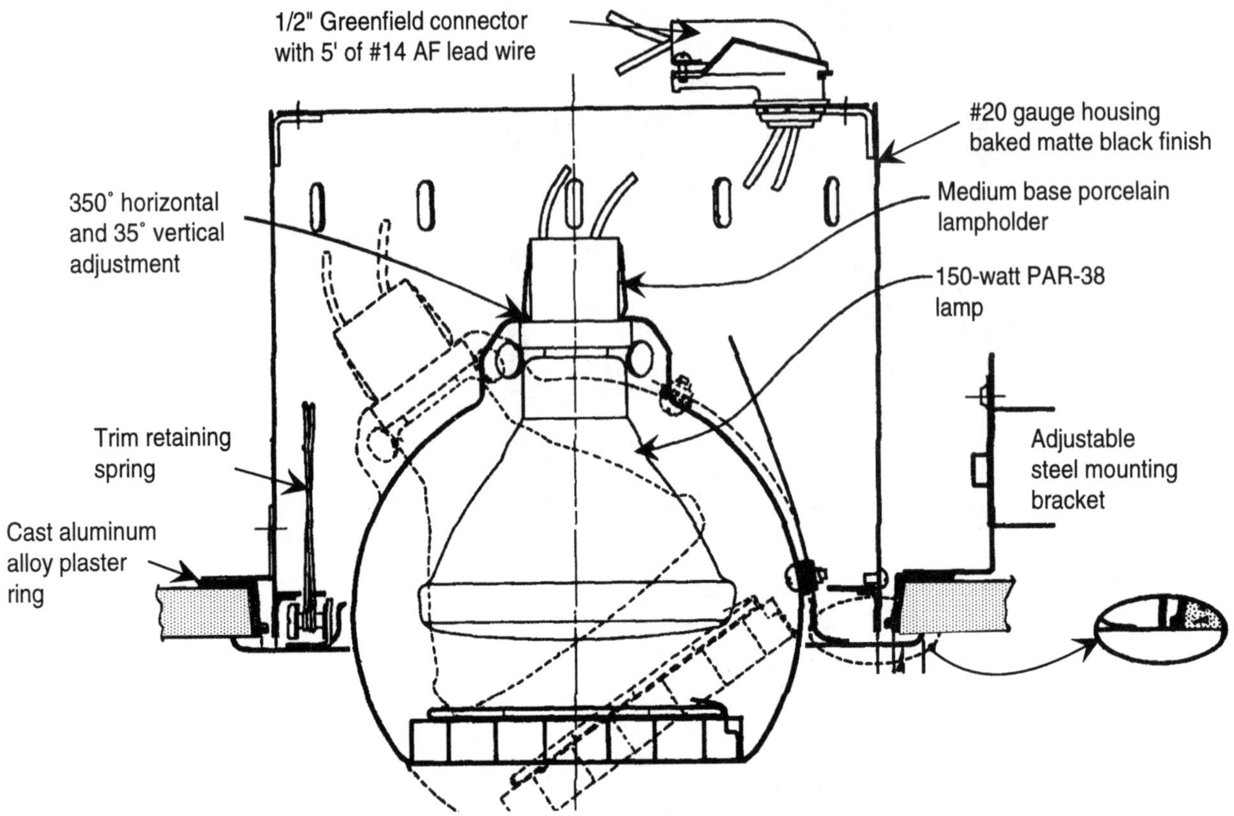
1/2" Greenfield connector
with 5' of #14 AF lead wire
#20 gauge housing
baked matte black finish
350° horizontal
and 35° vertical
adjustment
Medium base porcelain
lampholder
150-watt PAR-38
lamp
Trim retaining
spring
Adjustable
steel mounting
bracket
Cast aluminum
alloy plaster
ring

Figure 8-37: Sectional view of a directional lighting fixture

A floor plan of a typical lighting layout for a family room is shown in Figure 8-38. In this arrangement, the homeowners desired to use table lamps for general illumination. Consequently, only one ceiling-mounted incandescent lighting fixture is shown. This fixture is designated by the symbol "7" inside of a triangle. A lighting-fixture schedule appearing on the working drawings gives the manufacturer, catalog number, number and size of lamps, and the mounting. In this case, the fixture is a surface-mounted fixture containing two 60-watt incandescent lamps. A white opal wraparound diffuser is used to curtail any glare or brightness.

Note that recessed fixtures on the drawing are designated by a circle within a circle — again with the identifying mark next to each fixture or group of fixtures. Junction boxes are provided for the two built-in medicine cabinets in the bathrooms.

ANALYZING EXISTING LIGHTING LAYOUTS

It was previously mentioned that there is always more than one lighting solution to any given application. The final one decided upon usually depends on the homeowner's tastes and the amount allotted in the lighting-fixture budget.

One of the best ways to learn which lighting fixtures to use (and where to use them) in a residence is to analyze existing lighting layouts — those that are actually working and doing a good job for the homeowners. Such drawings follow — beginning with Figure 8-39. Note that Figure 8-39 is a floor plan of the residence used as a sample throughout this book, and the completed lighting layout is shown along with its related circuitry. This is an actual design used in a tenant house for a farm near Culpeper, Virginia. The layout is simple, yet functions well.

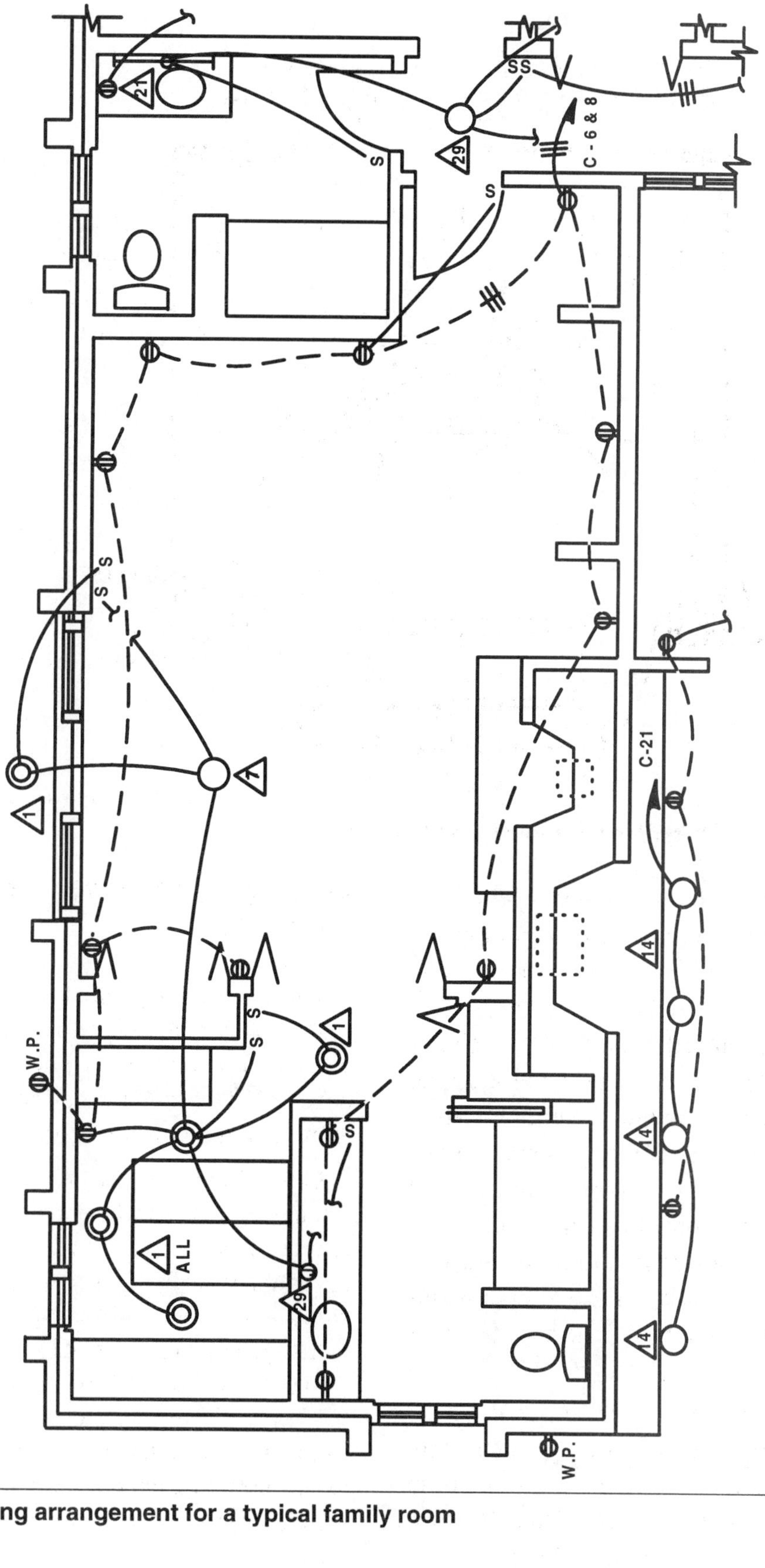
SS
C - 6 & 8
21
29
S
S
S
S
1
1
W.P.
C-21
14
14
14
1
1
ALL
29
S
S
S
W.P.

Figure 8-38: Lighting arrangement for a typical family room

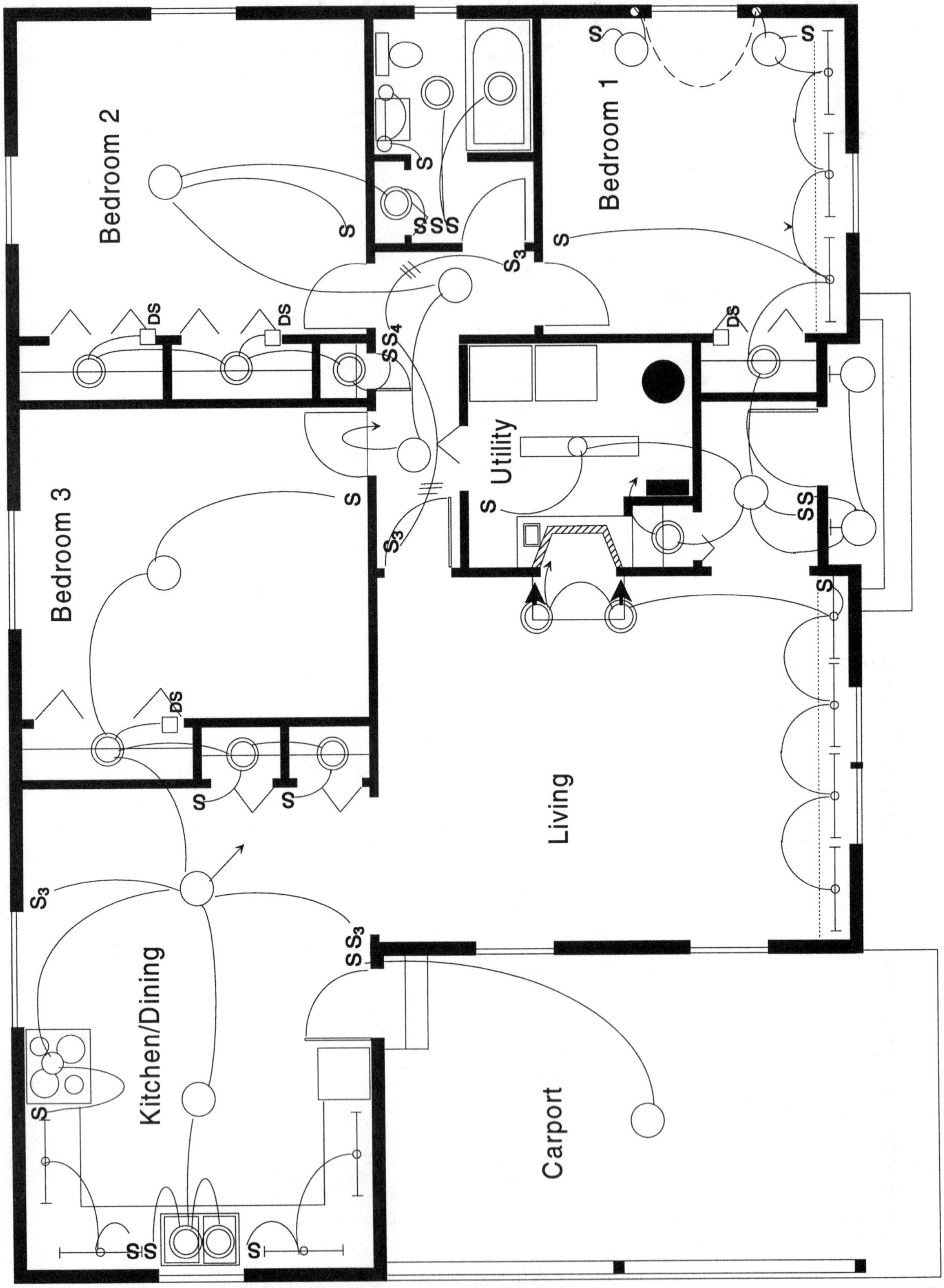
Bedroom 2
Bedroom 1
Bedroom 3
Utility
Living
Kitchen/Dining
Carport

Figure 8-39: Lighting layout for the sample residence used in this book

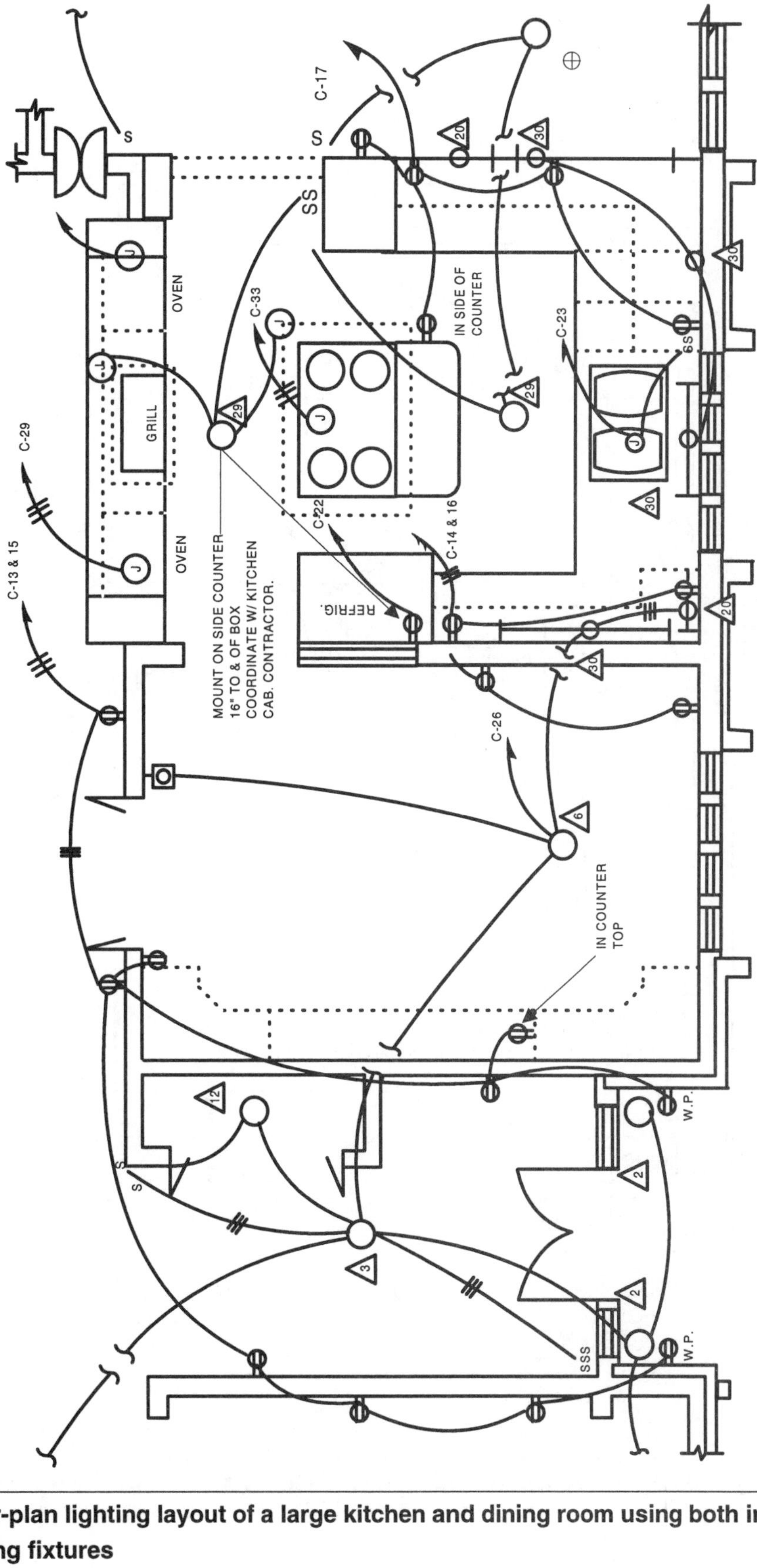
C-17
C-29
C-13 & 15
C-33
C-23
C-22
C-14 & 16
C-26
OVEN
GRILL
OVEN
REFRIG.
IN SIDE OF COUNTER
MOUNT ON SIDE COUNTER 16" TO & OF BOX COORDINATE W/ KITCHEN CAB. CONTRACTOR.
IN COUNTER TOP
W.P.
W.P.
SSS

Figure 8-40: Floor-plan lighting layout of a large kitchen and dining room using both incandescent and fluorescent lighting fixtures

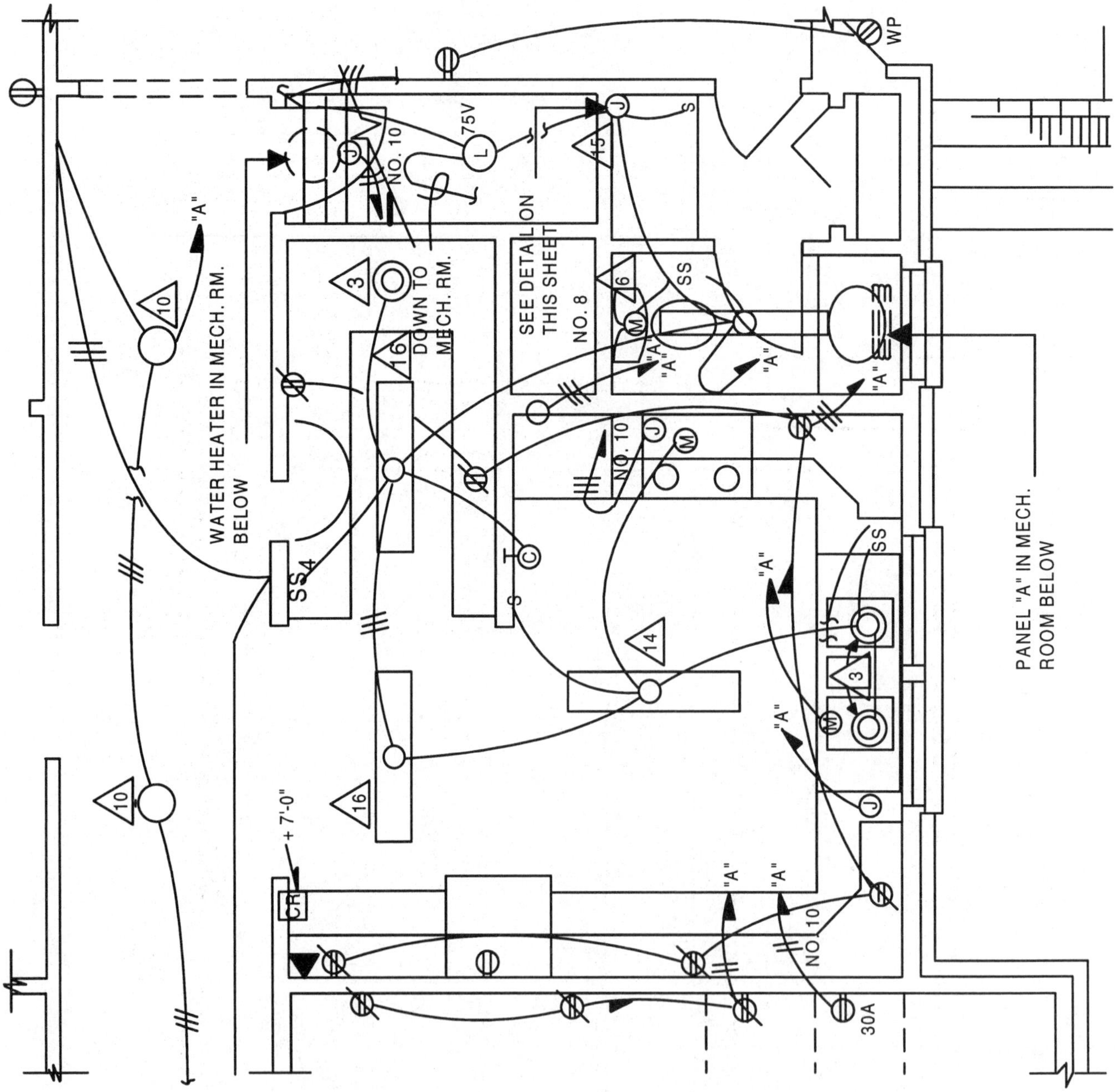
WATER HEATER IN MECH. RM. BELOW
DOWN TO MECH. RM.
SEE DETAIL NO. 8 THIS SHEET
PANEL "A" IN MECH. ROOM BELOW
WP
75V
30A
+ 7'-0"
NO. 10
"A"

Figure 8-41: Lighting and power layout of a modern kitchen utilizing both incandescent and fluorescent fixtures

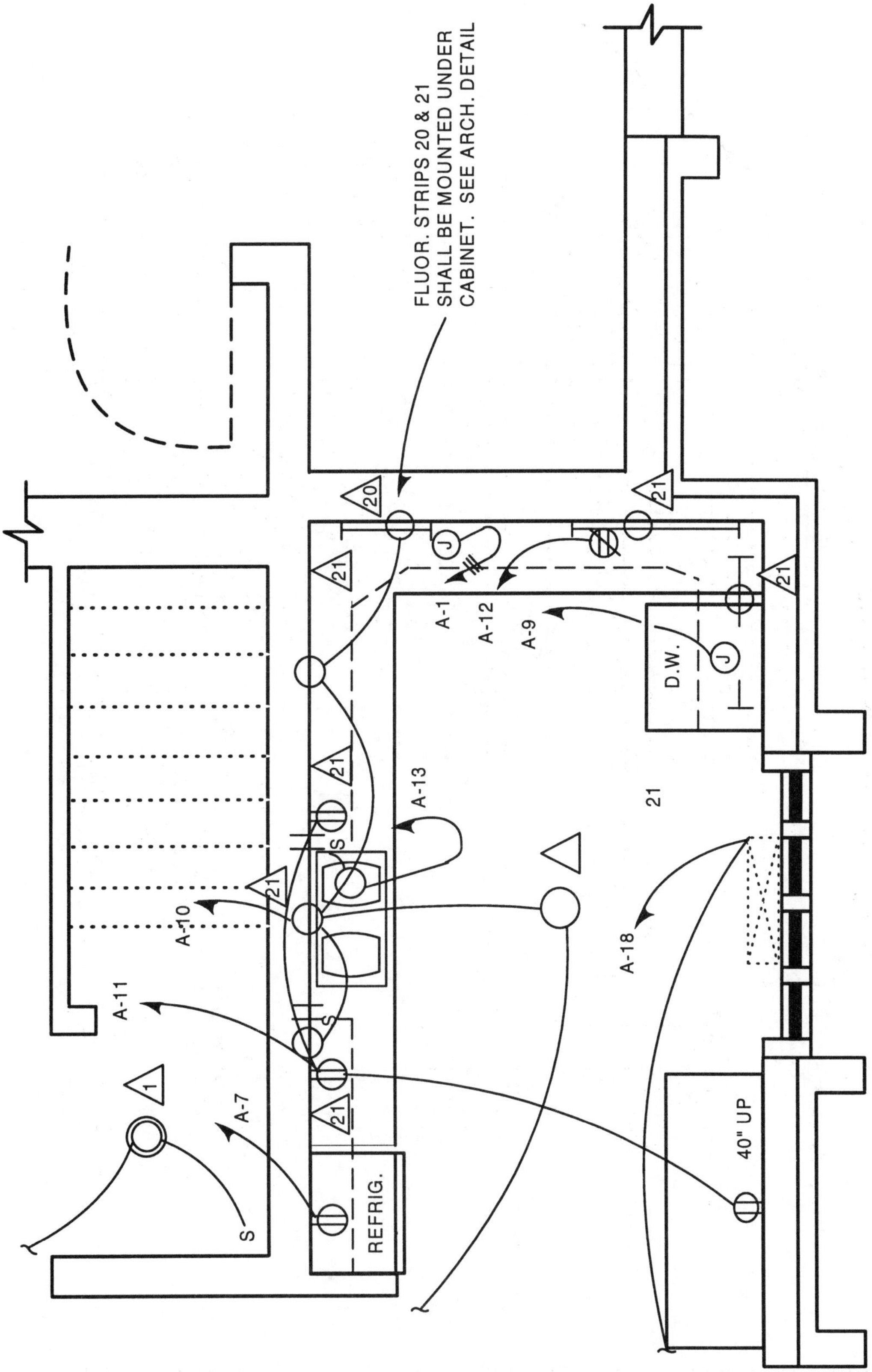
FLUOR. STRIPS 20 & 21
SHALL BE MOUNTED UNDER
CABINET. SEE ARCH. DETAIL
20
21
J
A-1
A-12
A-9
D.W.
A-13
A-10
A-11
A-18
A-7
S
1
REFRIG.
40" UP

Figure 8-42: A small residential kitchen using one incandescent lighting fixture and fluorescent fixtures under the cabinets

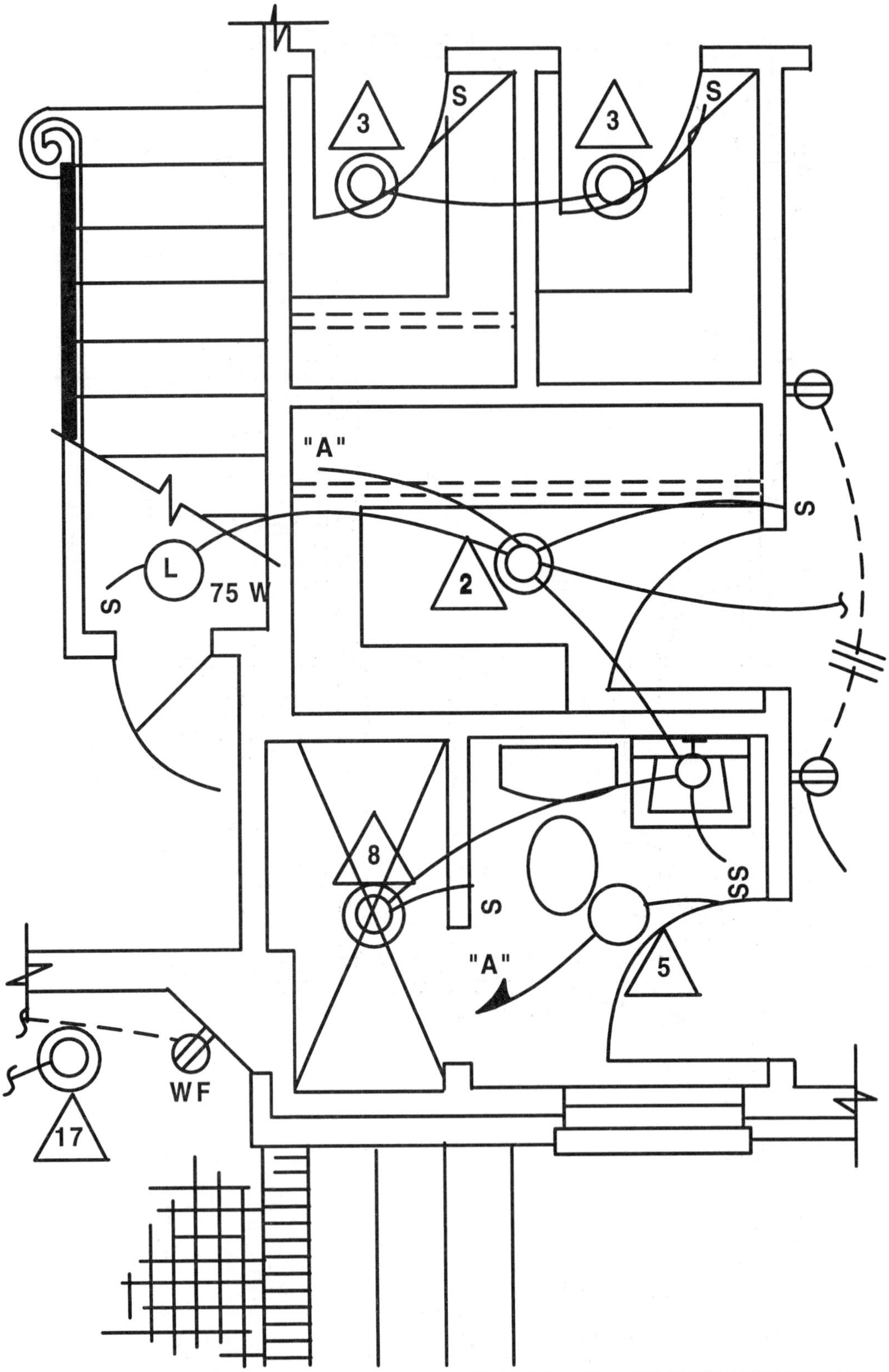

3
S
3
S
"A"
S
L
75 W
S
2
8
S
SS
"A"
5
WF
17

Figure 8-43: Plan for lighting a closet and a small bathroom; the junction box over the lavatory is for a built-in medicine cabinet containing its own lights

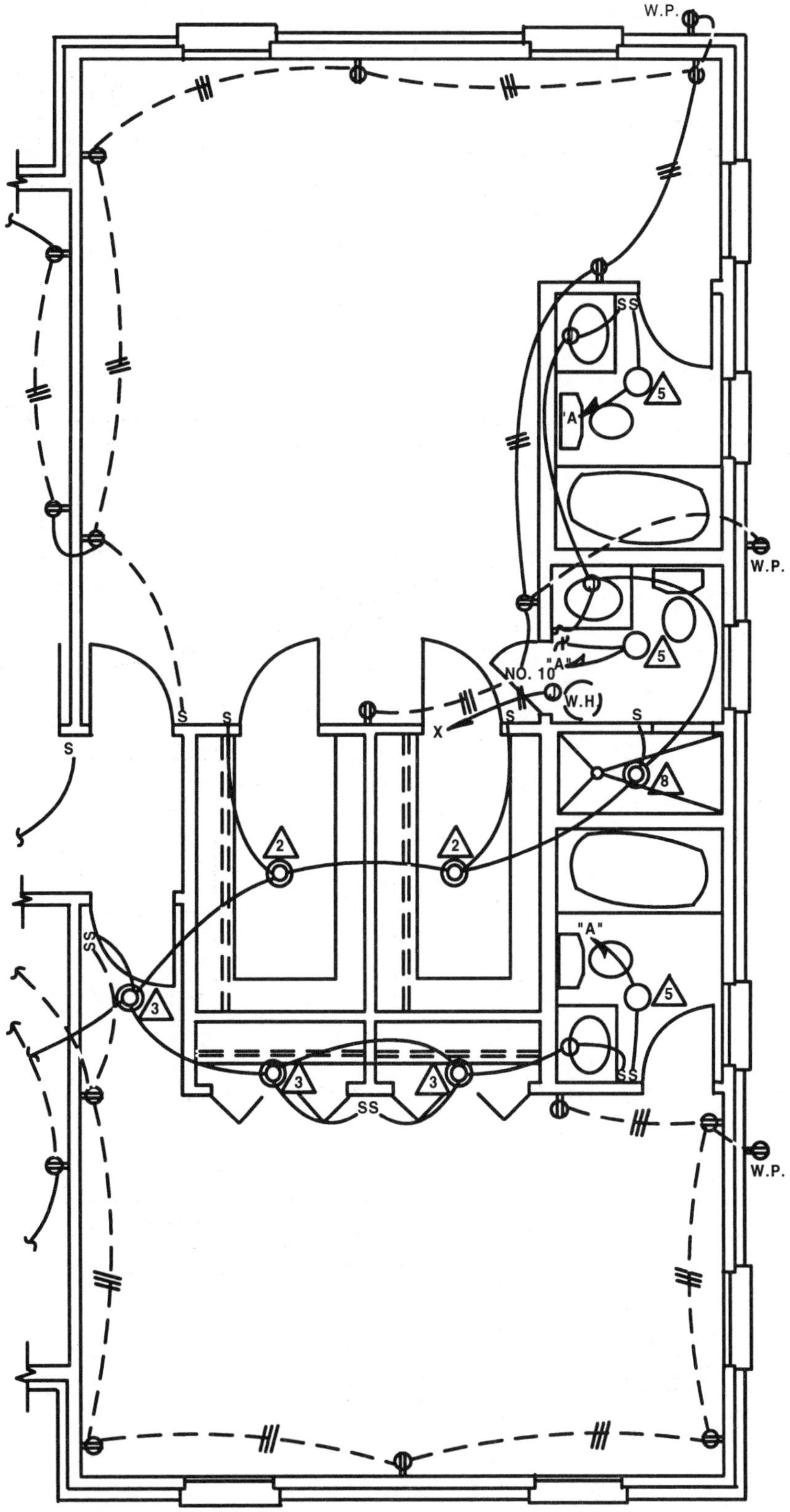
W.P.
W.P.
W.P.
"A"
"A"
"A"
NO. 10
W.H.
S
SS
X
2
3
5
8

Figure 8-44: Plan view of two bedrooms, walk-in closets, and adjoining bathrooms with the lighting outlets and related circuits

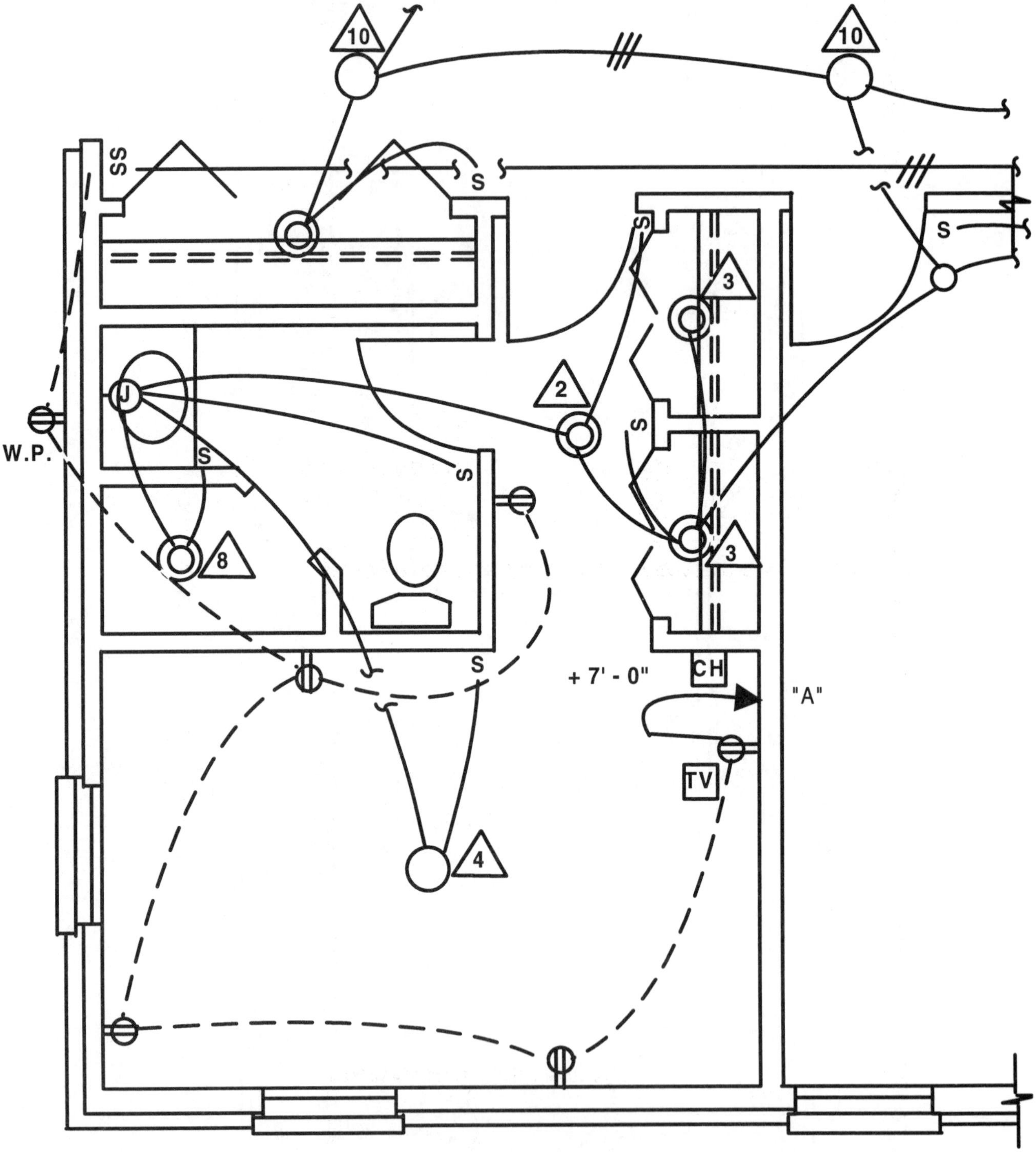
10
10
SS
S
S
3
2
3
8
W.P.
S
S
S
S
S
+ 7' - 0"
CH
"A"
TV
4

Figure 8-45: Lighting layout for a residence showing bedroom, closets, and adjoining bath

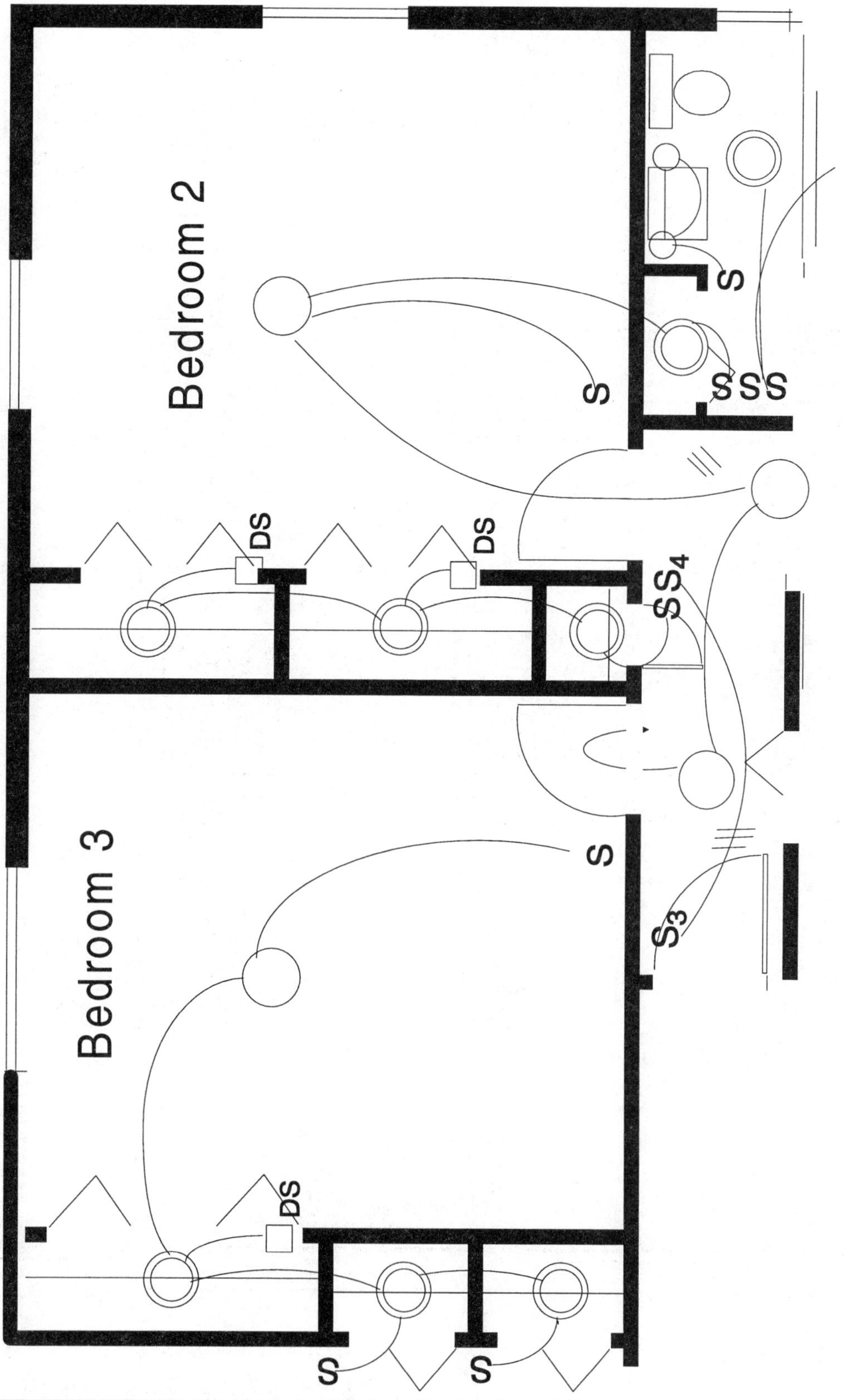
Bedroom 2
Bedroom 3
DS
DS
DS
S
SSS
S
SS_4
S
S_3
S
S

Figure 8-46: Lighting layout for bedrooms, closets, and bath

When designing a lighting layout for any project, remember not to "over-kill." Selecting lighting fixtures to match the style of the building or the home's decor should be the first consideration, while the amount of effective illumination should be second. Once this goal has been reached, stop! All that remains is to decide on the most adequate and convenient lighting-control system.

Lighting Catalogs

Manufacturers of lighting fixtures have colorful brochures and catalogs that can be extremely helpful to the electrician involved in residential lighting applications. Besides showing the types of fixtures available, and listing them by catalog number, examples of their use are also given. When a new project is encountered, a glance through a dozen or so of these brochures and catalogs will often give you some excellent ideas on how to proceed with your project.

Some of the larger lighting catalogs also provide design data for their fixtures that will prove helpful on the more sophisticated projects.

Brochures are normally available from lighting-fixture manufacturers at little or no charge. Your local lighting-fixture dealers will probably have free literature in their places of business. Visiting these dealers will also give you a chance to see the various types of fixtures on display.

Lighting Fixture Manufacturers

There are hundreds of lighting manufacturers in the United States. Names and addresses for most of these can be found in the CEE News Buyers' Guide, available for about $20 from Intertec Publishing Corp, 9221 Quivira Road, Overland Park, KS 66215.

UNDERWATER LIGHTING

Many homes throughout the United States incorporate water fountains into their landscaping designs. This is especially true of multi-family dwellings where a water fountain becomes a centerpiece for a veranda or courtyard. Swimming pools are also becoming more popular for private homes. New designs with vinyl pool liners have made private swimming pools more affordable than they were a couple of decades ago. Consequently, swimming pools are found in thousands of residential backyards.

This section discusses the fundamentals of designing underwater lighting for fountains and swimming pools. Such lighting is used mainly for decorative purposes and can be compared to a painting or any other work of art. While the designer's artistic ability is very important in lighting designs of this type, there are certain basic rules which may be followed in order to produce good lighting layouts. *NE Code* installation requirements also tightly govern the installation methods of underwater lighting.

Fountain Lighting Design

A fountain is utilized for one or more of the following reasons:

- Sheer fascination of visual and sound effects.
- To create product or trademark identification.
- A decorative feature piece.
- For animation.
- Air conditioning reject heat load.
- Enhance the surroundings of outstanding architectural structures.

Considerations for fountain lighting should include:

- The type of water effect to be lighted.
- Color selection.
- Maximum height to be illuminated.
- Selecting the type and number of fixtures and lamps.

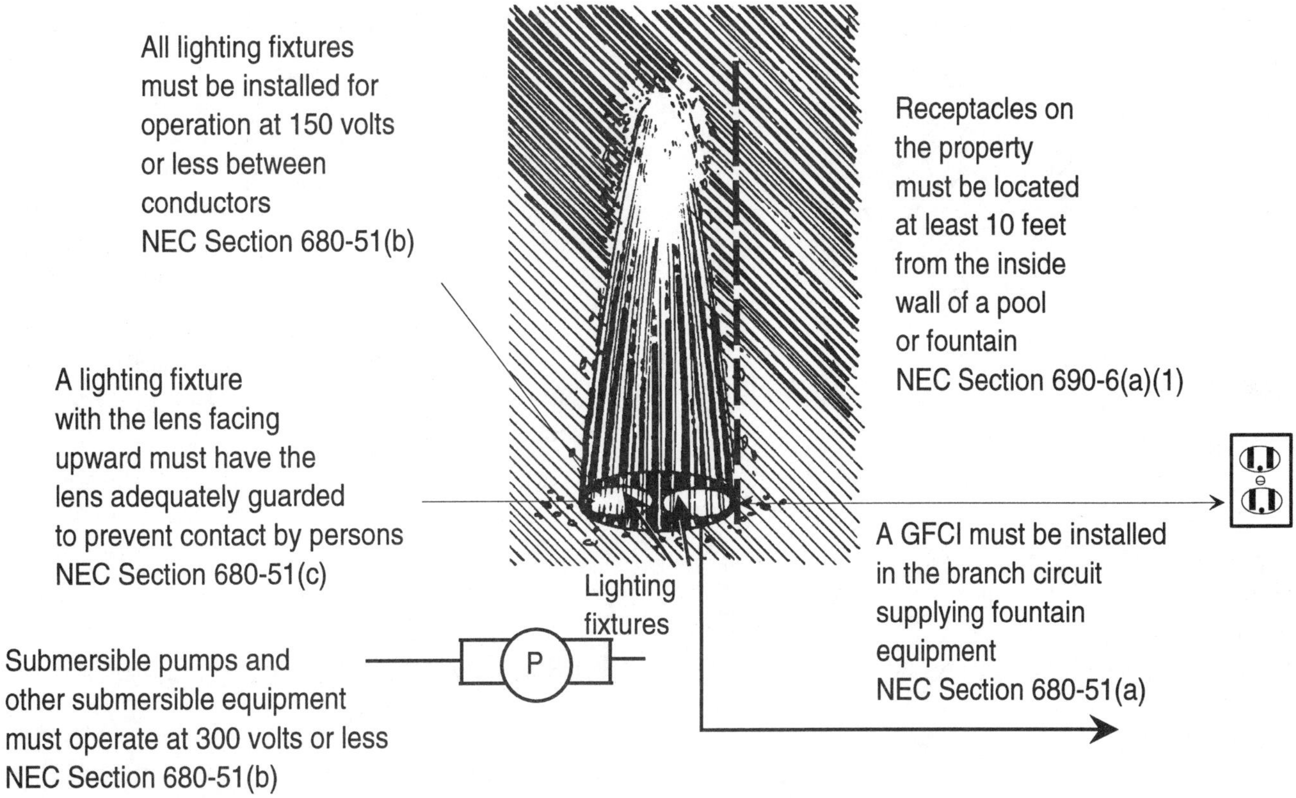
All lighting fixtures
must be installed for
operation at 150 volts
or less between
conductors
NEC Section 680-51(b)
Receptacles on
the property
must be located
at least 10 feet
from the inside
wall of a pool
or fountain
NEC Section 690-6(a)(1)
A lighting fixture
with the lens facing
upward must have the
lens adequately guarded
to prevent contact by persons
NEC Section 680-51(c)
A GFCI must be installed
in the branch circuit
supplying fountain
equipment
NEC Section 680-51(a)
Lighting
fixtures
Submersible pumps and
other submersible equipment
must operate at 300 volts or less
NEC Section 680-51(b)
P

Figure 8-47: A single-nozzle fountain with *NE Code* installation requirements

Types of Water Effect to Be Lighted

Single-nozzle fountains are normally lighted with two lighting fixtures with spot lamps, such as illustrated in Figure 8-47.

The smaller spray ring fountains may be lighted with one fixture, such as illustrated in Figure 8-48. The larger spray ring fountains are normally lighted with one fixture every 4 to 5 feet for illuminating heights up to 15 feet. Wide-angle flood lamps are used. For heights over 15 feet, use one fixture every 2 to 3 feet with medium-spread flood lamps. See example in Figure 8-49. This procedure also applies to waterfalls and weirs.

Color selection: The selection of colors is a subjective matter and can vary as much as opinions of designers or owners of buildings. However, the following may be used as a guide in selecting colors of lamps.

- Colors directly affect the selection of fixtures and lamps, inasmuch as various colors require differing candlepower to achieve decorative effects of light.
- Amber and turquoise lamps require 50% more candlepower than a clear lens for the same level of illumination.
- Red lamps require 100% more candlepower than a clear lens for the same level of illumination.
- Blue and green lamps require 250% more candlepower than a clear lens for the same level of illumination.
- Where high levels of illumination surround the fountain, use caution when selecting colors; the surrounding light will tend to wash out the colored light.

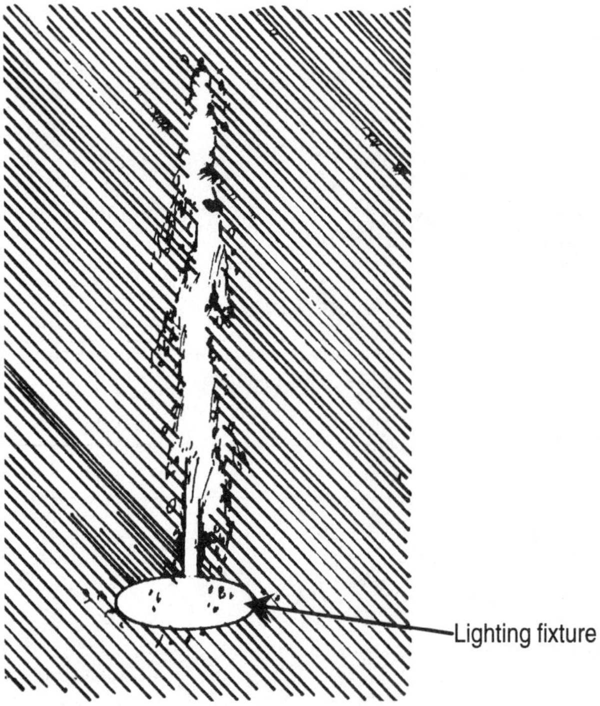
Lighting fixture

Figure 8-48: A spray-ring fountain

Heights to Be Illuminated

The table in Figure 8-50 may be used as a guide in determining the minimum beam candlepower required for a given height in order to achieve a reasonable balance of decorative effect colors. The values are based upon the use of standard lenses as manufactured by Kim Lighting & Manufacturing Co., Inc.

The total footcandle requirements for a specific lighting layout would be contingent upon the perimeter of the layout.

A tabulation of standard available lamps and their rated candlepower can be found in Figure 8-51.

Typical Layout

The illustration in Figure 8-49 shows a typical fountain layout. In the first part of the table in Figure 8-50, the twelve fountain lights surrounding the jet nozzle are located on a $2\frac{1}{2}$-foot radius. In the second part of the table, the outside ring is lighted by 24 fountain lights located on a 6-foot radius.

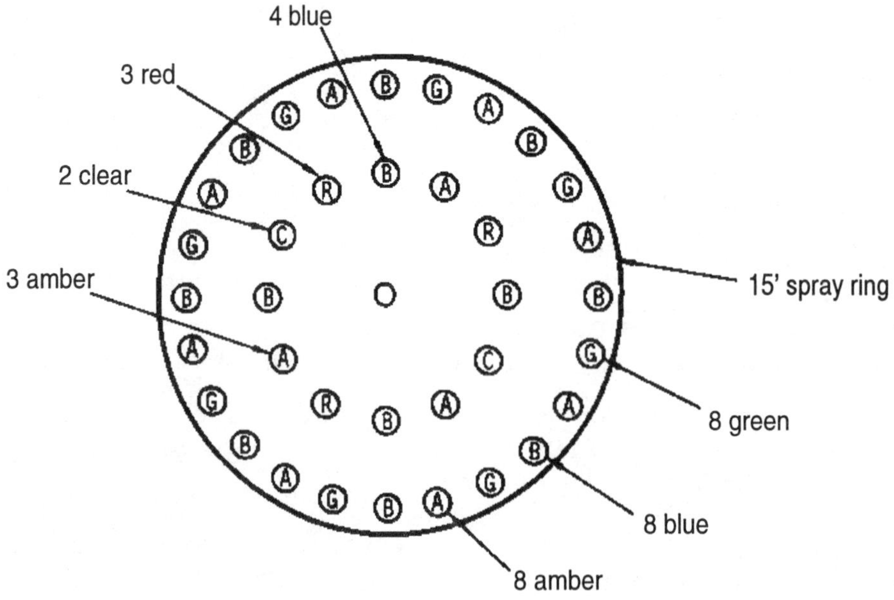
4 blue
3 red
2 clear
3 amber
15' spray ring
8 green
8 blue
8 amber

Figure 8-49: A large spray-ring fountain

Height of Water Effect (ft)	Clear	Amber & Turquoise	Red	Blue & Green
5	4,000	6,000	8,000	14.000
10	11,000	16,000	22,000	38,000
l5	21,000	31,000	42,000	73,000
20	34,000	51,000	68,000	119,000
25	50,000	75,000	100,000	175,000
30	69,000	103,000	138,000	241,000
35	91,000	136,000	182,000	378,000
40	115,000	174,000	230,000	406,000

Figure 8-50: Minimum desirable beam candlepower

Watts	Bulb	Ordering Code	Beam Type	Beam Spread (degrees)	Initial Average Maximum Beam Candlepower	Approx Life (hours)
150	PAR-38	150PAR/SP	Spot	30 × 30	10,500	2000
150	PAR-38	150PAR/FL	Flood	60 × 60	3500	2000
150	R-40	150R/SP	Spot	40	6300	2000
150	R-40	150R/FL	Flood	110	1300	2000
250	PAR-38	Q250 PAR38SP	Spot	26	34,000	4000
250	PAR-38	Q250PAR38FL	Flood	60	6000	4000
300	PAR-56	300PAR56 /NSP	Spot	15 × 20	70,000	2000
300	PAR-56	300PAR56/MFL	Med-Flood	20 × 35	22,000	2000
300	PAR-56	300PAR56/WFL	Wide-Flood	30 × 60	10,000	2000
500	PAR-56	Q500PAR56/NSP	Spot	15 × 32	90,000	4000
500	PAR-56	Q500PAR56/MFL	Med-Flood	20 × 42	49,000	4000
500	PAR-56	Q500PAR56/WFL	Wide-Flood	34 × 66	18,000	4000
1000	PAR-64	Q1000PAR64NSP	Spot	14 × 31	160,000	4000
1000	PAR-64	Q1000PAR64MFL	Med-Flood	22 × 45	60,000	4000
1000	PAR-64	Q1000PAR64WFL	Wide-Flood	45 × 72	27,000	4000

Figure 8-51: Standard lamps and their rated candlepower

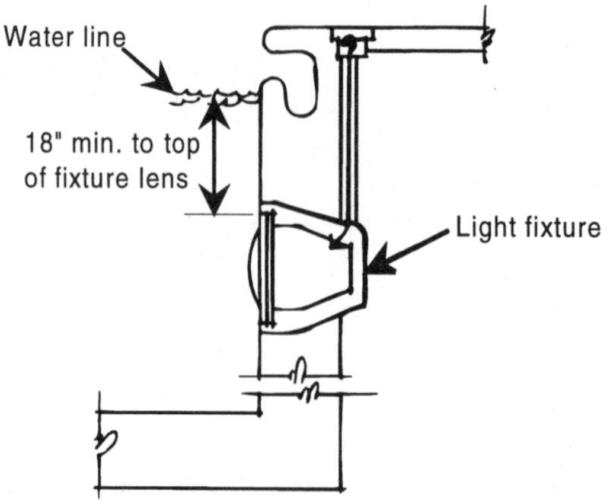
Water line
18" min. to top
of fixture lens
Light fixture

Figure 8-52: Cross-sectional view of a typical swimming-pool lighting fixture

SWIMMING POOL LIGHTING

There is a large selection of underwater lights, from small 75-watt units to large 1000-watt units, to meet practically every underwater lighting requirement.

In selecting equipment for underwater lighting, the greatest economy is achieved through the selection of a high quality fixture designed to give years of dependable service. By their very nature, underwater lighting fixtures are subjected to the forces most destructive to electrical fixtures — water, chemicals, and neglect.

The illustrations in Figures 8-52 and 8-53 show some recommended lighting layouts for swimming pools.

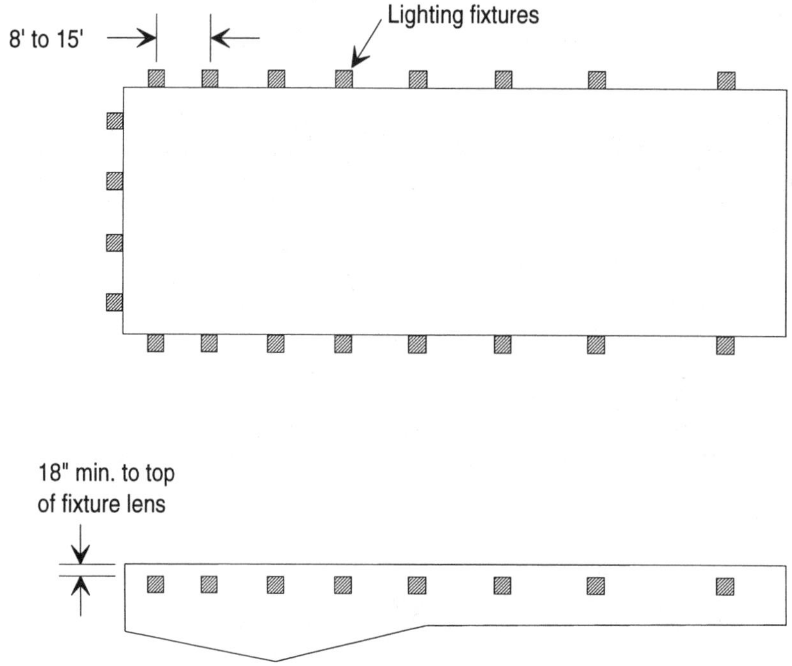
8' to 15'
Lighting fixtures
18" min. to top
of fixture lens

Figure 8-53: Recommended lighting layout for swimming pools

NE Code Requirements for Swimming Pools

The *NE Code* recognizes the potential danger of electric shock to persons in swimming pools, wading pools, and therapeutic pools or near decorative pools or fountains. This shock could occur from electric potential in the water itself or as a result of a person in the water or a wet area touching an enclosure that is not at ground potential. Accordingly, the *NE Code* provides rules for the safe installation of electrical equipment and wiring in or adjacent to swimming pools and similar locations. *NE Code* Article 680 covers the specific rules governing the installation and maintenance of swimming pools and similar installations.

Installation procedures for swimming pool electrical systems are too vast to be covered in detail in this chapter. However, the general requirements for the installation of outlets, overhead conductors and other equipment are summarized in Figures 8-54 and 8-55.

Besides *NE Code* Article 680, another good source for learning more about electrical installations in and around swimming pools is from manufacturers of swimming pool equipment, including those who manufacture and distribute underwater lighting fixtures. Many of these manufacturers offer pamphlets detailing the installation of their equipment with helpful illustrations, code explanations, and similar details. This literature is usually available at little or no cost to qualified personnel. You can write directly to manufacturers to request information about available literature, or contact your local electrical supplier or contractors who specialize in installing residential swimming pools.

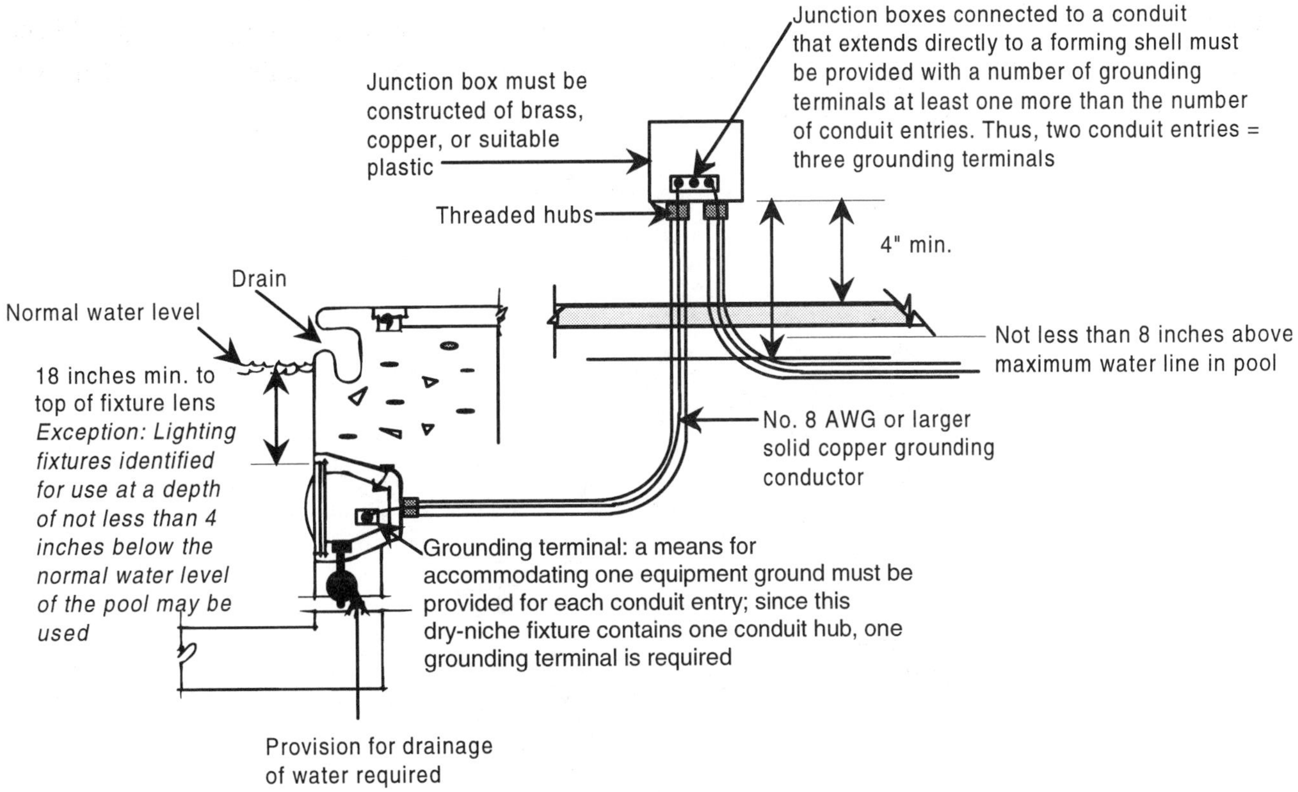
Junction boxes connected to a conduit that extends directly to a forming shell must be provided with a number of grounding terminals at least one more than the number of conduit entries. Thus, two conduit entries = three grounding terminals
Junction box must be constructed of brass, copper, or suitable plastic
Threaded hubs
4" min.
Drain
Normal water level
Not less than 8 inches above maximum water line in pool
18 inches min. to top of fixture lens
Exception: Lighting fixtures identified for use at a depth of not less than 4 inches below the normal water level of the pool may be used
No. 8 AWG or larger solid copper grounding conductor
Grounding terminal: a means for accommodating one equipment ground must be provided for each conduit entry; since this dry-niche fixture contains one conduit hub, one grounding terminal is required
Provision for drainage of water required

Figure 8-54: *NE Code* installations for pool lights and related components

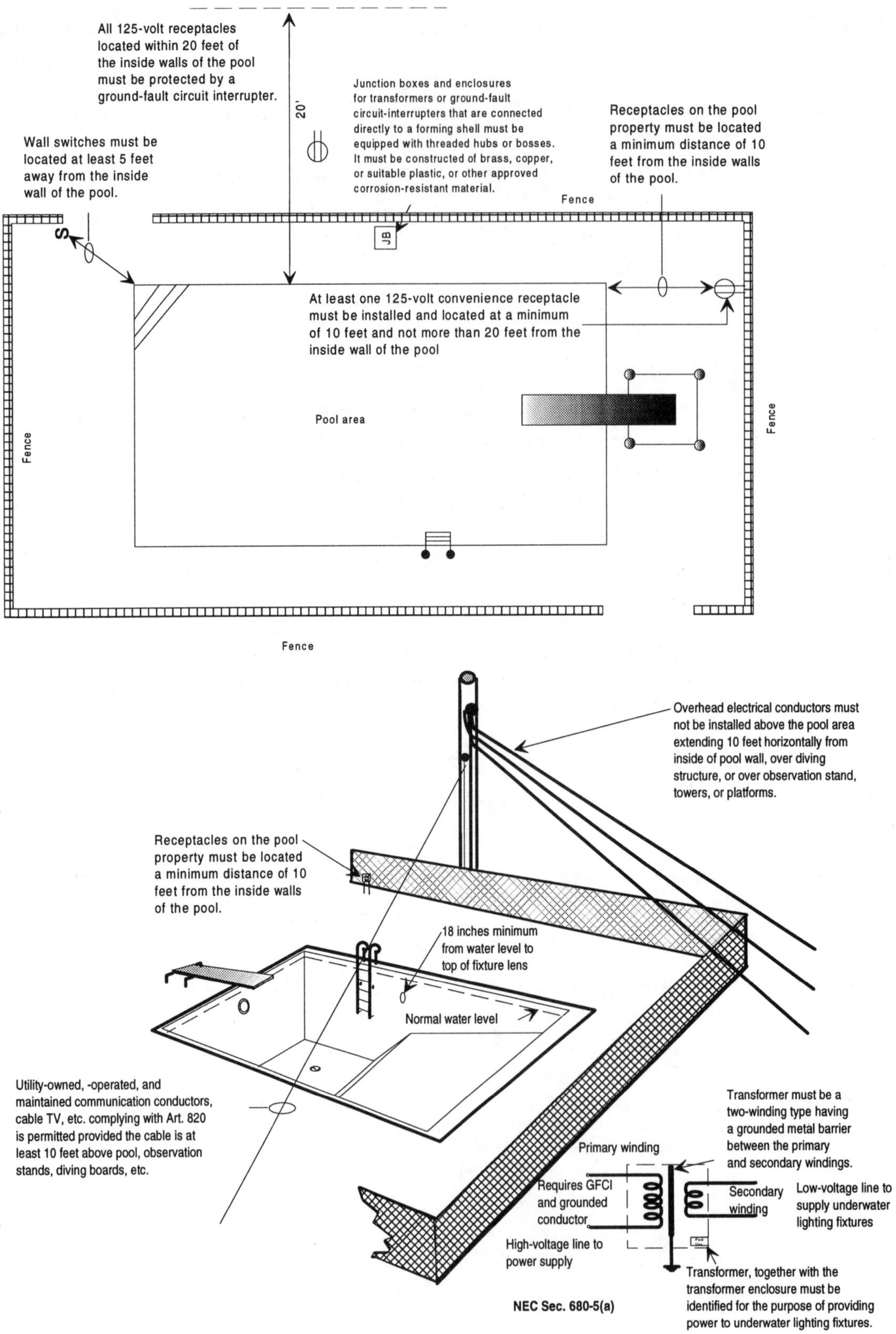
All 125-volt receptacles located within 20 feet of the inside walls of the pool must be protected by a ground-fault circuit interrupter.
20'
Junction boxes and enclosures for transformers or ground-fault circuit-interrupters that are connected directly to a forming shell must be equipped with threaded hubs or bosses. It must be constructed of brass, copper, or suitable plastic, or other approved corrosion-resistant material.
Receptacles on the pool property must be located a minimum distance of 10 feet from the inside walls of the pool.
Wall switches must be located at least 5 feet away from the inside wall of the pool.
Fence
S
JB
At least one 125-volt convenience receptacle must be installed and located at a minimum of 10 feet and not more than 20 feet from the inside wall of the pool
Pool area
Fence
Fence
Fence
Overhead electrical conductors must not be installed above the pool area extending 10 feet horizontally from inside of pool wall, over diving structure, or over observation stand, towers, or platforms.
Receptacles on the pool property must be located a minimum distance of 10 feet from the inside walls of the pool.
18 inches minimum from water level to top of fixture lens
Normal water level
Utility-owned, -operated, and maintained communication conductors, cable TV, etc. complying with Art. 820 is permitted provided the cable is at least 10 feet above pool, observation stands, diving boards, etc.
Transformer must be a two-winding type having a grounded metal barrier between the primary and secondary windings.
Primary winding
Requires GFCI and grounded conductor
Secondary winding
Low-voltage line to supply underwater lighting fixtures
High-voltage line to power supply
Transformer, together with the transformer enclosure must be identified for the purpose of providing power to underwater lighting fixtures.
NEC Sec. 680-5(a)

Figure 8-55: General *NE Code* requirements for pool electrical installations

Application	Requirements	*NE Code* Reference
Receptacles	Receptacles must be at least 10 feet from the inside wall of the pool. At least one receptacle must be located in an area 10 feet to 20 feet from the inside wall of a pool permanently installed at a residential occupancy. All receptacles in an area 10 feet to 20 feet from the inside wall of the pool must have a GFCI installed to protect the circuit.	680-6(a)
Lighting Fixtures	Lighting fixtures over pool or the area 5 feet from the pool must be 12 feet above water level, unless otherwise excepted in the *NE Code*. A cord-connected lighting fixture within 16 feet of the water surface must meet the requirements of other cord- and plug-connected equipment.	680-6(b)
Switching Devices	Switching devices must be at least 5 feet from the inside wall of the pool unless separated from the pool by a permanent barrier.	680-6(c)
Cord- and Plug-Connected Equipment	Fixed or stationary equipment rated 20 amperes or less may be cord and plug connected.	680-7
Clearance of Overhead Conductors	Overhead conductors must at least 10 feet above water level, diving boards, towers, and other structures.	680-8
Pool Heaters	Electric water heaters for pools must not exceed 48 amperes in rating or be protected at more than 60 amperes.	680-9
Underground Wiring	Underground wiring is not permitted under the pool or under the area within 5 feet from the pool with some exceptions.	680-10

Figure 8-56: Summary of *NE Code* installation rules

Chapter 9
Outdoor Lighting

Outdoor lighting is a partner in modern residential living. When well planned, it creates a total home environment, combining maximum aesthetic appeal with efficiency. It welcomes guests and lights their way to the house entrance; it creates a hospitable look and turns the area surrounding the home into an extra living area in warm months; reveals the beauty of gardens, trees, and foliage; expands the hospitality and comfort of patios and porches; stretches the hours for outdoor recreation or work; provides sure seeing to safeguard persons against accidents at night; and protects the home from prowlers.

This chapter is designed to show the reader how to create appealing outdoor lighting and will cover the following:

- Capturing the mystery and subtle qualities of outdoor lighting.
- Lighting ground contours and focal points.
- Creating silhouetted forms and shadow patterns.
- Using colored light.
- Using outdoor lighting to make the home more attractive, safe, and fun to live in.

Unlike some types of indoor lighting, designing outdoor lighting is mainly the process of using techniques gained from experience.

ENTRANCE LIGHTING

Entrance lighting makes a visitor's first impression a good one. At the same time, good entrance lighting flatters the home. Even with the modest residence covered in this book, the two wall-bracket lighting fixtures mounted on each side of the entrance door create a festive and somewhat luxurious look.

The two low-level lighting fixtures along the walk from the driveway to the front door clear a path through the darkness and see visitors safely to the front door. When used in combination with the wall-bracket lights, these two lights create a grand setting for the front lawn, making it seem larger and more beautiful. Figure 9-1 shows how these fixtures appear on the plan.

The post light at the driveway entrance lights a small sign giving the owners' name and house number — a convenience first-time visitors to the home will appreciate.

Since our residence under consideration is a modest one, the outdoor lighting is not too elaborate. The four lighting fixtures previously described are all that were used for entrance lighting. However, in larger and more luxurious homes other outdoor entrance lighting may be appropriate:

- Lighting fixtures recessed in the soffit of the front of the house will act as wall-wash fixtures, lighting a stone

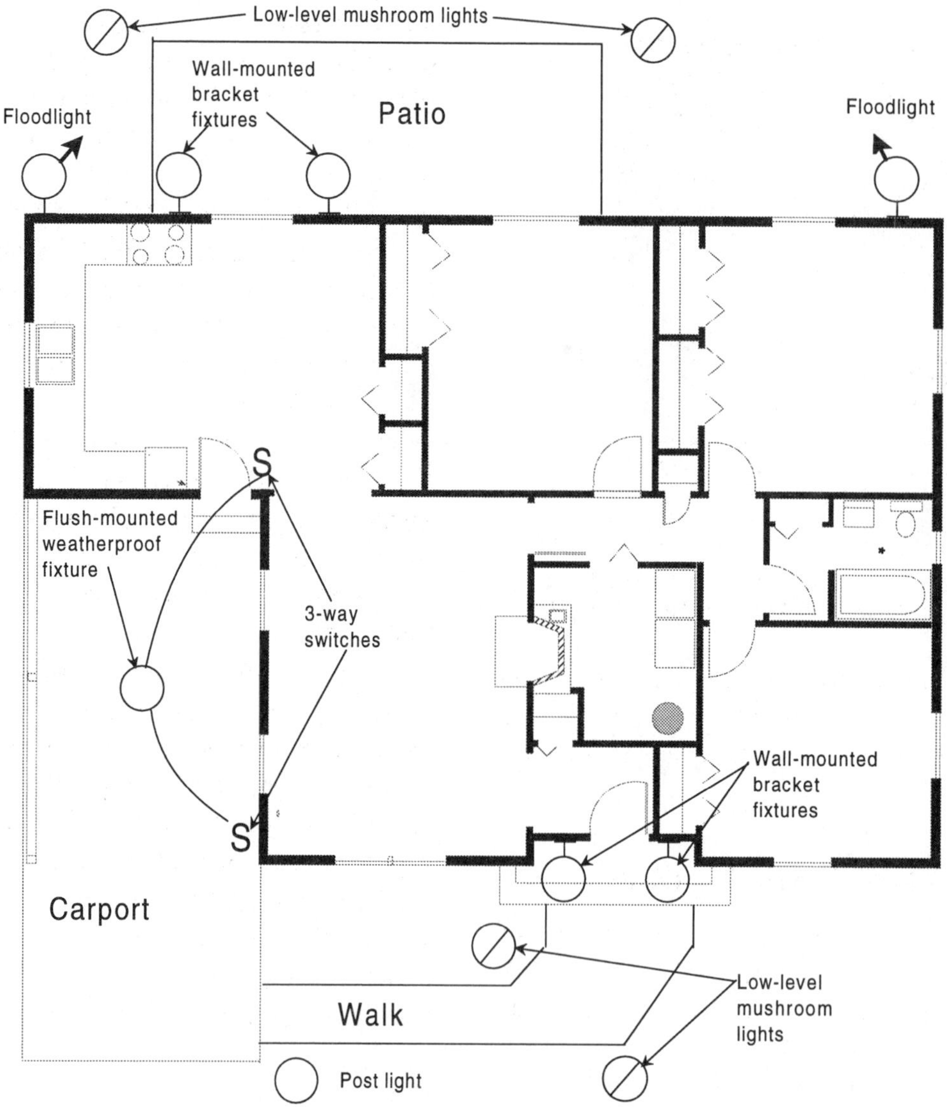
Low-level mushroom lights
Wall-mounted
bracket
fixtures
Patio
Floodlight
Floodlight
Flush-mounted
weatherproof
fixture
S
3-way
switches
S
Wall-mounted
bracket
fixtures
Carport
Low-level
mushroom
lights
Walk
Post light

Figure 9-1: Outdoor lighting layout for the residence under consideration in this book

facade or bringing out the texture of beautiful brick.

- Ground-mounted up-lights installed under trees on the lawn shining up through the tree branches will add elegance to any lawn and home.
- Spotlights recessed into the roof overhang can accent the house finish, light walks near the house, accent the entrance door, and dramatize painting or architectural detail — again extending a friendly welcome.

This same outdoor lighting that gives the home a proper introduction to friends also chases away shadows where danger could lurk. The lighting that enhances the architecture and landscape around the sides and rear of the home also forms a protective ring of light. Therefore, properly planned outdoor lighting tightens the security of any home at no additional cost. It is recommended that some or all of the outdoor lighting be controlled by either a photocell or a time switch, or a combination of both. Details of outdoor, as well as other types of lighting control are covered in Chapters 10 and 11.

Figure 9-2: Example of wide-beam uplighting

Figures 9-2 through 9-7 give examples of various types of outdoor lighting; accompanying plans show how these same illustrations appear on electrical working drawings.

PORCHES, PATIOS, AND TERRACES

The carport of the residence under consideration has one ceiling-mounted lighting fixture installed in the center of the area. A three-way switch with a weatherproof cover is located on the outside wall of the house, about where a person would get out of the car. This makes it convenient to turn the light on when the owners are coming home at night. The light may be turned off inside the house by means of another three-way switch.

The outdoor patio in back of our residence is lighted by a combination of wall-bracket fixtures and floodlights; low-level "mushroom" lighting fixtures are used along the outside edge of the patio for added flexibility.

This arrangement of outdoor lighting pushes back the barriers of darkness and opens up wide vistas of family fun and fulfillment. The owners of this house now enjoy the relaxing qualities of night in the comfort and convenience of a family recreation room.

A patio or terrace can be lighted in several ways, such as with a group of carefully positioned lighting fixtures installed under the roof overhang to evenly flood the terrace or patio with illumination. A pendant-type fixture hung from the terrace ceiling over a table on the terrace to provide glare-free light for card games, eating, or just engaging in conversation with friends.

When friends are being entertained on a patio or screened-in porch, a higher level of illumination

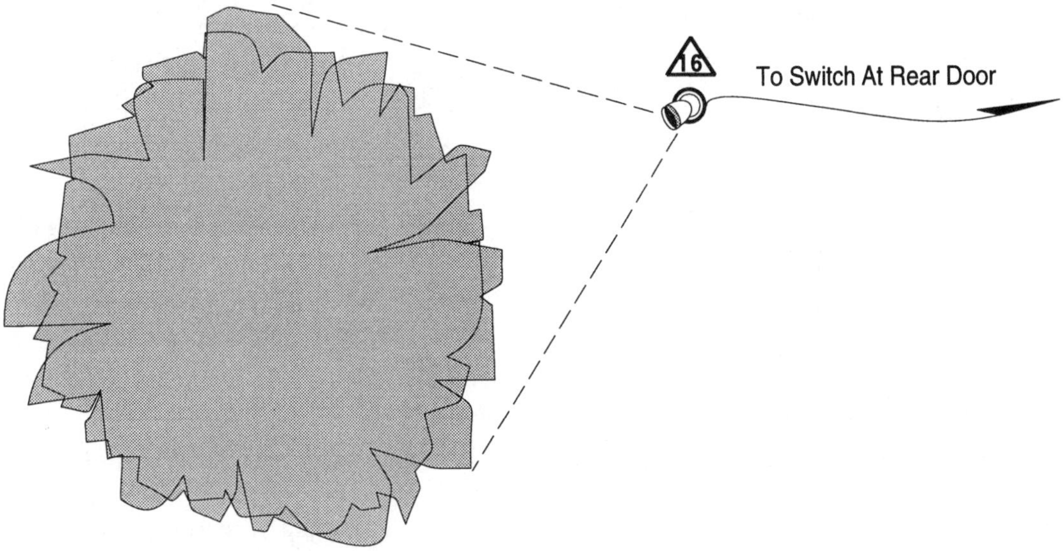
16
To Switch At Rear Door

Figure 9-3: Plan view of Figure 9-2

Figure 9-4: Example of controlled wide-beam uplighting

on colorful shrubs and flowers than that used for the porch will make the porch seem more spacious besides showing off the floral growth in the yard.

In choosing outdoor lighting for any home, the designer should coordinate the fixtures, matching them in styling to the architectural character of the home. This will enhance the home's appearance and value.

Wall-bracket fixtures should be used on both sides of the main entrance. The outlet boxes should be located approximately 66 inches above the top landing of the entrance steps. The lighting fixtures should be shielded with 8-inch or larger enclosures and contain a 50-watt or larger lamp in each fixture. Other entrances require only one fixture mounted at the same height and on the lock side of the door. Sometimes, however, it is desirable to have a fixture on each side of the door at other entrances, as in the case of our residence discussed earlier.

When it is not feasible to use two wall-bracket fixtures at the front door, one large fixture can be located directly above the door, or recessed fixtures can be mounted in the porch ceiling or roof overhang, as close to the door as possible. Each of these fixtures should have a minimum lamp size of 100 watts.

In addition to doorway lighting, a ceiling-mounted lighting fixture should be located at the center of all breezeway and porch ceilings. Again, this fixture should be shielded with ceramic-enameled or opal glass and be a minimum of 10 inches in diameter. The fixture should contain lamps that are rated at a minimum of 150 watts total. If the ceiling is over a small portico, an 8-inch diameter fixture with a 60-watt bulb will suffice.

A post lantern should be located at the main entrance walk to mark the driveway or the sidewalk to the main entrance. The shielded fixture should be a minimum of 12 inches square or 12 inches in diameter. Avoid clear glass designs and exposed high-wattage lamps because the after-image caused by viewing these will hamper seeing and can cause accidents.

A single-car garage should have two 8-inch diameter fixtures or porcelain lampholders mounted on each side of the car and about 6 feet back from the front bumper to light passageways at the sides of the car. Three such fixtures will give better illumination for two-car garages.

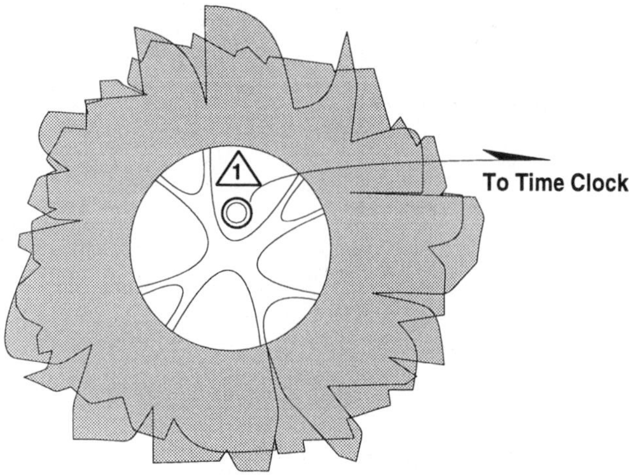
1
To Time Clock

Figure 9-5: Plan view of Figure 9-4

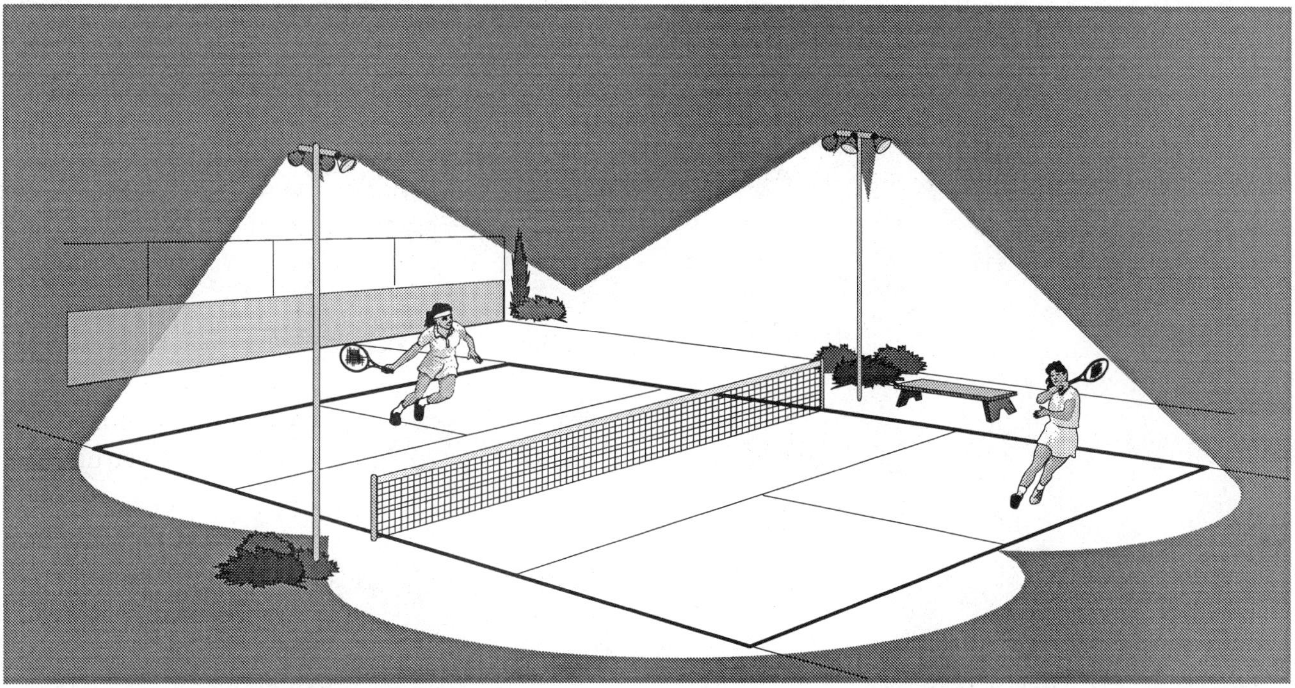

Figure 9-6: Example of wide-beam downlighting

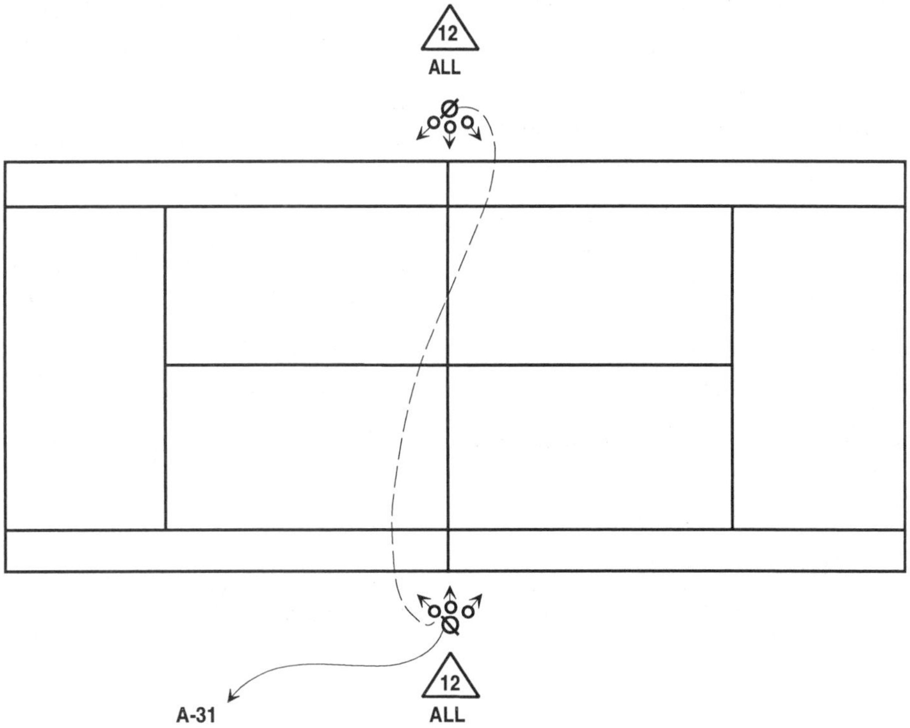
12
ALL
12
ALL
A-31

Figure 9-7: Plan view of Figure 9-6

Figure 9-8: Example of controlled-beam downlighting

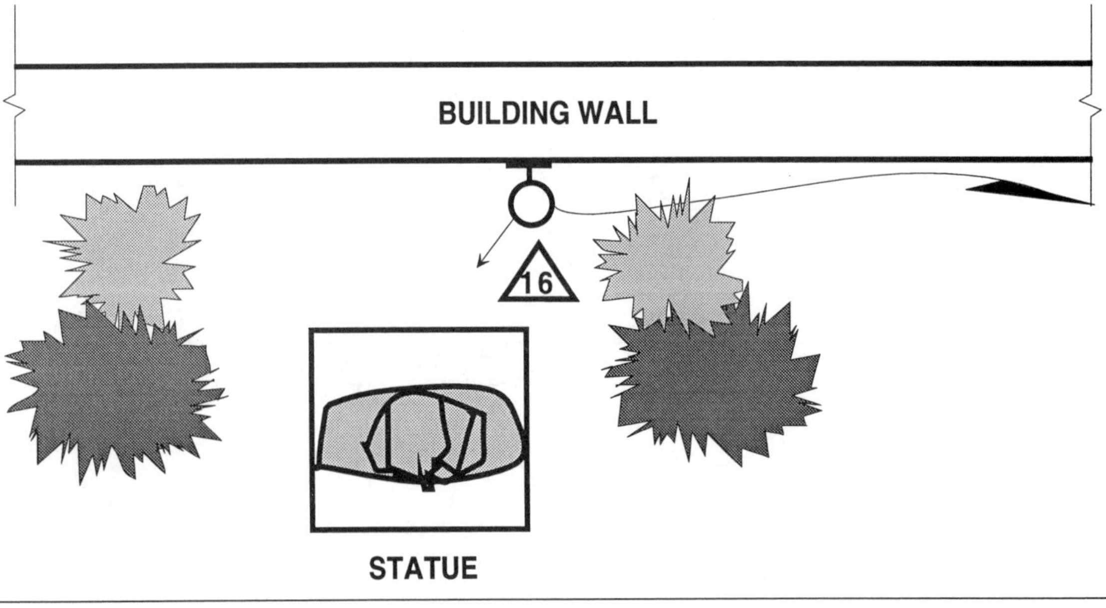
BUILDING WALL
16
STATUE

Figure 9-9: Plan view of Figure 9-8

Figure 9-10: Example of low horizontal light distribution

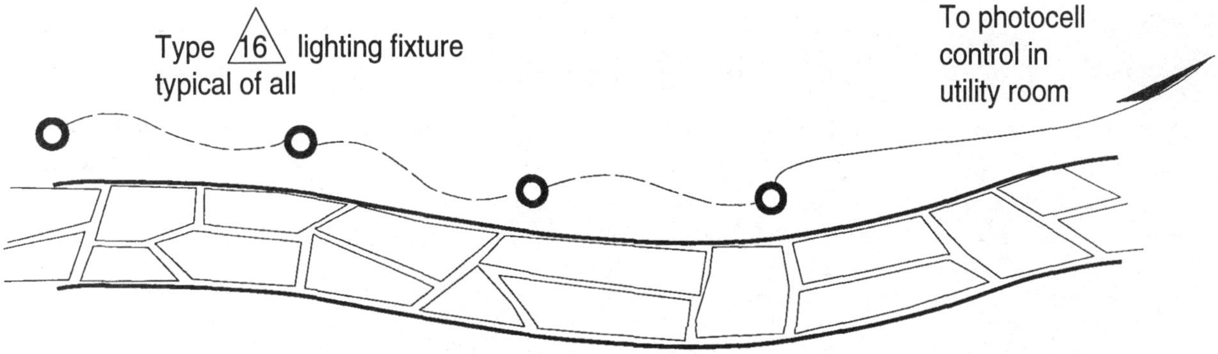
Type 16 lighting fixture
typical of all
To photocell
control in
utility room

Figure 9-11: Plan view of Figure 9-10

Figure 9-12: Example of general diffused lighting

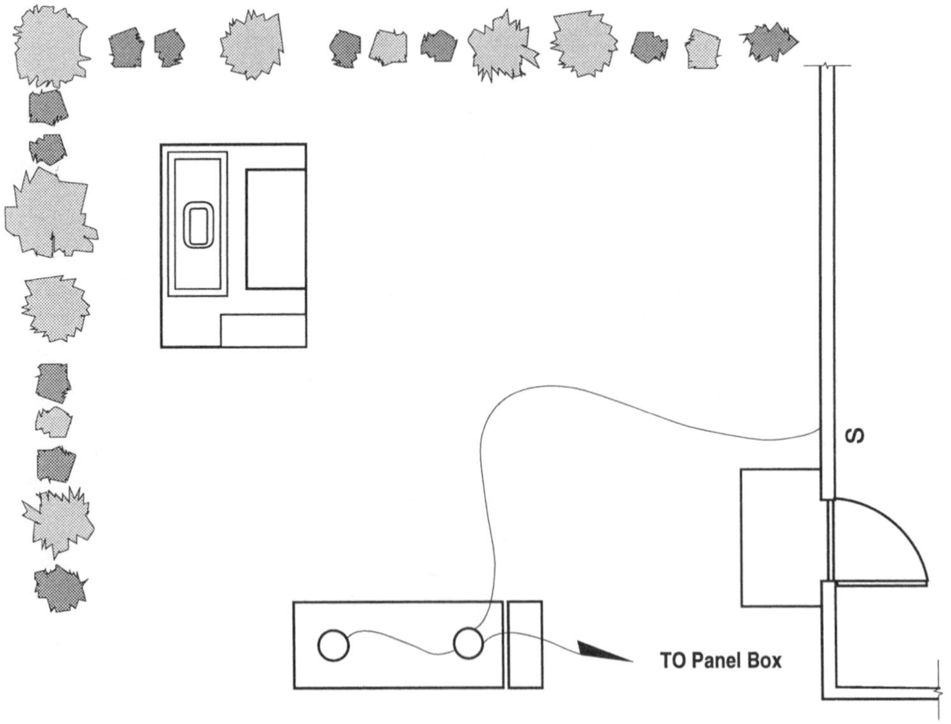
S
TO Panel Box

Figure 9-13: Plan view of Figure 9-12

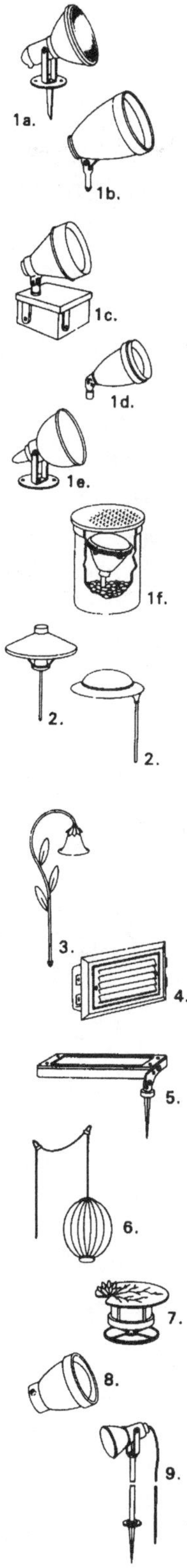
1a.
1b.
1c.
1d.
1e.
1f.
2.
2.
3.
4.
5.
6.
7.
8.
9.

1. Adjustable holders — for PAR 38 projector lamps and others with spike for ground placement, cover plates for outlet boxes and attachment clamps for use on tree or pole. USES: for all areas described in this chapter.

a. for PAR lamps. Color glass covers, louver and shield clip on bulb.

b. of metal, offers deep shielding, better appearance.

c. mercury floodlights — adjustable units mounted on enclosed ballast.

d. for R20 floodlights. Cover lens protects bulb.

e. enclosed floodlights — often called "handy floodlights." Use up to 300 watts. Cover glass protects regular household and reflector bulbs.

f. flush mounted fixtures for projector lamps. Specific housings available for 150 PAR 38 up to 500 PAR 64 and for mercury lamps. USE: in open areas without available natural shielding.

2. Mushroom unit — use with any height stem or post. Both side-suspended and center stem type available. Wide range of reflector width, depth and contour result in differences in complete bulb shielding. USES: general lighting on terrace with 4- to 5-foot stems; most visual tasks with 100 watt bulb, placed at rear corner of a chair; circulation areas; flowers and plants of all heights. Bulb wattage: as desired. (Many variations of the basic mushroom design are available).

3. Bell-type reflector — suspended from fixed height stem with lamp base up. USES: flowers and plants in small area. Garden steps.

4. Recessed units with lens or louver control. Directs light down about 45° below horizontal. Bulb wattage: from 6 to 25 watts depending on unit. Locate from 4 inches to 24 inches above ground. USES: Paths and walks near buildings — 8 to 12 feet apart, steps — mounted in risers or adjacent building; terraces to light floor — 6 to 10 feet apart.

5. Weatherproof fluorescent units — wattage depends on length of unit. USES: where line of light is desired.

6. Diffusing plastic shade attached to suspended socket. Various sizes shapes. USES: general lighting for terraces with roof or overhang — 40 to 75 watts in 10-inch diameter units — 100 to 150 watts in larger sizes. Decorative: with 10 to 40 watt colored or white bulbs.

7. Underwater fixtures — lily pad shield attached to glass enclosed housing. Use 25 to 60 watt bulb.

8. Underwater fixtures — provide controlled light. Bulb size depends on unit. Available in both low voltage and 120 volts, depending on unit.

9. Telescopic poles for PAR 38 and enclosed floodlights. Fit into pipe sleeves driven into ground. USES: sports and area floodlighting.

Figure 9-14: Examples of outdoor lighting equipment

For wide-coverage yard lighting, single or double weatherproof, adjustable floodlight units should be mounted under the eaves or roof overhang and aimed at the desired areas. By using 150-watt PAR-38 lamps, a relatively large area can be covered with light at a very reasonable cost.

At least one switch-controlled weatherproof receptacle should be located on the outside near the front entrance for holiday lighting. However, remember that *NEC* Section 210-8(a) states:

For residential occupancies, all 120-volt, single phase, 15- and 20-ampere receptacle outlets installed outdoors shall have approved ground-fault circuit protection for personnel . . . .

OUTDOOR LIGHTING EQUIPMENT

A wide variety of outdoor lighting equipment is readily available. Many factors related to the mechanical and electrical design are important, but the lighting effects desired are the primary considerations in selecting the equipment for outdoor lighting. Both daytime and nighttime appearance of the lighting fixtures should be considered very carefully. Also, make certain that fixture lamps and all electrical wiring are concealed from direct view.

The resistance to weather is another important consideration in selecting outdoor lighting fixtures, since the equipment will be subjected to sun, rain, wind, and snow. Aluminum, brass, copper, stainless steel, and even plastic are the materials most used in outdoor lighting fixtures.

Durability is a prime concern in selecting outdoor lighting equipment, and the cost of the fixtures — like almost everything else — is usually directly proportional to durability. The equipment must be able to stand up under a great amount of abuse throughout the year — not only against the weather but also the destructiveness of lawn mowers, snow throwers, and humans. Fixture design and color normally should blend with the landscape in the case of ground-mounted lawn or garden lighting fixtures. These should also be as inconspicuous as possible in the daytime as well as nighttime. Fixtures mounted on the house should follow the general architectural theme.

Personal taste is the final factor to be considered in the process of selecting outdoor lighting equipment. Figure 9-14 gives typical examples of outdoor lighting fixtures.

FLOODLIGHTING

Electrical designers who anticipate doing much work involving outdoor lighting should study the principles of floodlighting. Many excellent references are available — often free of charge — from manufacturers of lighting fixtures and lamps. Such literature is usually available from electrical-equipment suppliers, or may be ordered directly from the manufacturers.

You will also want to obtain recommendations from the Illuminating Engineering Society and the Institute of Electrical Engineers. Addresses are listed in Appendix II of this book.

Chapter 10

Lighting Control

Many lighting-control devices have been developed since Edison's first lamp. They have been designed to make the best use of the lighting equipment provided by the lighting industry. These include:

- Automatic timing devices for outdoor lighting
- Dimmers for residential lighting
- The common single-pole, 3-way, and 4-way switches.

SWITCHES

A switch, for our purposes, is a device used on branch circuits to control lighting. Switches fall into the following basic categories:

- Snap-action switches
- Mercury switches
- Quiet switches

Snap-action switches: A single-pole snap-action switch consists of a device containing two stationary current-carrying elements, a moving current-carrying element, a toggle handle, a spring, and a housing. When the contacts are open, as in Figure 10-1, the circuit is "broken" and no current flows. When the moving element is closed, by manually flipping the toggle handle, the contacts complete the circuit and the lamp will be energized. See Figure 10-2.

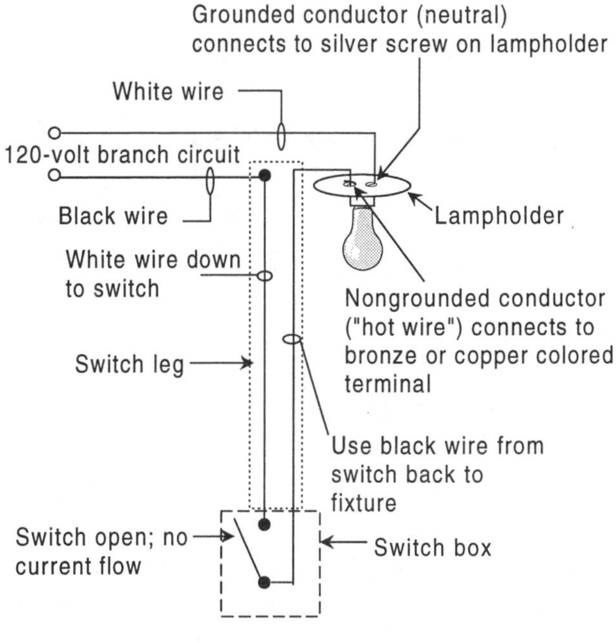

Grounded conductor (neutral) connects to silver screw on lampholder
White wire
120-volt branch circuit
Black wire
Lampholder
White wire down to switch
Nongrounded conductor ("hot wire") connects to bronze or copper colored terminal
Switch leg
Use black wire from switch back to fixture
Switch open; no current flow
Switch box

Figure 10-1: Switch in open (OFF) position

Mercury switches: Mercury switches consist of a sealed capsule containing mercury, as illustrated in Figure 10-3. Inside the capsule are contacting surfaces "A" and "B," which may be part of the wall of the capsule. The switch is operated by means of a handle which moves the capsule.

As shown in Figure 10-3, the capsule is tilted so that the mercury "C" has collected at one end of the capsule. Here, it bridges two contact points,

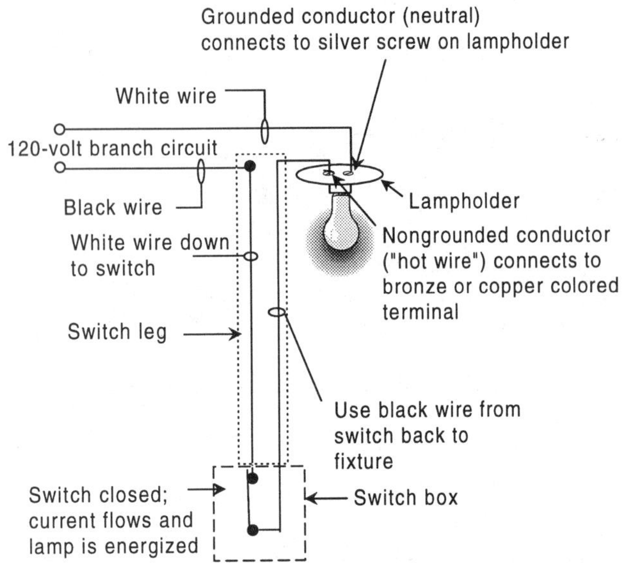
Grounded conductor (neutral) connects to silver screw on lampholder
White wire
120-volt branch circuit
Black wire
White wire down to switch
Switch leg
Lampholder
Nongrounded conductor ("hot wire") connects to bronze or copper colored terminal
Use black wire from switch back to fixture
Switch closed; current flows and lamp is energized
Switch box

Figure 10-2: Switch in closed (ON) position

"A" and "B," to complete the circuit and light the lamp. However, if the capsule is tilted the opposite way, the circuit between contacts "A" and "B" will not be completed, and the lamp will be de-energized (be turned off). Mercury switches offer the ultimate in silent operation and are recommended where the "clicking" of a light switch may be annoying.

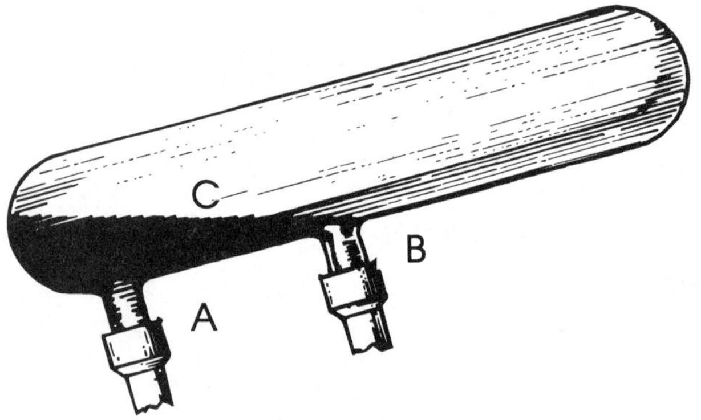
C
A
B

Figure 10-3: Principles of mercury-switch operation

Quiet switch: The quiet switch is a compromise between the snap-action switch and the mercury switch. Its operation is much quieter than the snap-action switch, yet it is not as expensive as the mercury switch.

The quiet switch consists of a stationary contact and a moving contact that are close together when the switch is open. Only a short, gentle movement is applied to open and close the switch, producing very little noise. This type of switch may be used only on alternating current, since the arc will not extinguish on direct current.

The quiet switch (Figure 10-4) is the most commonly used switch for modern lighting practice. These switches are common for loads from 10 to 20 amperes, in single-pole, three-way, four-way, or other configurations.

Many other types of switches are available for lighting control. One type of switch used mainly in residential occupancies is the door-actuated type which is generally installed in the door jamb of a closet to control a light inside the closet. When the door is open, the light comes on; when the door is closed, the light goes out. Refrigerator and oven lights are usually controlled by door switches.

The Despard switch is another special switch. Due to its small size, up to three may be mounted in a standard single-gang switch box. Weatherproof switches are made for outdoor use. Combination switch-indicator light assemblies are also available for use where the light cannot be seen from the switch location, such as an attic or garage. Switches are also made with small neon lamps in the handle that light when the switch is off. These low-current-consuming lamps make the switches easy to find in the dark.

Three-Way Switches

Three-way switches are used to control one or more lamps from two different locations, such as at the top and bottom of stairways, in a room that has two entrances, etc. A typical three-way switch is shown in Figure 10-5. Note that there are no ON/OFF markings on the handle. Furthermore, a

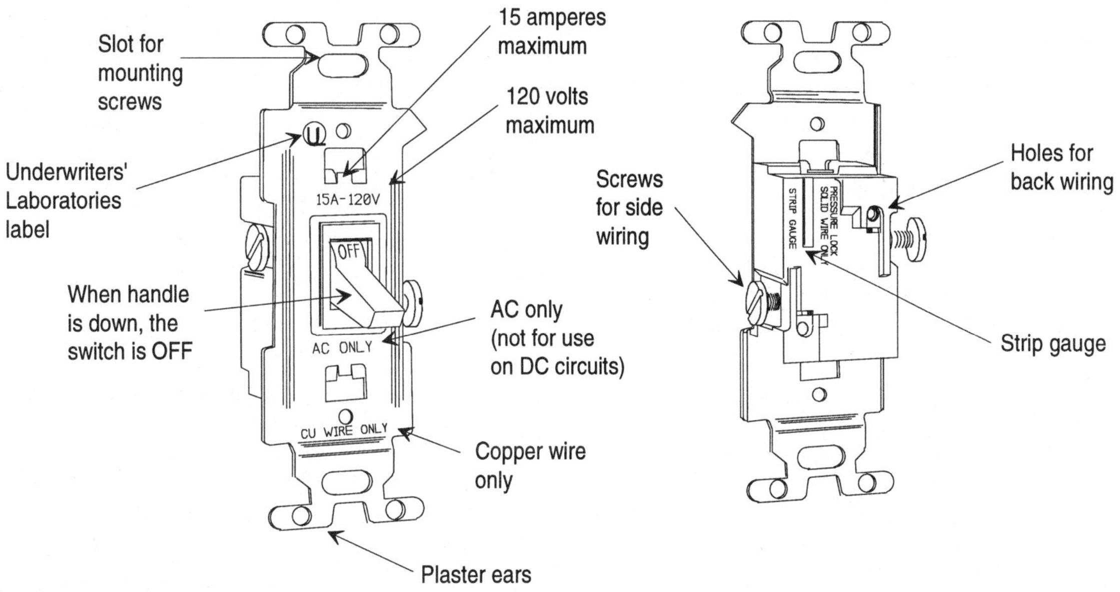
Slot for mounting screws
Underwriters' Laboratories label
When handle is down, the switch is OFF
15 amperes maximum
120 volts maximum
15A-120V
OFF
AC ONLY
CU WIRE ONLY
AC only (not for use on DC circuits)
Copper wire only
Plaster ears
Screws for side wiring
PRESSURE LOCK SOLID WIRE ONLY
STRIP GAUGE
Holes for back wiring
Strip gauge

Figure 10-4: Characteristics of a single-pole quiet switch

three-way switch has three terminals. The single terminal at one end of the switch is called the *common* (sometimes *hinge point*). This terminal is easily identified because it is darker than the two other terminals. The feeder ("hot" wire) or switch leg is always connected to the common dark or black terminal. The two remaining terminals are called *traveler terminals.* These terminals are used to connect three-way switches together.

The connection of three-way switches is shown in Figure 10-6. By means of the two three-way switches, it is possible to control the lamp from two locations. By tracing the circuit, it may be seen how these three-way switches operate.

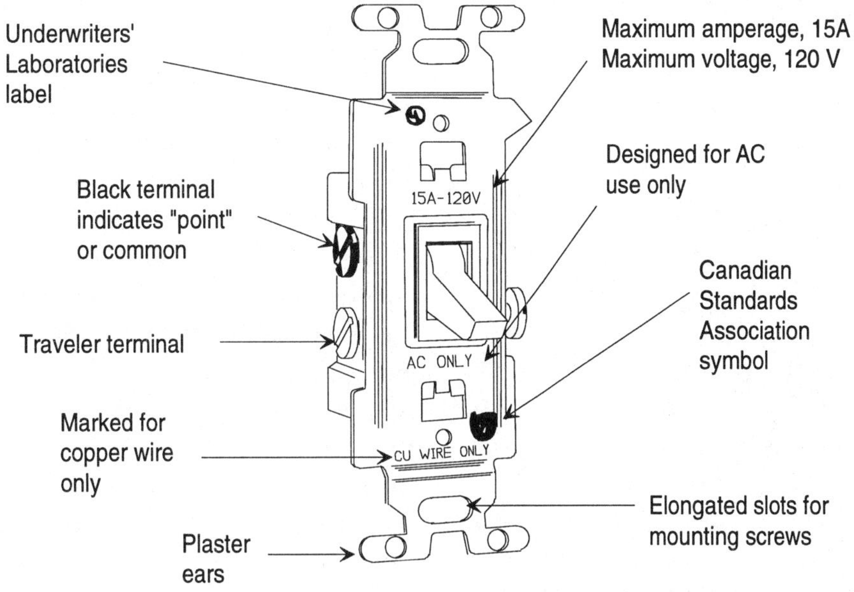
Underwriters' Laboratories label
Black terminal indicates "point" or common
Traveler terminal
Marked for copper wire only
Plaster ears
15A-120V
AC ONLY
CU WIRE ONLY
Maximum amperage, 15A
Maximum voltage, 120 V
Designed for AC use only
Canadian Standards Association symbol
Elongated slots for mounting screws

Figure 10-5: Characteristics of a three-way quiet switch

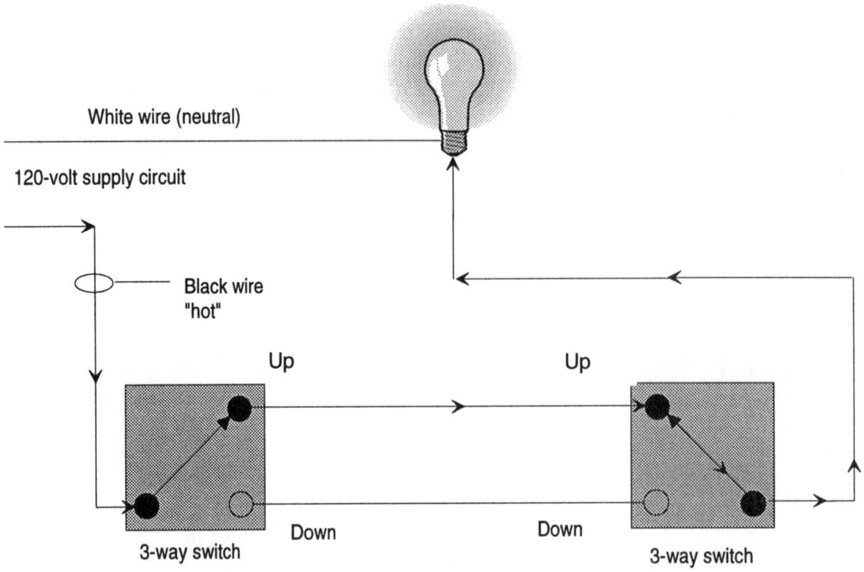
White wire (neutral)
120-volt supply circuit
Black wire
"hot"
Up
Up
Down
Down
3-way switch
3-way switch

Figure 10-6: Three-way switches in ON position; both handles up

A 120-volt circuit emerges from the left side of the drawing. The white or neutral wire connects directly to the neutral terminal of the lamp. The "hot" wire carries current, in the direction of the arrows, to the common terminal of the three-way switch on the left. Since the handle is in the up position, the current continues to the top traveler terminal and is carried by this traveler to the other three-way switch. Note that the handle is also in the up position on this switch; this picks up the current flow and carries it to the common point, which, in turn, continues on to the ungrounded terminal of the lamp to make a complete circuit. The lamp is energized.

Moving the handle to a different position on either three-way switch will break the circuit, which in turn, de-energizes the lamp. For example, let's say a person leaves the room at the point of the three-way switch on the left. The switch handle is flipped down, giving a condition as shown in Figure 10-7. Note that the current flow is now directed to the bottom traveler terminal, but since the handle of the three-way switch on the right is still in the up position, no current will flow to the lamp.

Another person enters the room at the location of the three-way switch on the right. The handle is flipped downward which gives the condition as shown in Figure 10-8. This change provides a complete circuit to the lamp which causes it to be energized. In this example, current flow is on the bottom traveler. Again, changing the position of the switch handle (pivot point) on either three-way switch will de-energize the lamp.

In actual practice, the exact wiring of the two three-way switches to control the operation of a lamp will be slightly different than the routing shown in these three diagrams. There are several ways that two three-way switches may be connected. One solution is shown in Figure 10-9. Here, 14/2 w/ground NM cable (Romex) is fed to the three-way switch on the left. The black or "hot" conductor is connected to the common terminal on the switch, while the white or neutral conductor is spliced to the white conductor of 14/3 w/ground NM cable leaving the switch. This 3-wire cable is necessary to carry the two travelers plus the neutral to the three-way switch on the right. At this point, the black and red wires connect to the two traveler terminals, respectively. The white or neutral wire is again spliced — this time to the white wire of another 14/2 w/ground NM cable. The neutral wire is never connected to the switch itself. The black

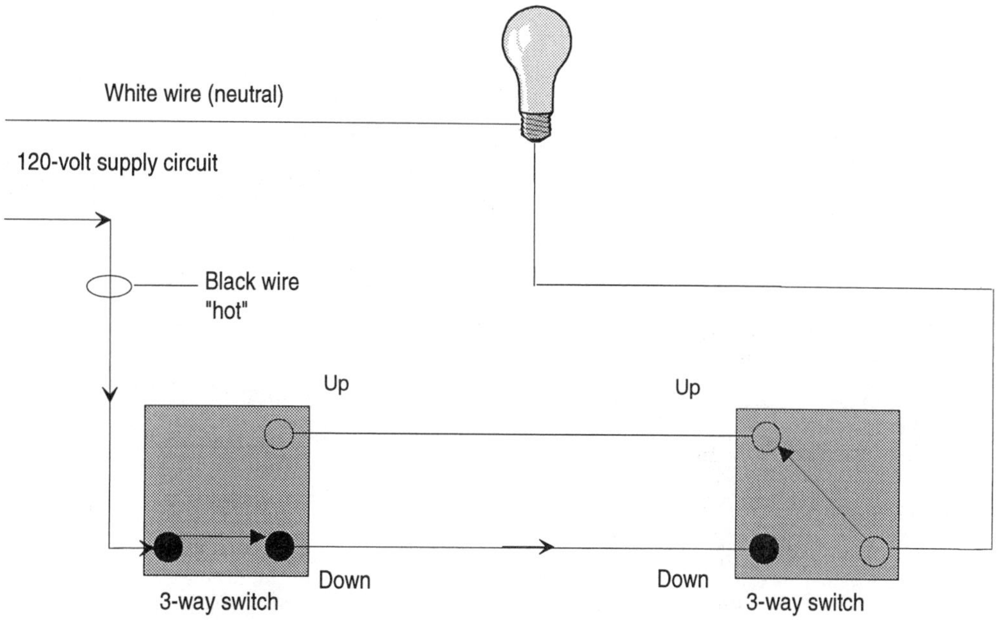
White wire (neutral)
120-volt supply circuit
Black wire
"hot"
Up
Up
Down
Down
3-way switch
3-way switch

Figurer 10-7: Three-way switches in OFF position; one handle down, one handle up

wire of the 14/2 w/ground NM cable connects to the common terminal on the three-way switch. This cable — carrying the "hot" and neutral conductors— is routed to the lighting fixture outlet for connection to the fixture.

Another solution is to feed the lighting-fixture outlet with two-wire cable. Run another two-wire cable — carrying the "hot" and neutral conductors — to one of the three-way switches. A three-wire cable is pulled between the two three-way switches, and then another two-wire cable is routed from the other three-way switch to the lighting-fixture outlet.

Some electricians use a short-cut method by eliminating one of the two-wire cables in the preceding method. Rather, a two-wire cable is run from the

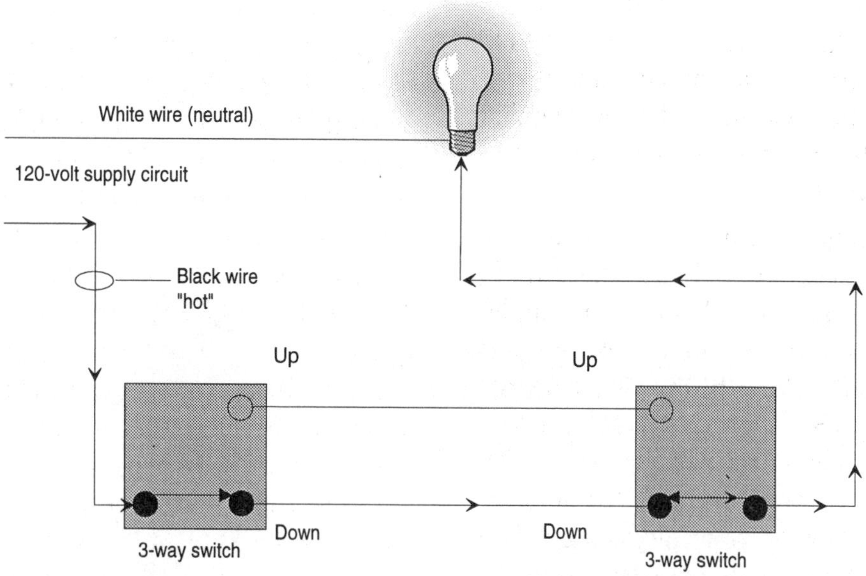
White wire (neutral)
120-volt supply circuit
Black wire
"hot"
Up
Up
Down
Down
3-way switch
3-way switch

Figure 10-8: Three-way switches with both handles down; light is energized

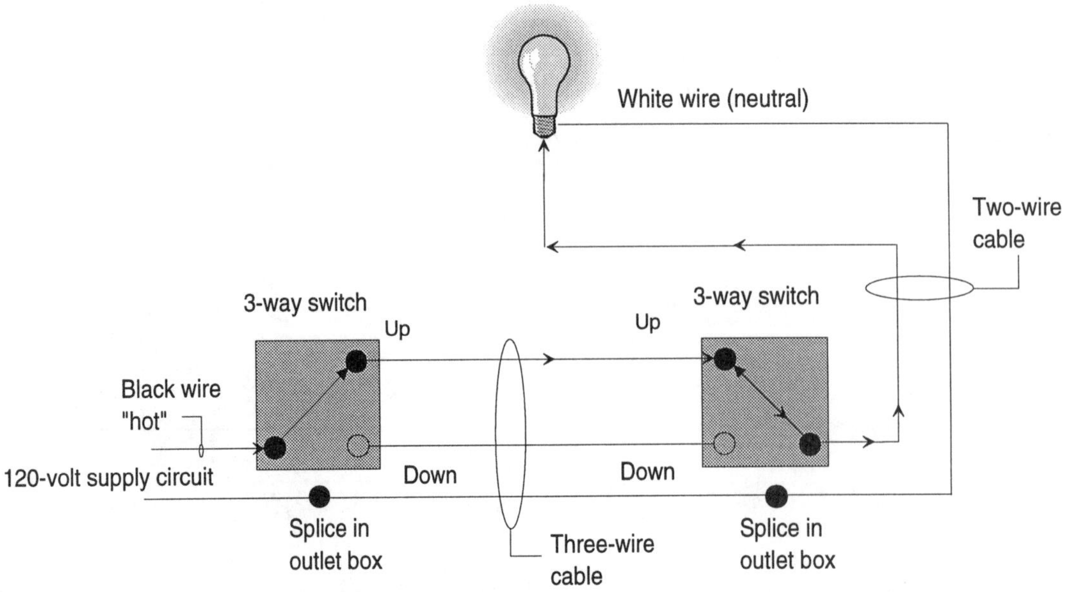
White wire (neutral)
Two-wire cable
3-way switch
Up
3-way switch
Up
Black wire "hot"
120-volt supply circuit
Down
Down
Splice in outlet box
Three-wire cable
Splice in outlet box

Figure 10-9: One way to connect a pair of three-way switches to control one lighting fixture or a group of lighting fixtures

lighting-fixture outlet to one three-way switch. Three-wire cable is pulled between the two three-way switches — two of the wires for travelers and the third for the common-point return. This method is shown in Figure 10-10, but should not be used with a metallic conduit system.

Four-Way Switches

Two three-way switches may be used in conjunction with any number of four-way switches to control a lamp, or a series of lamps, from any number of positions. When connected correctly, the actuation of any one of these switches will

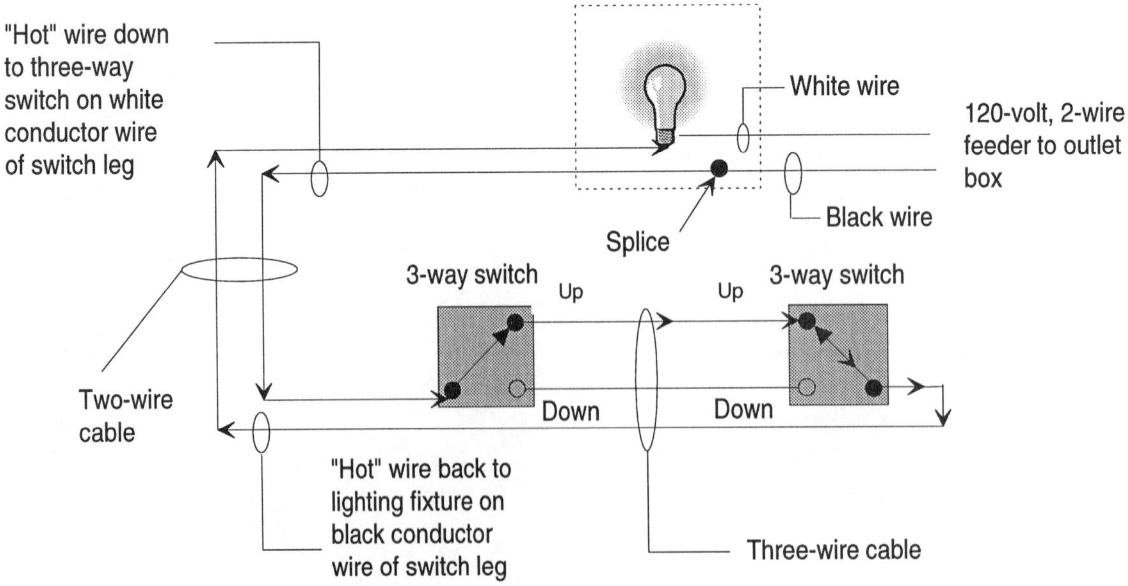
"Hot" wire down to three-way switch on white conductor wire of switch leg
White wire
120-volt, 2-wire feeder to outlet box
Black wire
Splice
3-way switch
Up
Up
3-way switch
Two-wire cable
Down
Down
"Hot" wire back to lighting fixture on black conductor wire of switch leg
Three-wire cable

Figure 10-10: Alternate method of connecting three-way switches

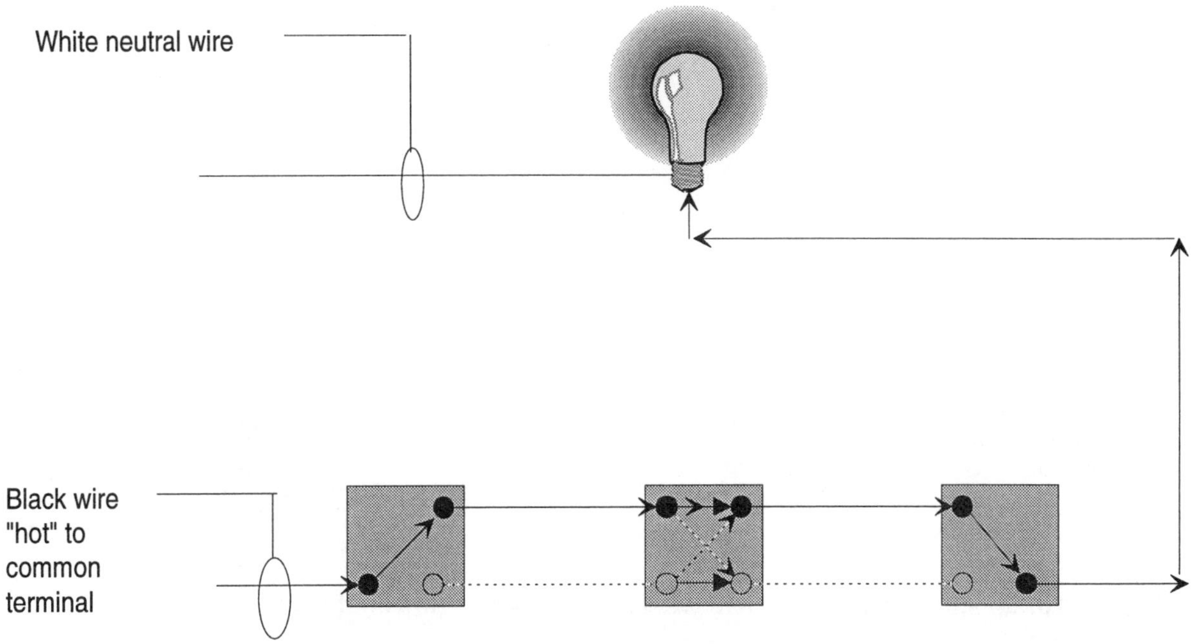
White neutral wire
Black wire
"hot" to
common
terminal

Figure 10-11: Three- and four-way switches used in combination; light is ON

change the operating condition of the lamp(s); that is, either turn the lamp on or off.

Figure 10-11 shows how a four-way switch may be used in combination with two three-way switches to control a device from three locations. In this example, note that the "hot" wire is connected to the common terminal on the three-way switch on the left. Current then travels to the top traveler terminal and continues on the top traveler conductor to the four-way switch. Since the handle is up on the four-way switch, current flows through the top terminals of the switch and onto the traveler conductor going to the other three-way switch. Again, the switch is in the up position. Therefore, current is carried from the top traveler terminal to the common terminal and on to the lighting fixture to energize it. Under this condition, if the position of any one of the three switch handles is changed, the circuit will be broken and no current will flow to the lamp.

For example, let's assume that the four-way switch handle is flipped downward. The circuit will now appear as shown in Figure 10-12, and the light will be out. The light would also go out if the position of either the right or left three-way switch handle was changed.

Remember, any number of four-way switches may be used in combination with two three-way switches, but two three-way switches are always necessary for the correct operation of one or more four-way switches.

Photoelectric Switches

The chief application of the photoswitch is to control outdoor lighting, especially the "dusk-to-dawn" lights found in suburban areas.

This interesting switch has an endless number of possible uses and is a great tool for electricians dealing with outdoor lighting situations.

RELAYS

Next to switches, relays play the most important part in the control of light. However, the design and application of relays is a study in itself, and far beyond the scope of this chapter. Still, brief mention of relays is necessary to round out your knowledge of lighting controls.

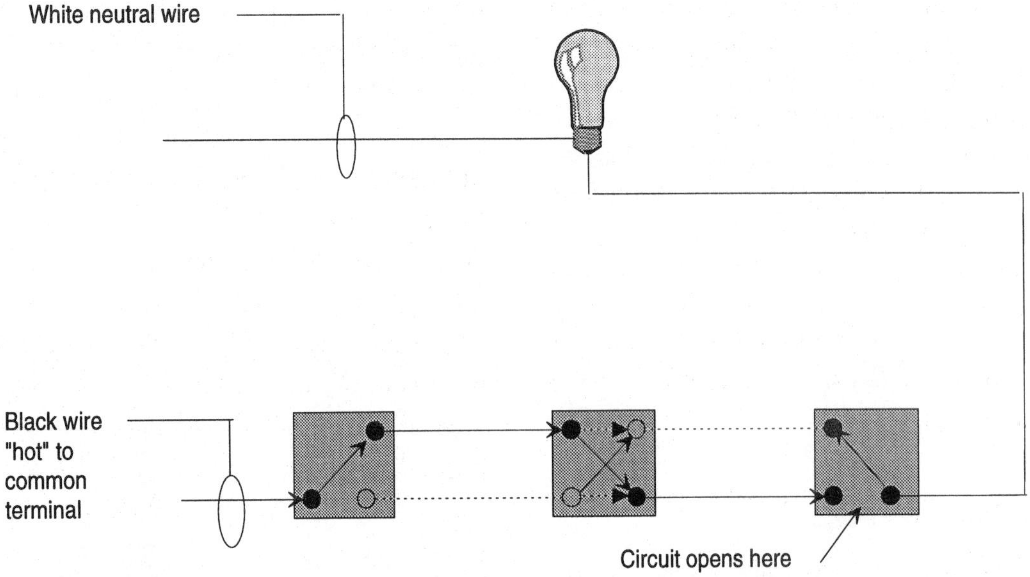
White neutral wire
Black wire "hot" to common terminal
Circuit opens here

Figure 10-12: Three- and four-way switches used in combination; the light is out

An electric relay is a device whereby an electric current causes the opening or closing of one or more pairs of contacts. These contacts are usually capable of controlling much more power than is necessary to operate the relay itself. This is one of the main advantages of relays.

One popular use of the relay in residential lighting systems is that of remote-control lighting. In this type of system, all relays are designed to operate on a 24-volt circuit and are used to control 120-volt lighting circuits. They are rated at 20 amperes which is sufficient to control the full load of a normal lighting branch circuit, if desired.

Remote-control switching makes it possible to install a switch wherever it is convenient and practical to do so or wherever there is an obvious need for having a switch — no matter how remote it is from the lamp or lamps it is to control. This method enables lighting designs to achieve new advances in lighting control convenience at a reasonable cost. Remote-control switching is also ideal for rewiring existing homes with finished walls and ceilings.

One relay is required for each fixture or for each group of fixtures that are controlled together. Switch locations for remote control follow the same rules as for conventional direct switching. However, since it is easy to add switches to control a given relay, no opportunities should be overlooked for adding a switch to improve the convenience of control.

Remote-controlled lighting also has the advantage of using selector switches at certain locations. For example, selector switches located in the master bedroom or in the kitchen of a home enable the owner to control every lighting fixture on the property from this location. The selector switch may turn on and off an outside or other light which customarily would be left on until bedtime and which might otherwise be forgotten.

Complete coverage of remote-control switching is covered in Chapter 11.

DIMMERS

Dimming a lighting system provides control of the quantity of illumination. It may be done to create various atmospheres and moods or to blend certain lights with others for various lighting effects.

For example, in homes with separate dining rooms, a chandelier mounted directly above the dining table and controlled by a dimmer switch adds versatility since it can set the mood of the activity — low brilliance (candlelight effect) for formal dining, medium brilliance for normal dining, or bright for a festive evening. When chandeliers with exposed lamps are used, the dimmer is essential to avoid a garish and uncomfortable atmosphere. The rating of the dimmer or rheostat must be sized for the load it is handling; that is, if the lighting load is, say, 1200 watts, the dimmer must be rated for at least this amount — 1500 watts would be better.

SWITCH LOCATIONS

Although the location of wall switches is usually provided for convenience, the *NE Code* also stipulates certain mandatory locations for lighting fixtures and wall switches. These locations are deemed necessary for added safety in the home — for both the occupants and service personnel who maintain the house systems.

In general, the *NE Code* requires adequate light in areas where HVAC equipment is placed. Furthermore, these lights must be conveniently controlled so that homeowners and service personnel do not have to enter the area where they might come in contact with dangerous equipment if the area is dark. Three-way switches are also required under certain conditions. The electrician should also become familiar with the regulations governing lighting fixtures in clothes closets, along with those governing weights of lighting fixtures that may be mounted directly to the outlet box without further support. Figure 10-13 summarizes *NE Code* requirements for light and switch placement in the home. For further details, refer to the appropriate sections in the *NE Code*.

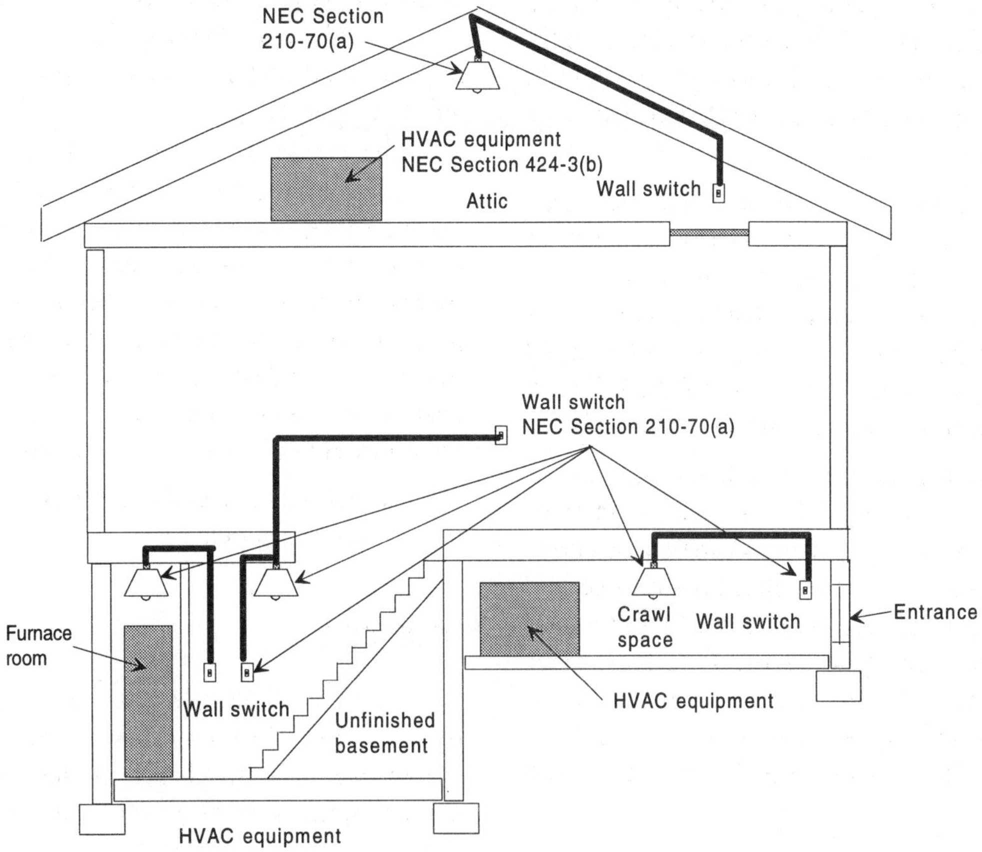
NEC Section
210-70(a)
HVAC equipment
NEC Section 424-3(b)
Attic
Wall switch
Wall switch
NEC Section 210-70(a)
Furnace
room
Wall switch
Unfinished
basement
Crawl
space
Wall switch
Entrance
HVAC equipment
HVAC equipment

Figure 10-13: Summary of *NE Code* requirements for lighting and lighting control

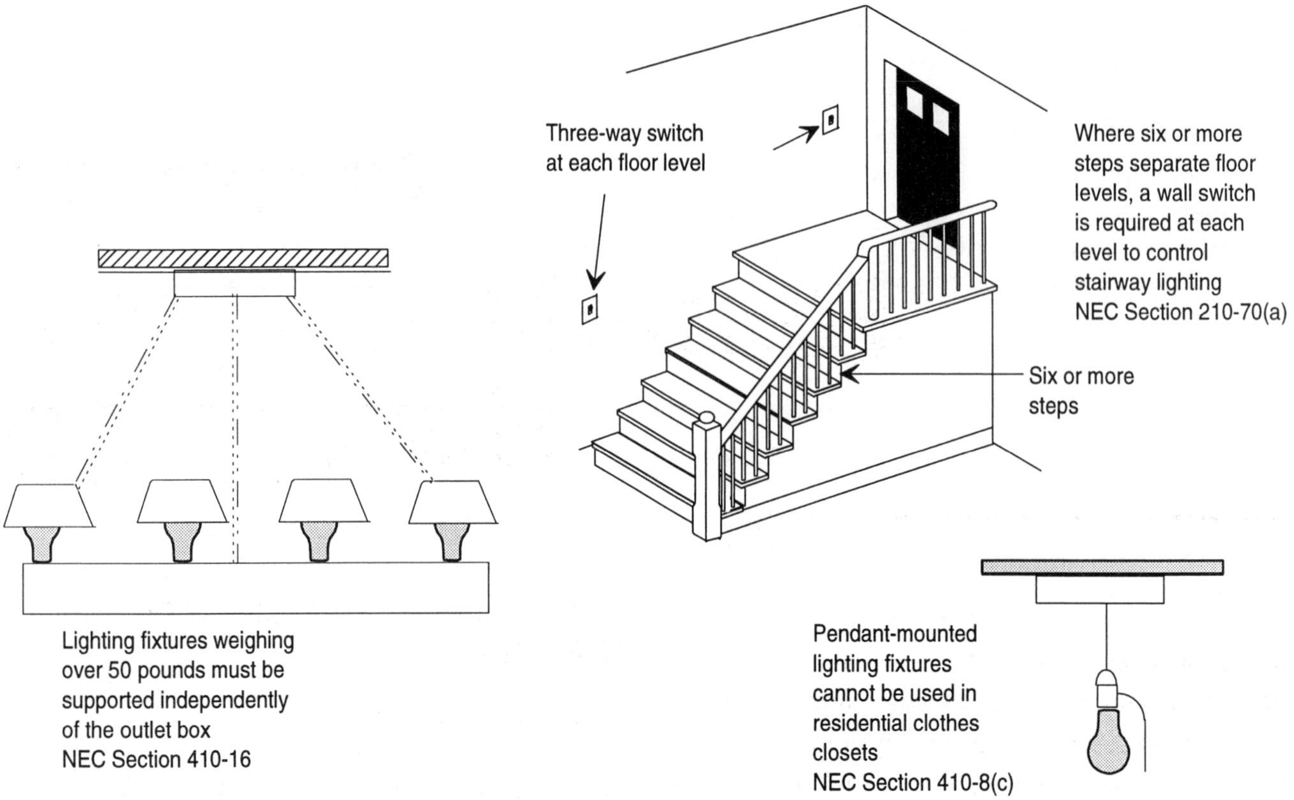
Three-way switch
at each floor level
Where six or more
steps separate floor
levels, a wall switch
is required at each
level to control
stairway lighting
NEC Section 210-70(a)
Six or more
steps
Lighting fixtures weighing
over 50 pounds must be
supported independently
of the outlet box
NEC Section 410-16
Pendant-mounted
lighting fixtures
cannot be used in
residential clothes
closets
NEC Section 410-8(c)

Figure 10-13: Summary of *NE Code* requirements for lighting and lighting control (Cont.)

Chapter 11

Remote-Control Switching

In applications where lighting must be controlled from several points, or where there is a complexity of lighting or power circuits, or where flexibility is desirable in certain systems, low-voltage remote-controlled relay systems have been applied. Basically, these systems use special low-voltage components, operated from a transformer, to switch relays which in turn control the standard line voltage circuits. Because the control wiring does not carry the line load directly, small lightweight cable can be used. It can be installed wherever and however convenient — placed behind moldings, stapled to woodwork, buried in shallow plaster channels, or installed in holes bored in wall studs.

A basic circuit of a remote-control switching system is shown in Figure 11-1 (page 166); another is shown in Figure 11-2 (page 167). In both circuits, the relay permits positive control for on and off. It can be located near the load or installed in centrally located distribution panel boxes, depending upon the application. Because no line voltage flows through the control circuits, and low voltage is used for all switch and relay wiring, it is possible to place the controls at a great distance from the source or load, thus offering many advantages.

ADVANTAGES OF REMOTE-CONTROL SWITCHING

Since branch circuits go directly to the loads in this type of system, no line-voltage switch legs are required. This saves costly larger-conductor runs through all switches and saves installation time and costs if multipoint switches are used.

Protection of the low-voltage conductors is not required by the *NE Code*; therefore, runs in open spaces above ceilings and through wall partitions usually can be made without further protection. Even an outlet box is not required at the switch locations. When the electrician roughs-in the wiring, he or she merely secures a plaster ring, of the correct depth, at each switch location and usually wraps the low-voltage cable around a nail driven behind the plate. The switch and its cover are then installed after the wall is finished, leaving a neat installation.

Low-voltage switching is especially useful in rewiring existing buildings since the small cables are as easy to run as telephone wires. They are easy to hide behind baseboards or even behind quarter-round molding. The cable can be run exposed without being very noticeable because of its small size. The small, flexible wires are easily "fished" in partitions. While not recommended for permanent installations, low-voltage wiring can be run

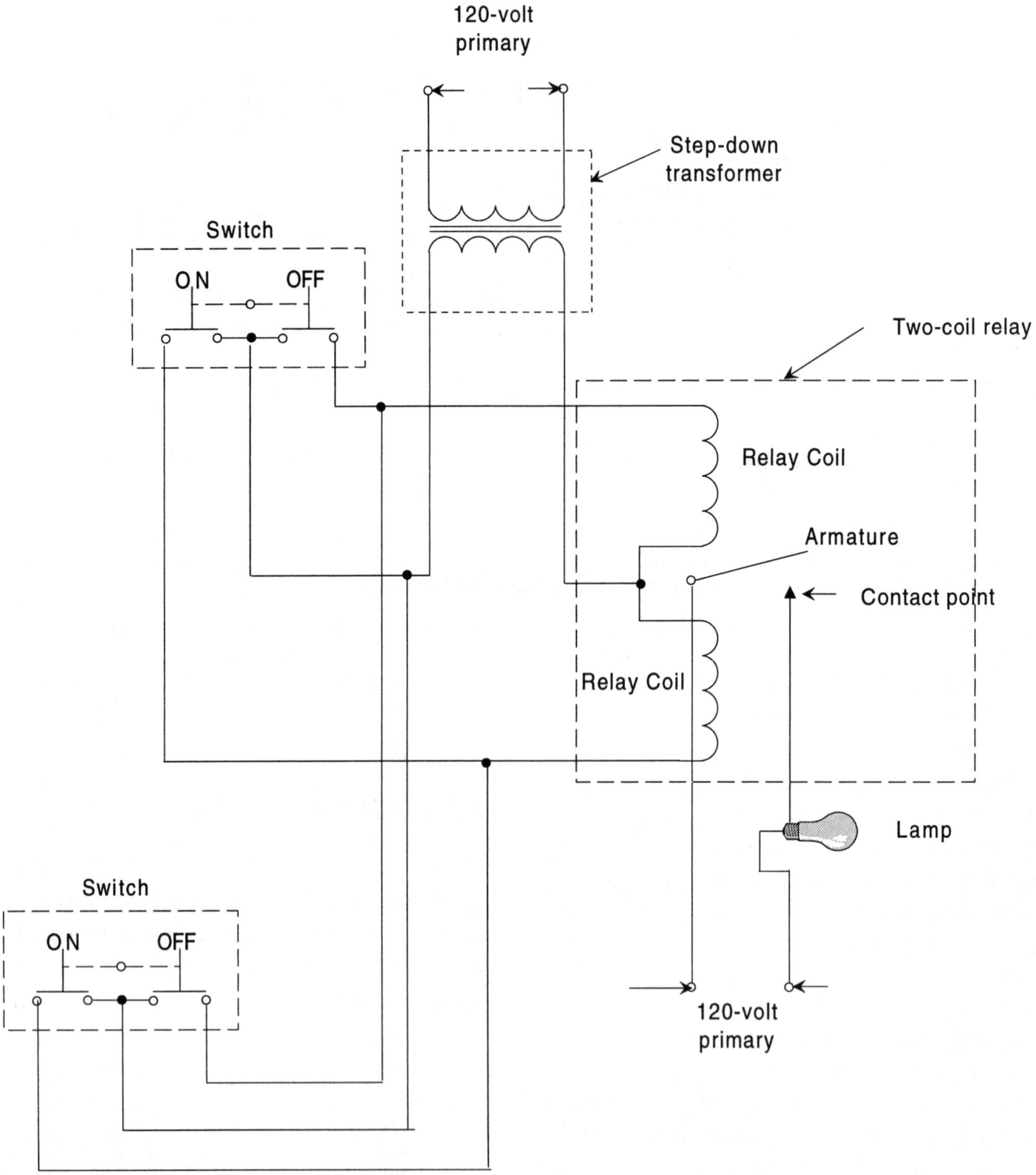
120-volt
primary
Step-down
transformer
Switch
ON
OFF
Two-coil relay
Relay Coil
Armature
Contact point
Relay Coil
Lamp
Switch
ON
OFF
120-volt
primary

Figure 11-1: Basic remote-control system

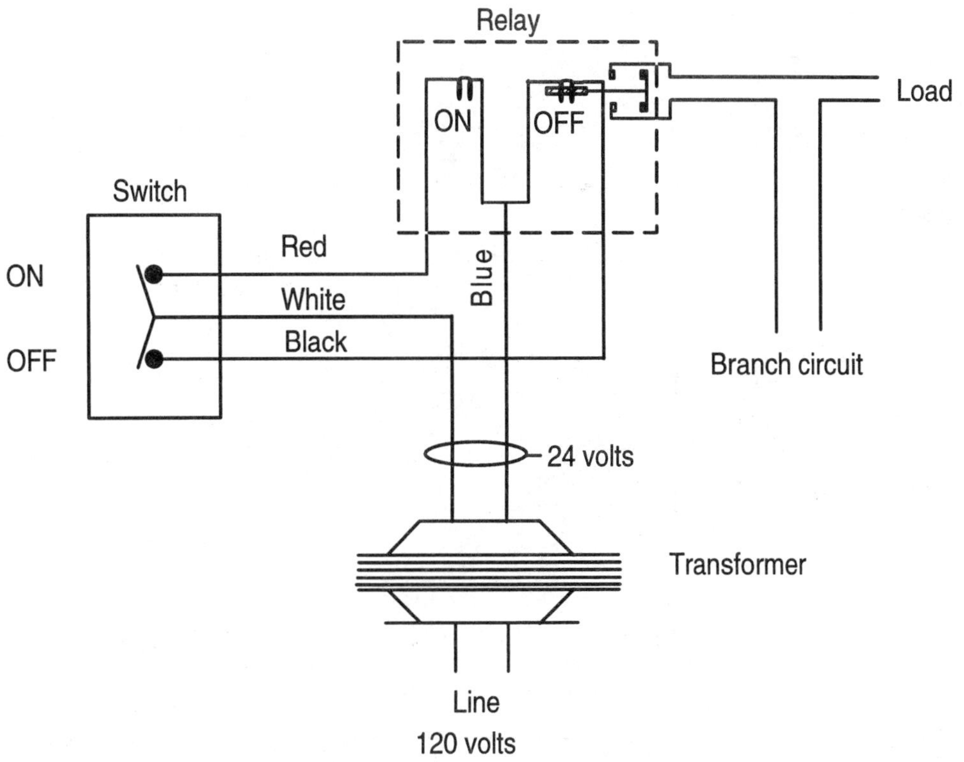
Relay
ON
OFF
Load
Switch
Red
ON
White
Blue
OFF
Black
Branch circuit
24 volts
Transformer
Line
120 volts

Figure 11-2: Basic layout of a remote-control switching system

under rugs to switches located on a table and used for switching a relay controlling a tv outlet, for example.

Figure 11-3 shows several types of remote-control circuits. In the circuits shown, any number of on-off switches can be connected to provide control from many remote points. In a typical installation, outside lights could be controlled from any area within the building without the need to run full-size cable for three- or four-way switch legs through the building. By running a low-voltage line from each relay to a central point, such as an exit door, all relays can be operated from one spot with a selector switch; this allows quick, convenient control of all exterior and interior lighting and power circuits.

A motor-driven rotary switch is available that allows turn-on or shutdown of 25 separate circuits or more with the push of one button. Dimming can also be accomplished by using the motorized control unit together with a modular incandescent dimming system.

SYSTEM COMPONENTS

The relay is the heart of a remote-control switching system. The relay employs a split low-voltage coil to move the line voltage contact armature to the ON (OFF) latched position. As illustrated in Figure 11-4, the ON coil moves the armature to the right when a 24 VAC control signal is impressed across its leads. This is similar to a magnet attracting the handle of a standard single-pole switch to the ON position when energized. The armature (handle) latches in the ON position and will remain there until the OFF coil is energized, drawing the armature into the OFF position. Figure 11-5 shows how a low-voltage relay is installed in an outlet box.

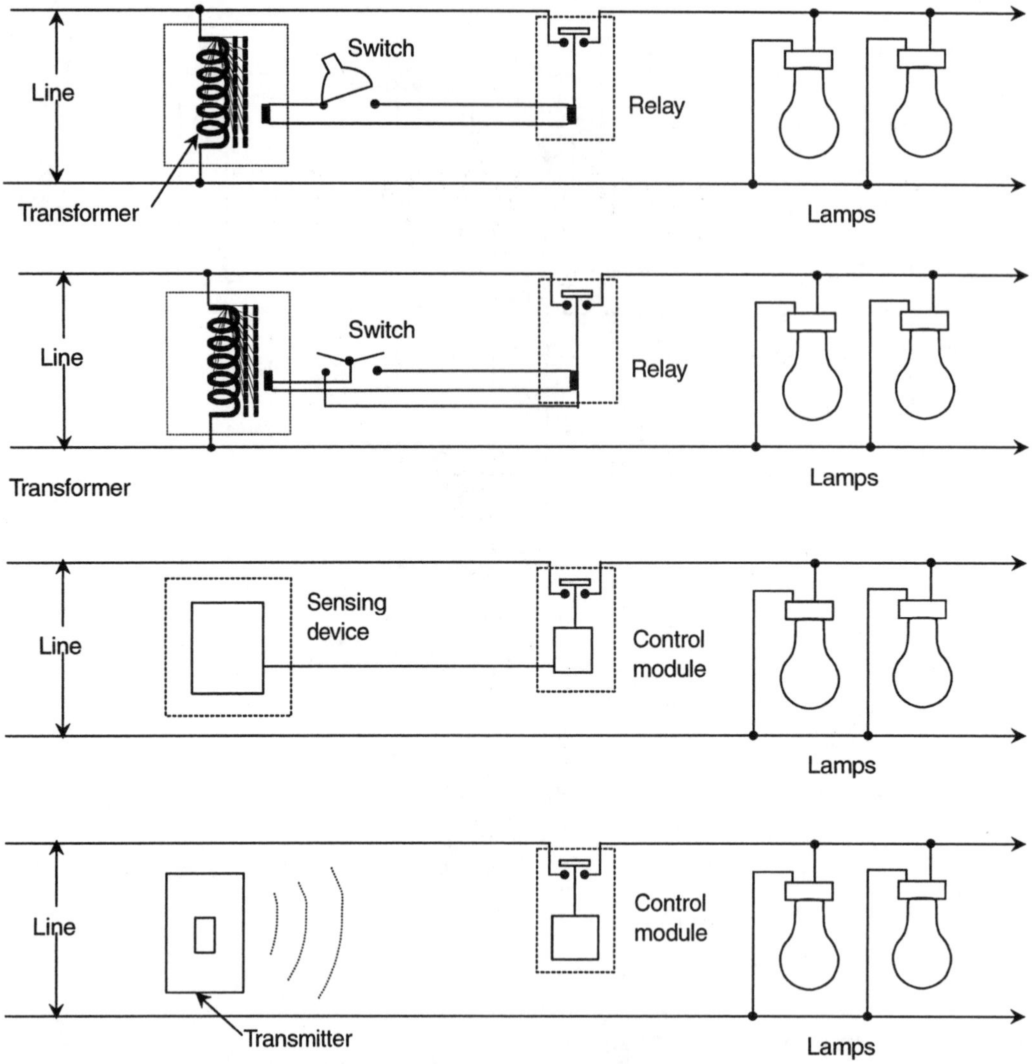
Switch
Line
Relay
Transformer
Lamps
Switch
Line
Relay
Transformer
Lamps
Sensing
device
Line
Control
module
Lamps
Line
Control
module
Transmitter
Lamps

Figure 11-3: Several types of remote-control circuits and their related components

Power Supplies

Transformers designed for use on remote-control switching systems supply 24 VAC power to operate the relays and their controls. The relay and pilot switch power is rectified to extend their life; electronic control component power is not rectified. The ac input voltage to the power supply can vary by manufacturer, with the more common voltages being 120, 208, 240, and 277 volts. The type of voltage used depends on the system. Figure 11-6 shows a control transformer along with connection details.

Switches

The standard low-voltage switch uses a rocker or two-button configuration to provide a momentary single-pole, double-throw action. Pushing the ON (OFF) button completes the circuit to the ON (OFF) coil of the relay, shifting the contact armature to the corresponding position. When the button is released, the relay remains in that position.

The pulse operation allows any number of switches to be wired in parallel as shown in Figure 11-7. A group of relays could also be wired for common switch control by paralleling their control leads. Those relays would operate as a group.

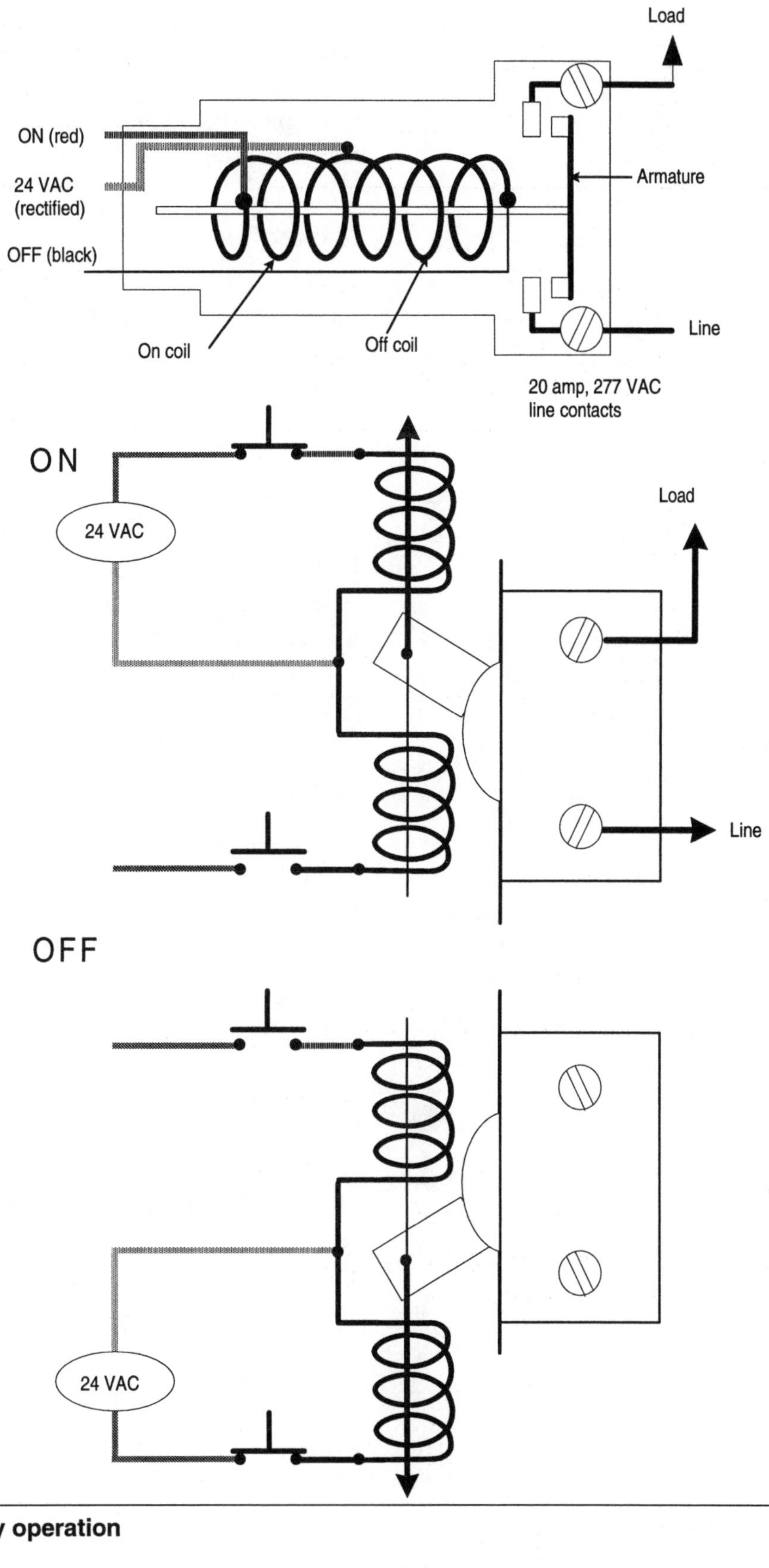
Load
ON (red)
24 VAC
(rectified)
OFF (black)
Armature
On coil
Off coil
Line
20 amp, 277 VAC
line contacts
ON
24 VAC
Load
Line
OFF
24 VAC

Figure 11-4: Relay operation

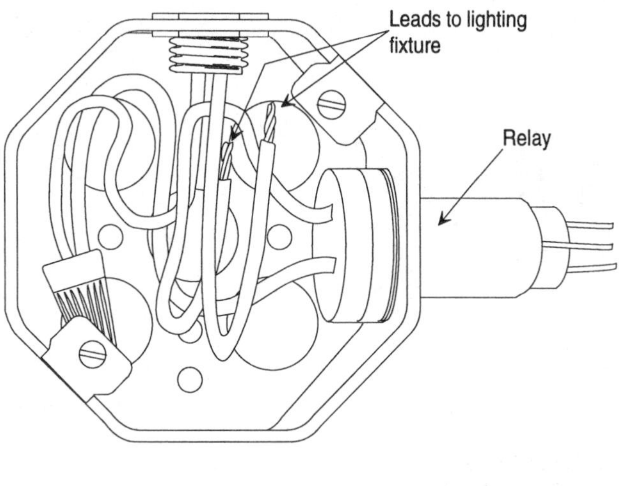
Leads to lighting fixture
Relay

Figure 11-5: Relay installed in outlet box

Pilot switches include a lamp wired between the switch common (white) and a pilot terminal. The auxiliary contact in the relay provides power to drive this lamp when the relay is ON.

CONTROLS

There are several variations of controls that are designed for use on low-voltage remote-control switching systems. A sampling of those available are described here. However, there are many more variations and it is recommended that you refer to manufacturers' catalogs for a listing of all types available. Such catalogs and supplements are ideal learning "tools."

Master Sequencer

The master sequencer allows relays to be controlled as a group while still allowing individual switch control for each. When the master switch is turned ON (OFF), the sequencer pulses each of its ON (OFF) relay outputs sequentially. A local switch can control an individual relay without affecting any other.

A second input channel allows timeclocks, building automation system outputs, photocells or other maintained contact devices to also control the sequencer. This provides simple automation coupled with local override of individual loads. See Figure 11-8.

Telephone Override Devices

Telephone override devices provide the same control functions as the sequencer; but in addition, it allows the occupant's Touchtone™ phone to be used in place of (or in addition to) hardwired

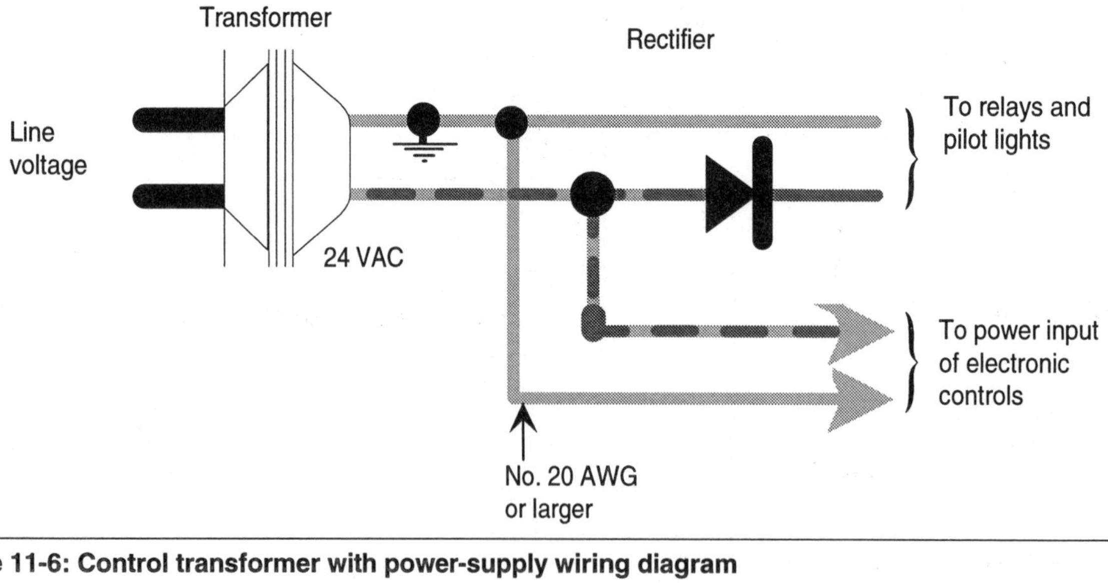
Transformer
Rectifier
Line voltage
To relays and pilot lights
24 VAC
To power input of electronic controls
No. 20 AWG or larger

Figure 11-6: Control transformer with power-supply wiring diagram

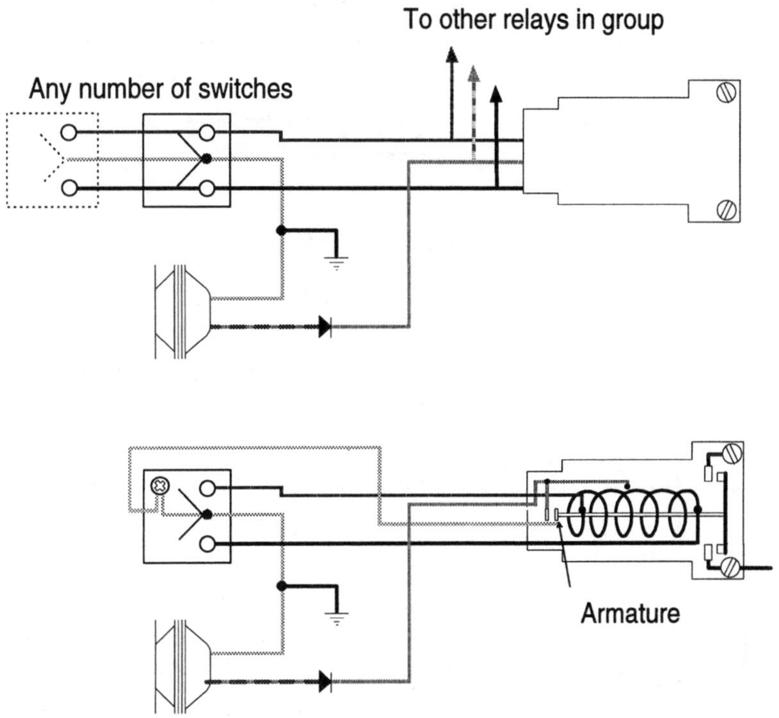
To other relays in group
Any number of switches
Armature

Figure 11-7: Switch operation

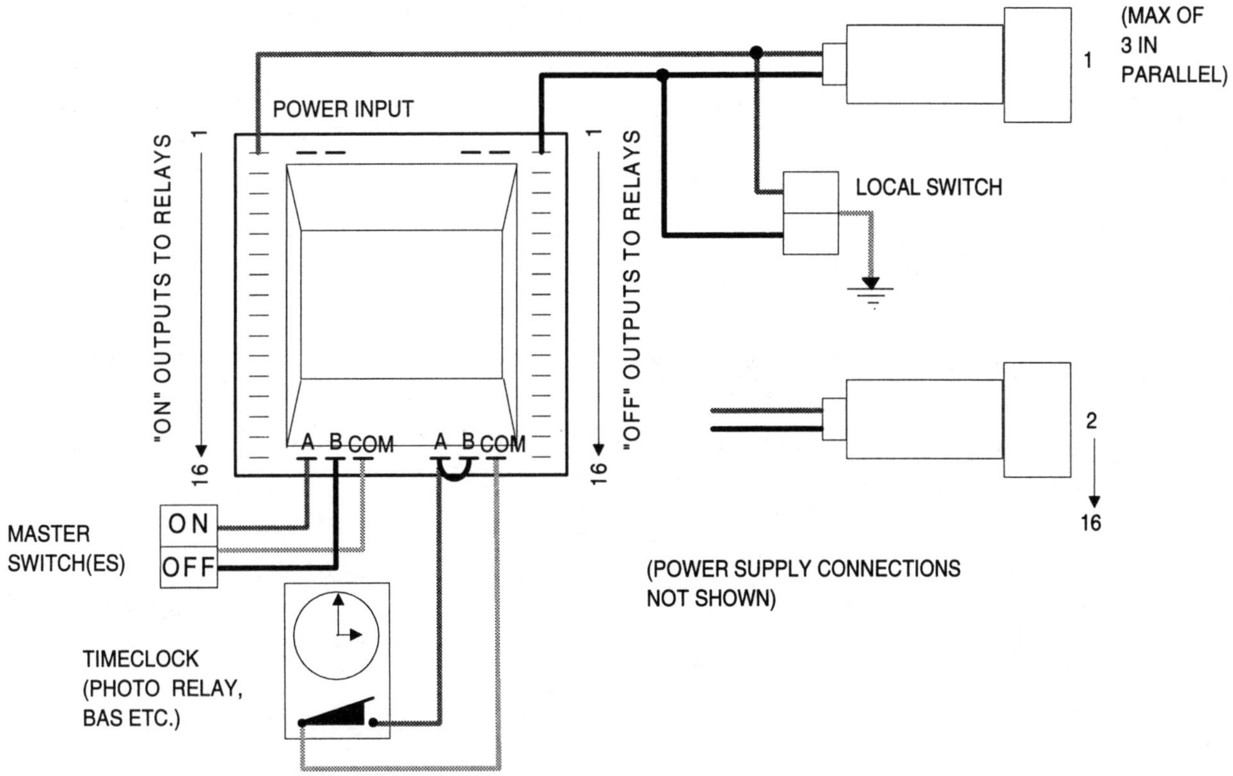
(MAX OF
3 IN
PARALLEL)
1
POWER INPUT
"ON" OUTPUTS TO RELAYS
1
16
"OFF" OUTPUTS TO RELAYS
1
16
LOCAL SWITCH
A B COM
A B COM
2
16
MASTER
SWITCH(ES)
ON
OFF
TIMECLOCK
(PHOTO RELAY,
BAS ETC.)
(POWER SUPPLY CONNECTIONS
NOT SHOWN)

Figure 11-8: Master sequencer connection

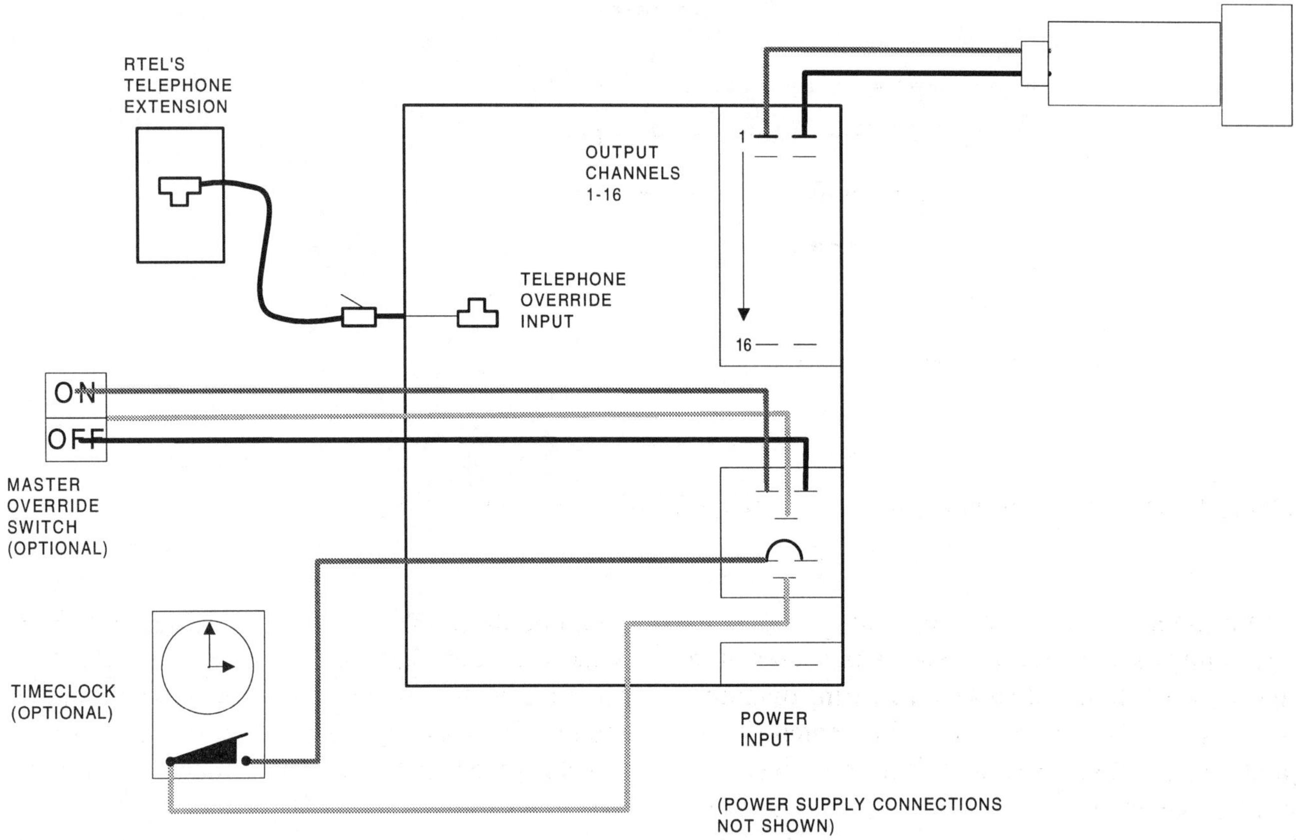
RTEL'S
TELEPHONE
EXTENSION
OUTPUT
CHANNELS
1-16
1
16
TELEPHONE
OVERRIDE
INPUT
ON
OFF
MASTER
OVERRIDE
SWITCH
(OPTIONAL)
TIMECLOCK
(OPTIONAL)
POWER
INPUT
(POWER SUPPLY CONNECTIONS
NOT SHOWN)

Figure 11-9: Telephone override circuit

switches as illustrated in Figure 11-9. Special function switches are available to allow the phone override function to be disabled or to be limited to ON overrides only.

BASIC OPERATION OF REMOTE-CONTROL SWITCHING

Figure 11-10 shows a single-switch control of a light or a group of lights in one area. This is the basic circuit of relay control and is similar to single-pole conventional switch circuits. The schematic wiring diagram of this circuit is shown in Figure 11-11.

The addition of a timeclock and/or photoelectric cell allows applications in which any number of control circuits are turned on or off at desired intervals. For example, the dusk-to-dawn control of outdoor lighting circuits.

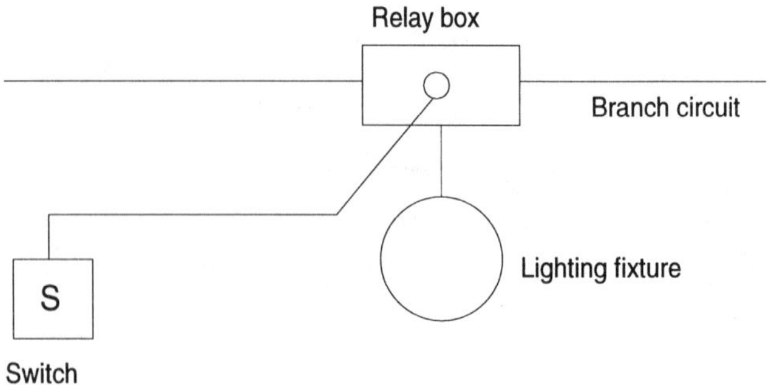
Relay box
Branch circuit
Lighting fixture
S
Switch

Figure 11-10: Single-switch control

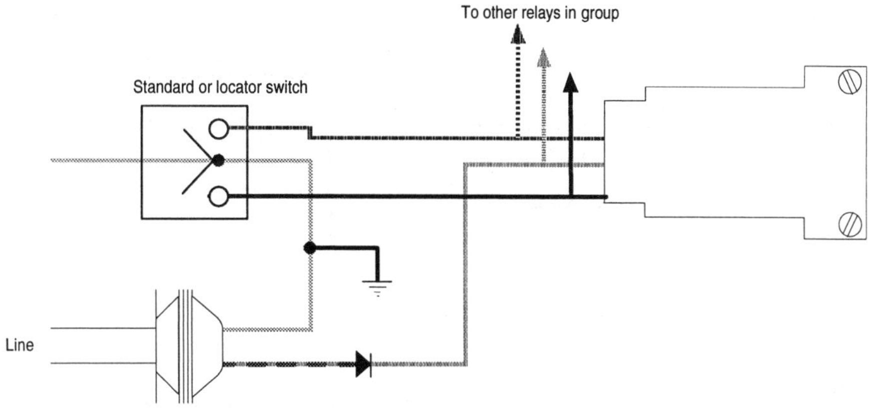
To other relays in group
Standard or locator switch
Line

Figure 11-11: Schematic wiring diagram of single switches controlling each circuit

Multipoint switching, or the switching of a single circuit from two or more switches, is shown in Figure 11-12 along with its related wiring diagram. In conventional wiring, multipoint switching requires costly three-way and four-way switches, plus the extension of switch legs and traveler wires through all switches. One of the greatest advantages of relay switching is the low cost of adding additional switch points.

Figure 11-13 shows many lighting circuits being controlled from one location. In conventional line-voltage switches, this is accomplished by ganging the individual switches together. The same procedure can be employed with remote-control switches. When more than six switches are required, it is usually desirable to install a master-selector switch that permits the control of individual circuits.

Master control of many individual and isolated circuits is possible. Where more than twelve circuits are to be controlled from a single location, and where the selection of individual circuits is not required, then a motor-master control unit automatically sweeps 25 circuits from either ON or OFF at the touch of a single "master" switch, which can be located at one or more locations.

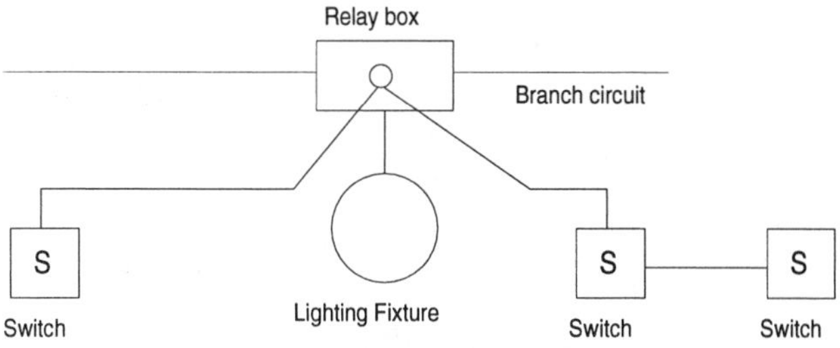
Relay box
Branch circuit
S
S
S
Switch
Lighting Fixture
Switch
Switch

Figure 11-12: Multipoint switching

PLANNING A REMOTE-CONTROL SWITCHING SYSTEM

Low-voltage lighting control systems can be installed in both residential and commercial installations. The systems installed in each type of location may differ. The manufacturer's instructions and recommendations should always be consulted and adhered to, as should the blueprint specifications.The first step in the installation procedure, whether the building is residential or commercial, is to develop a plan.When laying out a remote-

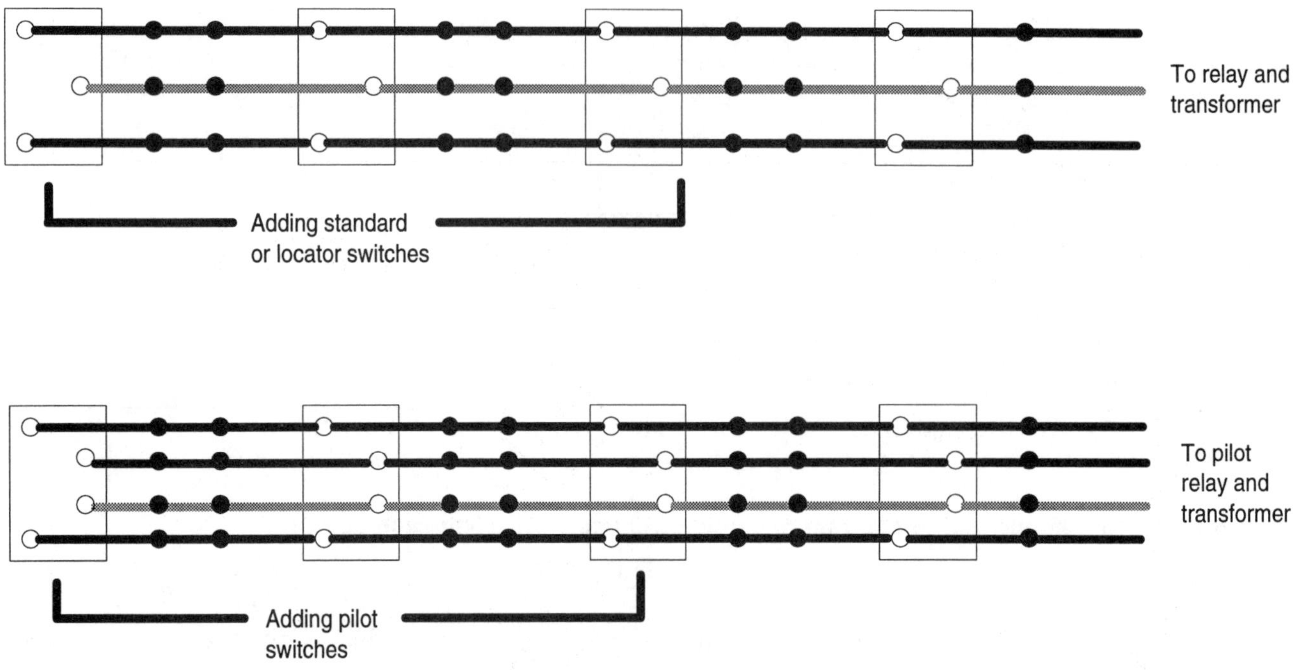
To relay and transformer
Adding standard or locator switches
To pilot relay and transformer
Adding pilot switches

Figure 11-13: Switching circuits being controlled from one location

control switching system for a building, drawing symbols such as the ones shown in Figure 11-14 are normally used to identify the various components. The drawing may be simple, requiring only the following:

- The location of each low-voltage control with a dashed line drawn from the switch to the outlet to be controlled.
- A schematic wiring diagram showing all components.
- The following note printed on the drawing:

"Furnish and install complete remote-control wiring system for control of lighting and other equipment as indicated on the drawings, diagrams, and schedules. System shall be complete with transformers, rectifiers, relays, switches, master-selector switches . . . wall plates, and wiring. All remote-control wiring components shall be of the same manufacture and installed in accordance with the recommendations of the manufacturer.

"Except where otherwise indicated, all remote-control wiring shall be installed in accordance with *NE Code* Article 725, Class 2."

DESIGN OF LOW-VOLTAGE SWITCHING

Many times the low-voltage lighting control system will be designed by architects, especially in new installations. However, in existing buildings, the electrician may be expected to design and install the system. If this situation occurs, the following steps, working from a set of blueprints, should be followed:

1. Assign each light or receptacle a specific relay. This can be accomplished by using the letters R1 for relay 1, R2 for relay 2, and so on.
2. Group together all the lights and outlets that are to be switched together.
3. Determine the number of relays that can be controlled by each control module.
4. Divide the number of relays by the number you computed in step 3. This is the

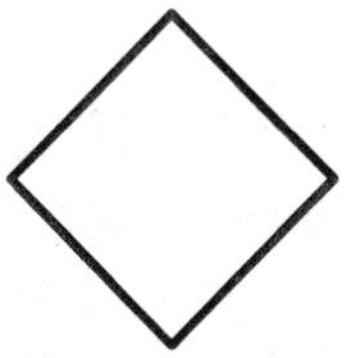

All appear within the diamond, a symbol widely recognized as denoting "low voltage."

Symbol	Description	Symbol	Description
— · · — ·	Remote-control low-voltage wire	S_T	Low-voltage switch, for interchangeable type bracket
T	Low-voltage transformer	S_X	Extension switch receptacle
	Rectifier, or rectifier assembly	MS 8	Master-selector push-button switch - 8 position
C A	Components cabinet - A refers to first cabinet	MS_P 8	Master-selector push-button switch with pilot lights - 8 position
R 1	Relay, standard type, spec. grade - 1 refers to number of relay	MM R	Motor master - red for ON
R_P 2	Relay with pilot switch, spec. grade - 2 refers to number of relay	MM B	Motor master - black for OFF
S	Low-voltage switch, spec. grade, std. type	P_L	Separate pilot light
S_P	Low-voltage switch, spec. grade, with pilot light	**Note:**	At controlled receptacles (ceiling or wall) use plain RA6 etc. which designates Relay, in conponent cabinet A, relay #6.

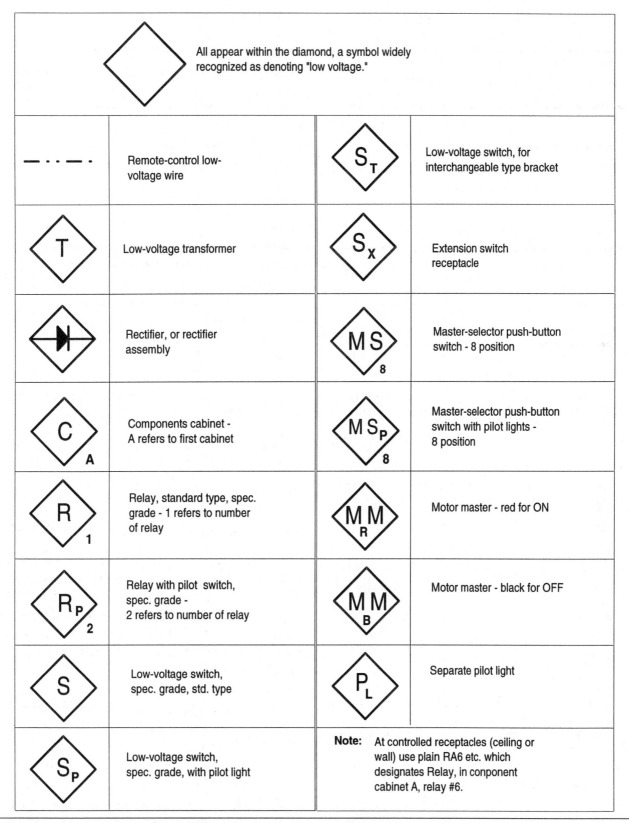

Figure 11-14: Recommended symbols for low-voltage drawings

number of control modules needed to control the relays.

5. Determine the location of the switch points on the building blueprints.
6. Indicate what lights and receptacles are being controlled by what relay. This step may be accomplished by indicating the group of lights on the blueprint. However, it is easier if a switch schedule is used.
7. Review the blueprints or schedule to determine the number and size of the switching stations.

Figure 11-15 shows the floor plan of a residential building. Symbols are used to show the location of all lighting fixtures just as they are used to show the location of conventional switches. Line-voltage circuits feeding the lighting outlets are indicated in the same way (by solid lines) that they are in conventional switching. All of the remote-control switches, however, are indicated by the symbol S_L, instead of S, S_3, S_4, etc. Lines from these remote-control switches to the lighting fixtures that they control are shown by lines from the switch location to the item controlled.

Another method of showing low-voltage remote-control switching on a residential floor plan is shown in Figure 11-16. Here the switches are shown by the manufacturer's catalog number (SF6, SF7, etc.) and each relay is numbered (R-1, R-2, etc.) When this type of layout arrangement is used a schedule is usually shown on the working drawings, or else placed in the written specifications to exactly identify each item.

A schematic wiring diagram, like the one in Figure 11-17, may sometimes be used for further details of all components and the related wiring connections. This wiring diagram aids the contractor or electrician in the installation of the system and leaves little doubt as to exactly what is required.

INSTALLATIONS

Before starting, consult both the *NE Code* and local codes for installation procedures. Once the layout plan has been developed and materials are acquired, the installation of the system can begin. The rough-in instructions presented here are only an example. It is important to follow the manufacturer's recommendations and blueprint specifications.

Enclosure: The enclosure should be placed in a location easily accessible but not hidden from sight. If at all possible, install a unit that is prewired. This will decrease installation time and also limit the possibility for mistakes. Note that, even though the panel may be prewired, the supply cables will still have to be installed. This should be done according to the manufacturer's recommendations.

Since heat will be produced by the enclosed control modules, vents should be placed around the enclosure to facilitate cooling. The location chosen for the enclosure should not hinder the cooling capabilities for which the panels are designed. The manufacturer may require that the high-voltage portion of the enclosure panel be mounted in the upper right corner. This is due to the *NE Code* requirement that is designed to keep low-voltage conductors apart from high-voltage conductors. The panel itself should be mounted at eye level between two studs.

The power supply should be mounted on the enclosure and the power connections made. The number of power supplies needed per system will depend on the size of the system and the capabilities of the power supply. Some power supplies are capable of supplying up to ten modules if the average number of LEDs per channel does not exceed 2 or 2.5. Be sure to check the manufacturer's specifications to ensure that the power supply will not be overloaded.

Switch points: The first step when installing the switch points is to mark their location. As with ordinary switches, the switch points should be located as specified by the plans or specifications.

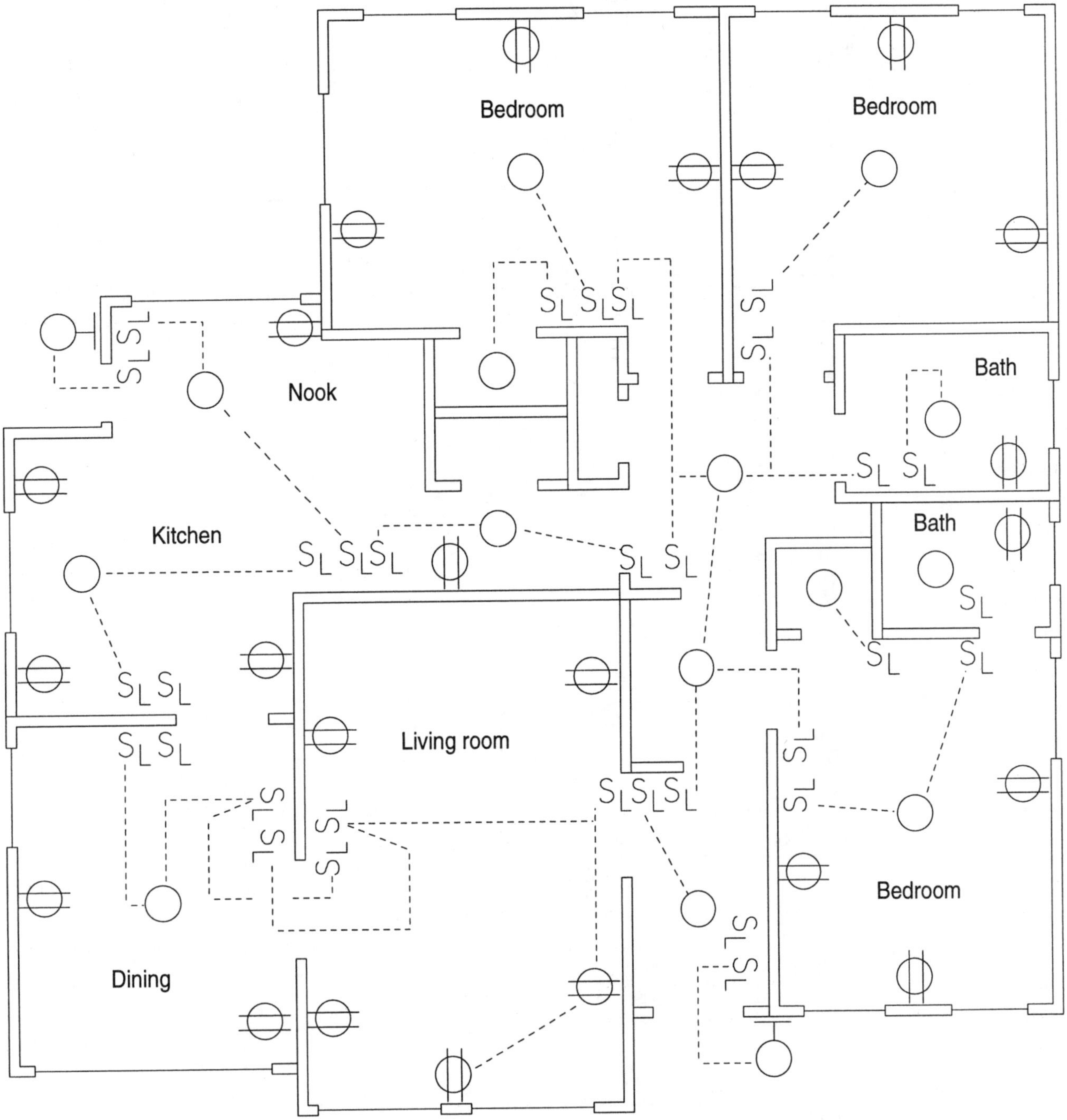
Bedroom
Bedroom
Nook
Bath
Kitchen
Bath
Living room
Bedroom
Dining

Figure 11-15: Low-voltage switching circuits laid out on a residential floor plan

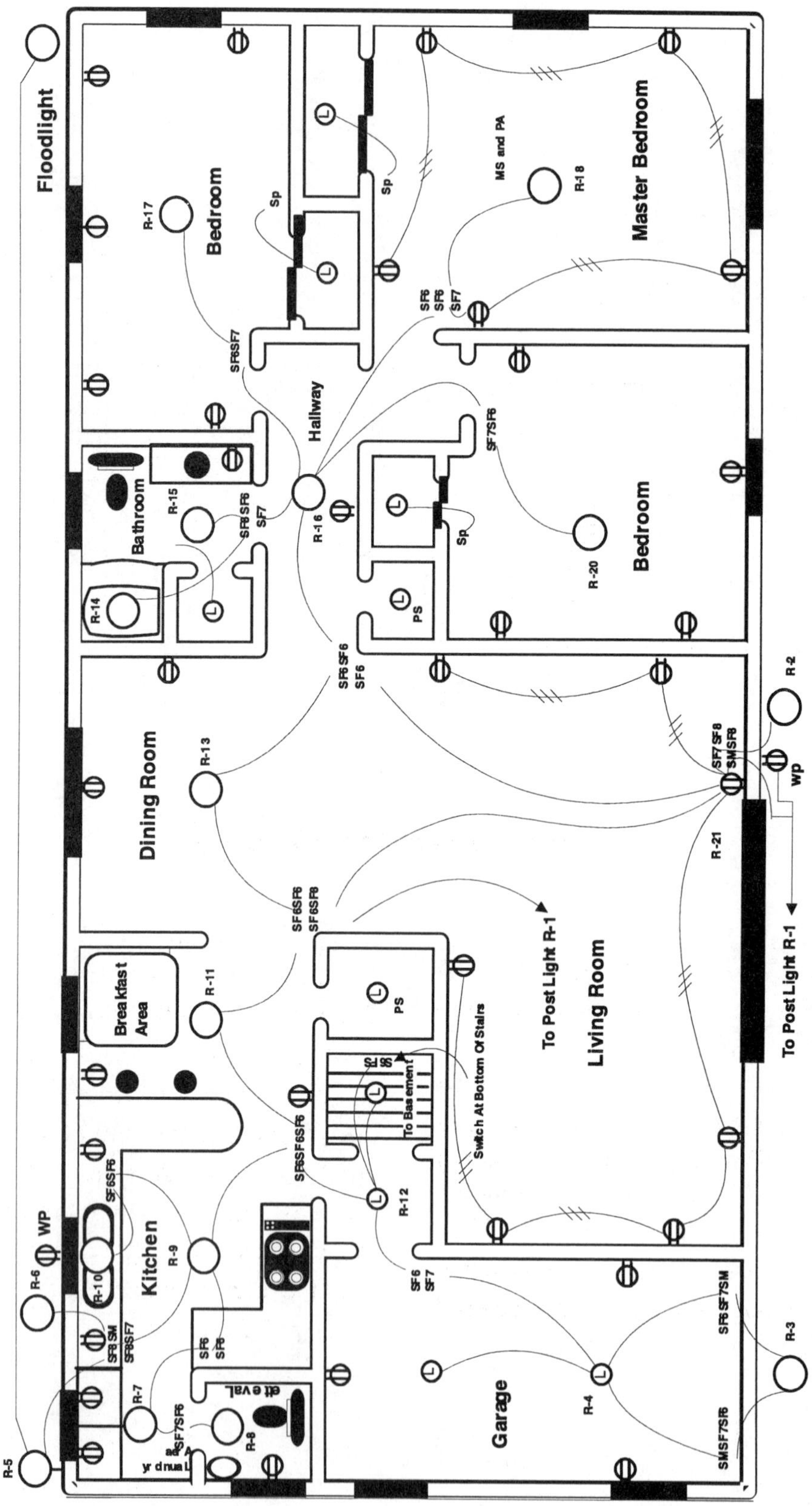
Floodlight
Bedroom
Master Bedroom
MS and PA
Hallway
Bathroom
Bedroom
Dining Room
Breakfast Area
Kitchen
Living Room
To Post Light R-1
To Post Light R-1
Switch At Bottom Of Stairs
To Basement
Garage
WP
WP
R-2
R-3
R-4
R-5
R-6
R-7
R-8
R-9
R-10
R-11
R-12
R-13
R-14
R-15
R-16
R-17
R-18
R-20
R-21

Figure 11-16: Alternate method of showing low-voltage switching on a residential floor plan

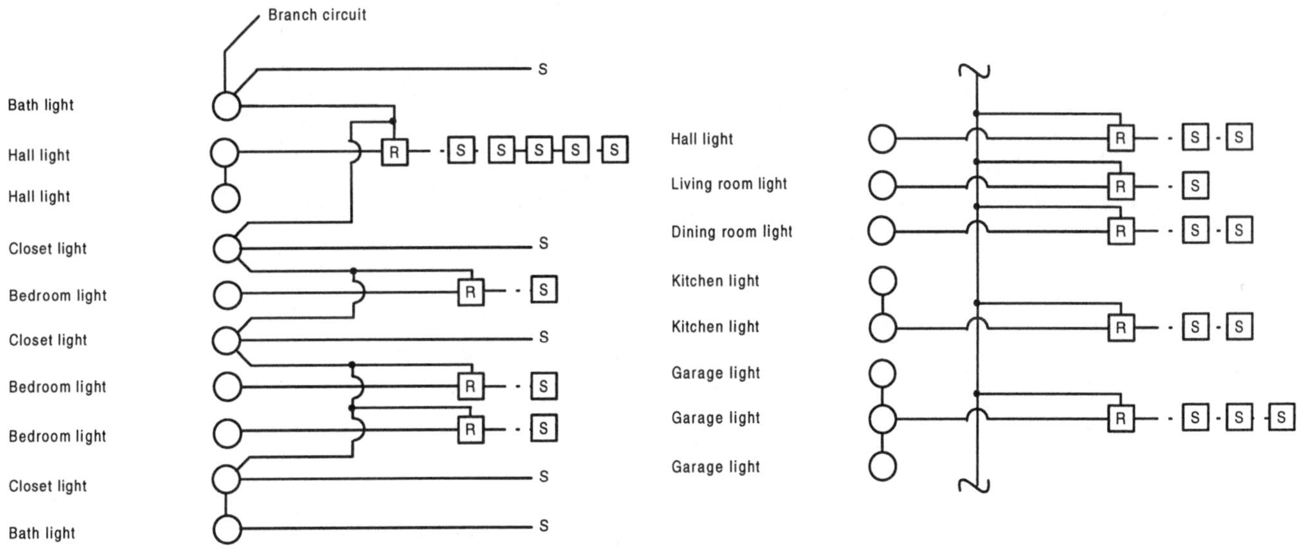
Branch circuit
S
Bath light
Hall light
Hall light
Closet light
Bedroom light
Closet light
Bedroom light
Bedroom light
Closet light
Bath light
R
S
Hall light
Living room light
Dining room light
Kitchen light
Kitchen light
Garage light
Garage light
Garage light

Figure 11-17: Schematic wiring diagram of a low-voltage switching system

Once the location has been determined, mount either a plaster ring or nonmetallic box. A box is not needed in most frame-construction installations.

A plaster ring or Romex staple should then be stapled part way in a stud, behind the plaster ring or nonmetallic box. The preparation for running the low-voltage wiring is the same as for running 120-volt wiring. If the edge of the hole is closer than $1\frac{1}{4}$ inches from the nearest edge of the stud, a $\frac{1}{16}$-inch steel nail plate or bushing must be used to protect the cable from future damage by nails or screws.

The wires are then pulled from the switch station to the panel enclosure. They should come from the left side if the manufacturer recommends that the enclosure be installed with the high-voltage section in the upper right corner. There should be approximately 10 to 12 inches of extra wire left at each switch location and 12 to 18 inches left at the panel enclosure. A wire-pulling technique that reduces installation time follows.

Multiconductor cable is frequently used due to simpler installation. Run one wire (low-voltage lead) from the enclosure to the farthest switching point and anchor the wire to the plaster ring. This wire will supply the low-voltage power to each of the switches. Working back towards the enclosure, loop the wire down and anchor it to each switch location. Take care to leave a sufficient amount of wire for connecting the switch. The low-voltage lead is then connected to the appropriate power-supply lead at the enclosure. The process is repeated for the dc common lead. When this wiring method is used, the dc positive and common lead will have the same color throughout the system and future connection mistakes will be minimized.

The next step is to select a multiple-switch circuit. Pull a different colored wire from the enclosure to the farthest switching location in this circuit. Working back towards the panel enclosure as before, loop the wire to the other switch locations in the multiple-switch circuit. Thus, there are two leads going to each switch with one of the leads common to all the switches in the system. However, if the switches contain a pilot light, two additional No. 18 wires must be run to the LED. The wires are connected to the appropriate pilot light transformer located in the panel enclosure.

If possible, use different colored wires for each run. This will limit confusion in the future when connecting the individual circuits to power. Wire schedules should be used to limit hook up mistakes and give a clearer picture of the system.

Remember that the colors for the control wiring should not be the same as the colors for the dc power and common conductors. Additional marking methods should also be used for future reference. Note that if the wires are marked using adhesive tabs or punch cards, they should be placed in a plastic bag and shoved behind the ring or secured into the plastic box. This will prevent the markings from being covered by paint or plaster when the walls are finished.

The low-voltage wiring should also be kept apart from the high-voltage wiring whenever possible. If the cables must be placed in parallel runs, they should be separated by at least 6 inches, otherwise the noise generated from the high-voltage conductors could seriously interfere with the operation of the low-voltage system.

Power and finish trim: Connecting the circuits to power is now a fairly simple task. Each wire from the switch is connected to a relay with the other side of the relay connected to the common of the dc supply.

When the walls of the building are finished, the switches are connected to the conductors located at each switch point location. Wall plates are then added. A common type of wall plate used with low-voltage lighting control systems requires no screws since it can be snapped into place.

RELAY TROUBLESHOOTING

Electromagnetic contactors and relays of various types are used to make and break circuits that control and protect the operation of such devices as electric motors, combustion and process controls, alarms and annunciators. Successful operation of a relay is dependent upon maintaining the proper interaction between its solenoid and its mechanical elements — springs, hinges, contacts, dashpots, and the like.

The operating coil is usually designed to operate the relay from 80% to 110% of rated voltage. At low voltage the magnet may not be strong enough to pull in the armature against the armature spring and gravity, while at high voltages excessive coil temperatures may develop. Increasing the armature spring force or the magnet gap will require higher pull-in voltage or current values and will result in higher drop-out values.

When sufficient voltage is applied to the operating coil, the magnetic field builds up, the armature is attracted and begins to close. The air gap is shortened, increasing the magnetic attraction and accelerating the closing action, so that the armature closes with a snap, closing normally opened contacts and opening normally closed contacts. Binding at the hinges or excessive armature spring force may cause a contactor or relay that normally has snap action to make and break sluggishly. This condition is often encountered in relays that are rarely operated. Contactors and relays should be operated by hand from time to time to make sure their parts are working freely with proper clearances and spring actions, but this procedure should be performed only by qualified personnel.

Chapter 12

Residential HVAC Systems

The use of electric heating in residential occupancies has risen tremendously over the past decade, and electric heating shows all signs of becoming the principal type of heat for residential construction in the very near future. This is due to the following advantages of electric heat which are not possible with most other heating systems:

- Electric heat is noncombustible and therefore safer than combustible fuels.
- It requires no storage space, fuel tanks, or chimneys.
- It requires little maintenance.
- The initial installation cost is low.
- The amount of heat may be easily controlled since each room may be controlled separately with its own thermostat.
- It is predicted that electricity will be more plentiful than other fuels in the future.

The purpose of this chapter is to discuss in detail the methods used to calculate heat loss and thus the wattage necessary to comfortably heat homes, the type of units available, and how to design and install electric heat systems according to the *NE Code*.

The goals of an electric heat installation are to obtain:

- Adequate, dependable, and trouble-free installation.
- Year-round comfort.
- Reasonable annual operating cost.
- Reasonable installation cost.
- Systems that are easy to service and maintain.

The electrical designer must make heat-loss calculations to ascertain that heating equipment of proper capacity has been selected and specified. Heat loss is expressed either in Btu per hour (which is abbreviated Btuh) or in watts. Both are measures of the rate at which heat is transferred and are easily converted from one to the other:

$$watts = \frac{Btuh}{3.4}$$

$$Btuh = watts \times 3.4$$

Basically, for the majority of residential projects, the calculation of heat loss through various structures and materials, such as walls, roofs, ceilings, windows, and floors, requires three simple steps:

Step 1. Determine the net area in square feet.

HEATING LOAD				
1. DESIGN CONDITIONS	DRY BULB (°F)		SPECIFIC HUMIDITY gr./lb.	
OUTSIDE				
INSIDE				
DIFFERENCE				
ITEMS				HEAT LOAD (BTU)
2. TRANSMISSION GAIN				
WINDOWS	SQ. FT. X	FACTOR X	DRY BULB TEMP. DIFF.	
		0.55		=
		0.55		=
WALLS		0.13		=
		0.13		=
		0.13		=
		0.13		=
ROOF		0.09		=
FLOOR		0.20		=
OTHER				=
3. VENTILATION OR INFILTRATION				
SENSIBLE LOAD	CFM X	DRY BULB TEMP. DIFF. X	FACTOR	
			1.08	=
HUMIDIFICATION LOAD	CFM X	SPECIFIC HUMIDITY DIFF. X		
			0.67	=
4. DUCT HEAT LOSS				=
HEAT LOSS ______X FACTOR FOR INSULATION THICKNESS ______X DUCT LENGTH (FT.)________ ÷ 100				=
5. TOTAL HEATING LOAD				=

Figure 12-1: Typical heating-calculation form

Step 2. Find the proper "heat-loss factor" from a table, or use the ones in the form.

Step 3. Multiply the area by the factor; the product will be expressed in Btuh. Since most electric-heat equipment is rated in watts rather than Btuh, divide this product by 3.4 to convert to watts.

Calculations of heat loss for any building may be made more quickly and more efficiently by using a prepared form, such as that shown in Figure 12-1. With spaces provided for all necessary data and calculations, the procedure becomes routine and simple. Note that this simple method is for residential construction only.

The heat load estimate is based on design conditions inside the building and outside in the atmosphere surrounding the building. Outside design conditions are the maximum extremes of temperature occurring in a specific locality. The inside design condition is the degree of temperature and humidity that will give optimum comfort.

Outside Design Conditions

The outside design dry-bulb temperatures for calculating heating loads are shown in Figure 12-2. For cities and localities not listed, use the design temperatures of listed cities that most closely approximate the local climate conditions, or refer to the ASHRAE Handbook and obtain specific humidity from a psychrometric chart, or use 80% average relative humidity for the outside design humidity.

Inside Design Conditions

The inside comfort design temperature for heating in winter is 72° to 75°F dry bulb. Homes where senior citizens reside may require an inside design temperature as high as 78°F dry bulb.

Heat-Load Calculations

First, calculate the area (in square feet) of all outside windows in the area under consideration and insert this figure in the proper space on the calculation form. Next, calculate the area of the outside walls (minus the area of any windows or doors in the wall) and insert this figure in its proper space. Continue by calculating the areas of the roof, floor, and doors, inserting all in their proper spaces on the form. Then refer to the calculation form and enter the figure in the appropriate locations. Finally, add all of the products to find the total heat loss for the area in question. The answer will be in Btuh. Now divide this answer by 3.4 to obtain the total watts required to maintain the inside design temperature.

To determine the required heating equipment for each area, calculate the remaining areas in the building in the same manner.

The space marked "infiltration" is used to determine the heat loss caused by air entering the house through window cracks and opened doors. There are several ways to determine the infiltration, but for residential structures the "air change" method is considered best. For use on the form, determine the cubic feet per minute (cfm) of infiltration air from the following formula:

$$\frac{\textit{(room) length} \times \textit{width} \times \textit{height}}{60} = \textit{total cfm}$$

Multiply by the dry-bulb difference and the factor given on the form to determine the sensible heat load.

Heating Calculations for Residence Under Study

From the table in Figure 12-2 we find that the design dry-bulb temperature for our residence located near Washington, DC is 0°F. Since the inside design temperature will be 72°F, the dry-bulb temperature difference will be 72°F.

We will not be concerned with the wet-bulb temperature difference for heating loads, and we will use the figure 80% for the outside relative humidity.

These inside and outside design conditions are entered on eight separate heating-calculation forms because there are eight different areas in the house that will need heat. (See Figure 12-3).

The areas in square feet of all windows, walls, roof, and floor are determined next and entered in the separate spaces provided. Remember that the wall area does not include windows.

Next, the appropriate factors for windows, walls, roof, and floor have already been determined and appear in their appropriate locations on the form.

Then the dry-bulb temperature difference is entered after each item. Next, the area in square feet is multiplied by the factor in the form times the

State	City	Winter (heating) dry-bulb °F	Summer (cooling) dry-bulb °F	Summer (cooling) wet-bulb °F
Alabama	Birmingham	10	95	78
	Mobile	15	95	80
Arizona	Flagstaff	-10	90	65
	Phoenix	25	105	76
	Yuma	30	110	78
Arkansas	Little Rock	5	95	78
California	Bakersfield	25	105	70
	El Centro	25	110	78
	Fresno	25	105	74
	Los Angeles	35	90	70
	San Diego	35	95	68
	San Francisco	35	85	65
Colorado	Denver	-10	95	64
	Pueblo	-20	95	65
Connecticut	Bridgeport	0	95	75
Delaware	Wilmington	0	95	78
District of Columbia	Washington	0	95	78
Florida	Jacksonville	25	95	78
	Miami	35	91	79
Georgia	Atlanta	10	95	76
	Savannah	20	95	78
Idaho	Boise	-10	95	65
Illinois	Chicago	-10	95	75
	Springfield	-10	98	77
Indiana	Indianapolis	-10	95	76
	Terre Haute	0	95	78
Iowa	Des Moines	-15	95	78
	Sioux City	-20	95	78
Kansas	Topeka	-10	100	78
Kentucky	Louisville	0	95	78

Figure 12-2: Recommended outside design conditions for localities in the contiguous United States

State	City	Winter (heating) dry-bulb °F	Summer (cooling) dry-bulb °F	Summer (cooling) wet-bulb °F
Louisiana	New Orleans	20	95	80
Maine	Augusta	-15	90	73
Maryland	Baltimore	0	95	78
Massachusetts	Boston	0	92	75
	Worcester	-5	93	75
Michigan	Detroit	-10	95	75
	Lansing	-10	95	75
Minnesota	Duluth	-25	93	73
	Minneapolis	-20	95	75
Mississippi	Vicksburg	10	95	78
Missouri	Kansas City	-10	100	76
	St. Louis	0	95	78
Montana	Butte	-20	. . .	. .
	Miles City	-35	. . .	. .
Nebraska	Omaha	-10	95	78
Nevada	Reno	-5	95	65
New Hampshire	Concord	-15	90	73
New Jersey	Newark	0	95	75
New Mexico	Albuquerque	0	95	70
New York	Buffalo	-5	93	73
	New York	0	95	75
	Syracuse	-10	93	75
North Carolina	Asheville	0	93	75
	Raleigh	10	95	78
North Dakota	Bismarck	-30	95	73
Ohio	Akron	-5	95	75
	Dayton	0	95	78
	Toledo	-10	95	75
Oklahoma	Tulsa	0	101	77

Figure 12-2: Recommended outside design conditions for localities in the contiguous United States

State	City	Winter (heating) dry-bulb °F	Summer (cooling) dry-bulb °F	Summer (cooling) wet-bulb °F
Oregon	Portland	10	90	68
Pennsylvania	Philadelphia	0	95	78
	Pittsburgh	0	95	75
Rhode Island	Providence	0	93	75
South Carolina	Charleston	15	95	78
South Dakota	Sioux Falls	-20	95	75
Tennessee	Nashville	0	95	78
Texas	Austin	20	100	78
	Dallas	0	100	78
	El Paso	10	100	69
	Houston	20	95	78
Utah	Salt Lake City	-10	95	65
Vermont	Burlington	-10	90	73
Virginia	Richmond	15	95	78
Washington	Seattle	15	85	65
	Spokane	-15	93	65
West Virginia	Charleston	0	95	75
	Wheeling	-5	95	75
Wisconsin	Milwaukee	-15	95	75
Wyoming	Cheyenne	-15	95	65

Figure 12-2: Recommended outside design conditions for localities in the contiguous United States

dry-bulb temperature difference. The answer in Btuh is entered in the heating-load column.

Infiltration is calculated next by using the equation given previously. The cfm figure is then multiplied by the dry-bulb temperature difference and by the factor on the form, which is 1.08. The resulting answer is the infiltration sensible heat load and should be entered in the heating-load column.

The transmission loads and the infiltration load are then added for each area, and the total result is entered in the space provided on the form. This final total figure is the total heating load in Btuh. Since most electric equipment is rated in watts, the total heating load should be divided by 3.4 to convert it from Btuh to watts.

The equipment selected must be capable of handling this load and maintaining the required temperature conditions inside the building.

The finished heat-loss forms for the residence shown in Figure 12-3 appear in Figures 12-4 through 12-11.

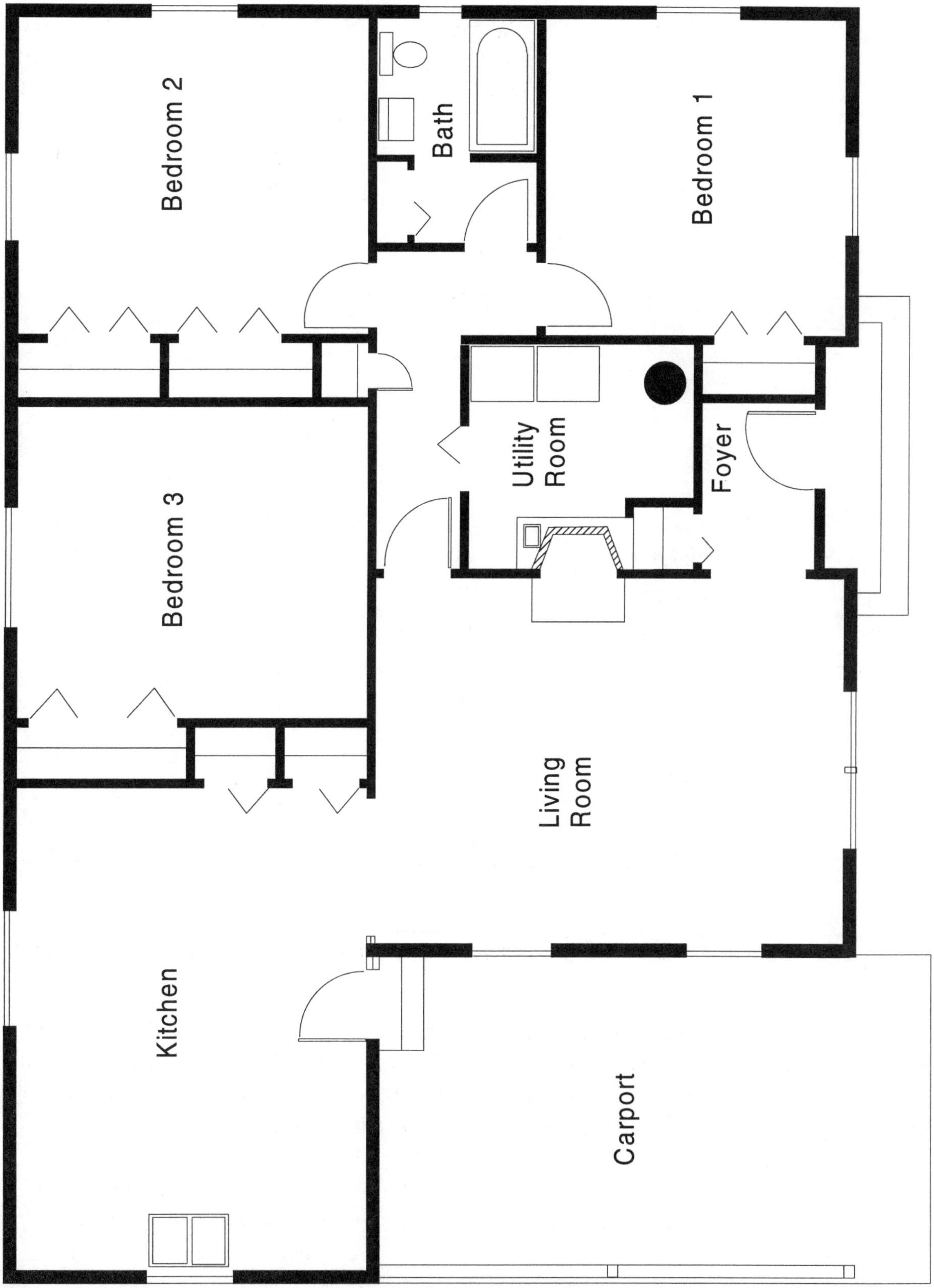
Bedroom 2
Bath
Bedroom 1
Utility Room
Foyer
Bedroom 3
Living Room
Kitchen
Carport

Figure 12-3: Floor plan of residence used for heating-calculation forms

HEATING LOAD		
1. DESIGN CONDITIONS	DRY BULB (°F)	SPECIFIC HUMIDITY gr./lb.
OUTSIDE	0°F	
INSIDE	70°F	
DIFFERENCE	70°F	

ITEMS					HEAT LOAD (BTU)
2. TRANSMISSION GAIN					
	SQ. FT. X	FACTOR X	DRY BULB TEMP. DIFF.		
WINDOWS	49	0.55	70	=	1886.5
		0.55		=	
WALLS	249	0.13	70	=	2265.9
		0.13		=	
		0.13		=	
		0.13		=	
ROOF	273	0.09	70	=	1719.9
FLOOR	273	0.20	70	=	3822.0
OTHER				=	
3. VENTILATION OR INFILTRATION					
SENSIBLE LOAD	CFM X	DRY BULB TEMP. DIFF. X	FACTOR		
	36.4	70	1.08	=	2751.8
HUMIDIFICATION LOAD	CFM X	SPECIFIC HUMIDITY DIFF. X			
			0.67	=	
4. DUCT HEAT LOSS				=	
HEAT LOSS _______X FACTOR FOR INSULATION THICKNESS _______ X					
DUCT LENGTH (FT.)_________ ÷ 100				=	
5. TOTAL HEATING LOAD	12,446.17/3.4			=	3660.63 watts

Figure 12-4: Heat-loss calculations for living room

HEATING LOAD

1. DESIGN CONDITIONS	DRY BULB (°F)	SPECIFIC HUMIDITY gr./lb.
OUTSIDE	0°F	
INSIDE	70°F	
DIFFERENCE	70°F	

ITEMS					HEAT LOAD (BTU)
2. TRANSMISSION GAIN					
WINDOWS	SQ. FT. X	FACTOR X	DRY BULB TEMP. DIFF.		
	43	0.55	70	=	1655.5
		0.55		=	
WALLS	293	0.13	70	=	2666.3
		0.13		=	
		0.13		=	
		0.13		=	
ROOF	227	0.09	70	=	1430.1
FLOOR	227	0.20	70	=	3178.0
OTHER				=	
3. VENTILATION OR INFILTRATION					
SENSIBLE LOAD	CFM X	DRY BULB TEMP. DIFF. X	FACTOR		
	30.33	70	1.08	=	2292.9
HUMIDIFICATION LOAD	CFM X	SPECIFIC HUMIDITY DIFF. X			
			0.67	=	
4. DUCT HEAT LOSS				=	
HEAT LOSS _______X FACTOR FOR INSULATION THICKNESS _______ X DUCT LENGTH (FT.)_________ ÷ 100				=	
5. TOTAL HEATING LOAD	11,222.84/3.4			=	3300.83 watts

Figure 12-5: Heat-loss calculations for kitchen

HEATING LOAD					
1. DESIGN CONDITIONS	DRY BULB (°F)			SPECIFIC HUMIDITY gr./lb.	
OUTSIDE	0°F				
INSIDE	70°F				
DIFFERENCE	70°F				
	ITEMS				HEAT LOAD (BTU)
2. TRANSMISSION GAIN					
	SQ. FT. X	FACTOR X	DRY BULB TEMP. DIFF.		
WINDOWS		0.55		=	
		0.55		=	
WALLS		0.13		=	
		0.13		=	
		0.13		=	
		0.13		=	
ROOF	54	0.09	70	=	340.2
FLOOR	54	0.20	70	=	756.0
OTHER				=	
3. VENTILATION OR INFILTRATION					
SENSIBLE LOAD	CFM X	DRY BULB TEMP. DIFF. X	FACTOR		
	7.6	70	1.08	=	574.6
HUMIDIFICATION LOAD	CFM X	SPECIFIC HUMIDITY DIFF. X			
			0.67	=	
4. DUCT HEAT LOSS				=	
HEAT LOSS ______ X FACTOR FOR INSULATION THICKNESS ______ X DUCT LENGTH (FT.)______ ÷ 100				=	
5. TOTAL HEATING LOAD	1670.8/3.4			=	491.41 watts

Figure 12-6: Heat-loss calculations for utility room

HEATING LOAD		
1. DESIGN CONDITIONS	DRY BULB (°F)	SPECIFIC HUMIDITY gr./lb.
OUTSIDE	0°F	
INSIDE	70°F	
DIFFERENCE	70°F	

	ITEMS				HEAT LOAD (BTU)
2. TRANSMISSION GAIN					
WINDOWS	SQ. FT. X	FACTOR X	DRY BULB TEMP. DIFF.		
	4	0.55	70	=	154.0
		0.55		=	
WALLS	48	0.13	70	=	436.8
		0.13		=	
		0.13		=	
		0.13		=	
ROOF	52	0.09	70	=	327.6
FLOOR	52	0.20	70	=	728.0
OTHER				=	
3. VENTILATION OR INFILTRATION					
SENSIBLE LOAD	CFM X	DRY BULB TEMP. DIFF. X	FACTOR		
	6.93	70	1.08	=	523.9
HUMIDIFICATION LOAD	CFM X	SPECIFIC HUMIDITY DIFF. X			
			0.67	=	
4. DUCT HEAT LOSS				=	
HEAT LOSS ______X FACTOR FOR INSULATION THICKNESS _______X DUCT LENGTH (FT.)_________ ÷ 100				=	
5. TOTAL HEATING LOAD	2170.3/3.4			=	638.32 watts

Figure 12-7: Heat-loss calculations for bath

HEATING LOAD						
1. DESIGN CONDITIONS	DRY BULB (°F)			SPECIFIC HUMIDITY gr./lb.		
OUTSIDE	0°F					
INSIDE	70°F					
DIFFERENCE	70°F					
	ITEMS					HEAT LOAD (BTU)
2. TRANSMISSION GAIN						
WINDOWS	SQ. FT. X	FACTOR X	DRY BULB TEMP. DIFF.			
	18	0.55	70		=	693.0
		0.55			=	
WALLS	30	0.13	70		=	273.0
		0.13			=	
		0.13			=	
		0.13			=	
ROOF	27	0.09	70		=	170.1
FLOOR	27	0.20	70		=	378.0
OTHER					=	
3. VENTILATION OR INFILTRATION						
SENSIBLE LOAD	CFM X	DRY BULB TEMP. DIFF. X	FACTOR			
	3.6	70	1.08		=	272.16
HUMIDIFICATION LOAD	CFM X	SPECIFIC HUMIDITY DIFF. X				
			0.67		=	
4. DUCT HEAT LOSS					=	
HEAT LOSS ______X FACTOR FOR INSULATION THICKNESS ______ X DUCT LENGTH (FT.)________ ÷ 100					=	
5. TOTAL HEATING LOAD	1786.26/3.4				=	525.37 watts

Figure 12-8: Heat-loss calculations for foyer

HEATING LOAD

1. DESIGN CONDITIONS	DRY BULB (°F)	SPECIFIC HUMIDITY gr./lb.
OUTSIDE	0°F	
INSIDE	70°F	
DIFFERENCE	70°F	

ITEMS					HEAT LOAD (BTU)
2. TRANSMISSION GAIN					
	SQ. FT.	X FACTOR	X DRY BULB TEMP. DIFF.		
WINDOWS	28	0.55	70	=	1078.0
		0.55		=	
WALLS	164	0.13	70	=	1492.4
		0.13		=	
		0.13		=	
		0.13		=	
ROOF	144	0.09	70	=	907.2
FLOOR	144	0.20	70	=	2016.0
OTHER				=	
3. VENTILATION OR INFILTRATION					
SENSIBLE LOAD	CFM	X DRY BULB TEMP. DIFF.	X FACTOR		
	19.2	70	1.08	=	1451.5
HUMIDIFICATION LOAD	CFM	X SPECIFIC HUMIDITY DIFF.	X		
			0.67	=	
4. DUCT HEAT LOSS				=	
HEAT LOSS ______X FACTOR FOR INSULATION THICKNESS ______X DUCT LENGTH (FT.)________ ÷ 100				=	
5. TOTAL HEATING LOAD	6945.1/3.4			=	2042.68 watts

Figure 12-9: Heat-loss calculations for bedroom 1

HEATING LOAD

1. DESIGN CONDITIONS	DRY BULB (°F)	SPECIFIC HUMIDITY gr./lb.
OUTSIDE	0°F	
INSIDE	70°F	
DIFFERENCE	70°F	

ITEMS					HEAT LOAD (BTU)
2. TRANSMISSION GAIN					
	SQ. FT. X	FACTOR X	DRY BULB TEMP. DIFF.		
WINDOWS	28	0.55	70	=	1078.0
		0.55		=	
WALLS	172	0.13	70	=	1565.2
		0.13		=	
		0.13		=	
		0.13		=	
ROOF	156	0.09	70	=	982.8
FLOOR	156	0.20	70	=	2184.0
OTHER				=	
3. VENTILATION OR INFILTRATION					
SENSIBLE LOAD	CFM X	DRY BULB TEMP. DIFF. X	FACTOR		
	20.82	70	1.08	=	1573.99
HUMIDIFICATION LOAD	CFM X	SPECIFIC HUMIDITY DIFF. X			
			0.67	=	
4. DUCT HEAT LOSS				=	
HEAT LOSS _______ X FACTOR FOR INSULATION THICKNESS _______ X DUCT LENGTH (FT.) _______ ÷ 100				=	
5. TOTAL HEATING LOAD	7383.99/3.4			=	2171.76 watts

Figure 12-10: Heat-loss calculations for bedroom 2

HEATING LOAD		
1. DESIGN CONDITIONS	DRY BULB (F)	SPECIFIC HUMIDITY gr./lb.
OUTSIDE	0°F	
INSIDE	70°F	
DIFFERENCE	70°F	

ITEMS					HEAT LOAD (BTU)
2. TRANSMISSION GAIN					
	SQ. FT. X	FACTOR X	DRY BULB TEMP. DIFF.		
WINDOWS	21	0.55	70	=	808.5
		0.55		=	
WALLS	75	0.13	70	=	682.5
		0.13		=	
		0.13		=	
		0.13		=	
ROOF	156	0.09	70	=	982.8
FLOOR	156	0.20	70	=	2184.0
OTHER				=	
3. VENTILATION OR INFILTRATION					
SENSIBLE LOAD	CFM X	DRY BULB TEMP. DIFF. X	FACTOR		
	20.82	70	1.08	=	1573.99
HUMIDIFICATION LOAD	CFM X	SPECIFIC HUMIDITY DIFF. X			
			0.67	=	
4. DUCT HEAT LOSS				=	
HEAT LOSS _____X FACTOR FOR INSULATION THICKNESS _____X DUCT LENGTH (FT.)_______ ÷ 100				=	
5. TOTAL HEATING LOAD	6231.79/3.4			=	1832.88 watts

Figure 12-11: Heat-loss calculations for bedroom 3

SELECTING HEATING EQUIPMENT

The type of electric heating system used for a given residence will usually depend on the structural conditions, the kind of room, and the purpose for which the room will be used. The owner's preference will also enter into the final decision.

Several types of electric heating units are available (see Figure 12-12) and a brief description of each is in order.

Electric baseboard heaters: Electric baseboard heaters are mounted on the floor along the baseboard, preferably on outside walls under windows for the most efficient operation. They are absolutely noiseless in operation and are the type most often used for heating residential occupancies and for use as supplemental heat in many commercial areas.

Electric baseboard heaters may be mounted on practically any surface (wood, plaster, drywall, and so on), but if polystyrene foam insulation is used near the unit, a ¾-inch (minimum) ventilated spacer strip must be used between the heater and the wall. In such cases, the heater should also be elevated above the floor or rug to allow ventilation to flow from the floor upward over the total heater space.

One complaint received over the years about this type of heater has been wall discoloration directly above the heating units. When this problem occurs, the reason is almost always traced to one or more of the following factors:

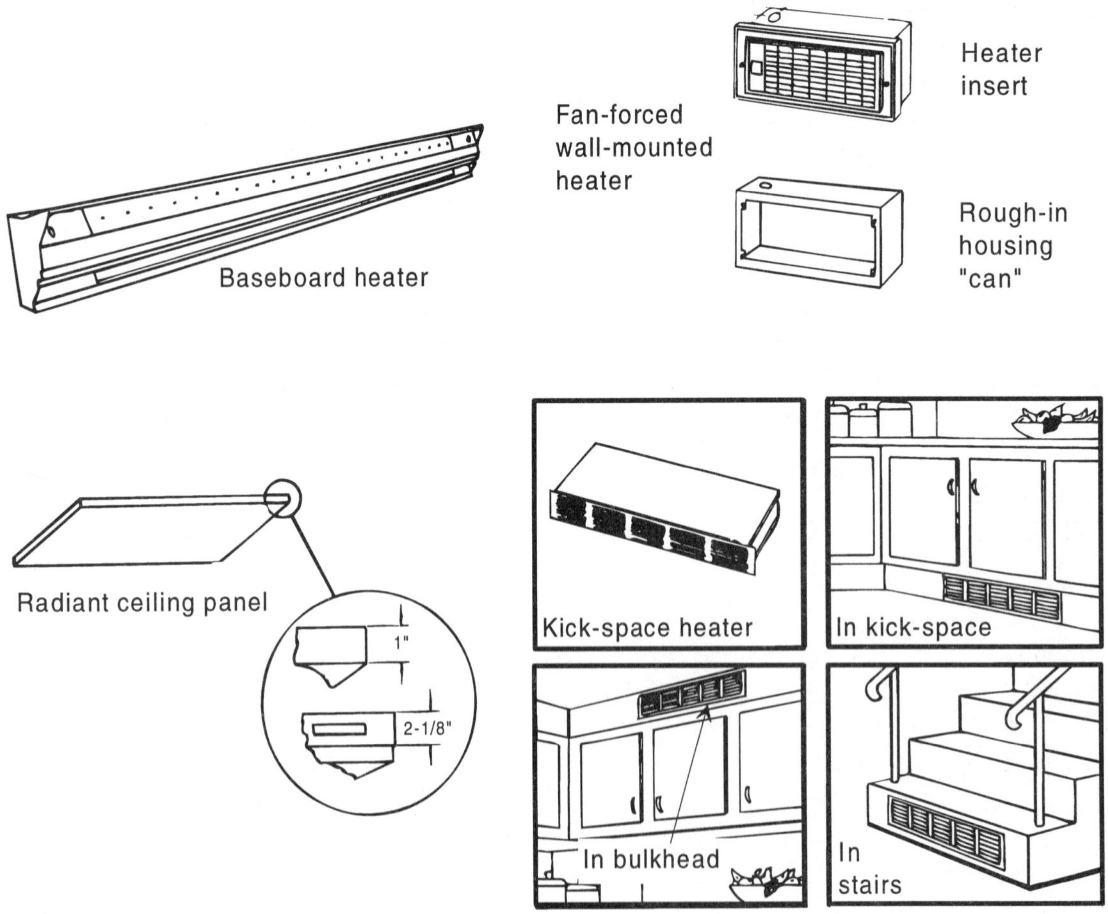
Baseboard heater
Fan-forced wall-mounted heater
Heater insert
Rough-in housing "can"
Radiant ceiling panel
1"
2-1/8"
Kick-space heater
In kick-space
In bulkhead
In stairs

Figure 12-12: Several types of electric-heating units

- High wattage per square foot of heating element.
- Heavy smoking by occupants.
- Poor housekeeping.

Radiant ceiling heaters: Radiant ceiling heaters are often used in bathrooms and similar areas so that the entire room does not have to be overheated to meet the need for extra warmth after a bath or shower. They are also used in larger areas, such as a garage or basement, or for spot-warming a person standing at a workbench.

Most of these units are rated from 800 to 1,500 watts (W) and normally operate on 120-volt circuits. As with most electric units, they may be controlled by a remote thermostat, but since they are usually used for supplemental heat, a conventional wall switch is often used. They are quickly and easily mounted on an outlet box in much the same way as conventional lighting fixtures. In fact, where very low wattage is used, ceiling heaters may often be installed by merely replacing the ceiling lighting fixture with a light/heater combination.

Radiant heating panels: Radiant heating panels are commonly manufactured in 2-feet by 4-feet sizes and are rated at 500 watts. They may be located on ceilings or walls to provide radiant heat that spreads evenly through the room. Each room may be controlled by its own thermostat. Since this type of heater may be mounted on the ceiling, its use allows complete freedom for room decor, furniture placement, and drapery arrangement. Most are finished in beige to blend in with nearly any room or furniture color.

Units mounted on the ceiling give the best results when located parallel to, and approximately 2 feet from the outside wall. However, this type of unit may also be mounted on walls.

Electric infrared heaters: Rays from infrared heaters do not heat the air through which they travel. Rather, they heat only persons and certain objects that they strike. Therefore, infrared heaters are designed to deliver heat into controlled areas for the efficient warming of people and surfaces both indoors and outdoors (such as to heat persons on a patio on a chilly night or around the perimeter of an outdoor swimming pool). This type of heater is excellent for heating a person standing at a workbench without heating the entire room, for melting snow from steps or porches, and for providing sunlike heat over outdoor areas. Some of the major advantages of infrared heat include:

- No warm-up period is required. Heat is immediate.
- Heat rays are confined to the desired areas.
- They are easy to install, as no ducts, vents, and so on, are required.

When installing this type of heating unit, never mount the heater closer than 24 inches from vertical walls unless the specific heating unit is designed for closer installation. Read the manufacturer's instructions carefully.

NOTE: Infrared quartz lamps provide some light in addition to heat.

Forced-air wall heaters: Forced-air wall heaters are designed to bring quick heat into an area where the sound of a quiet fan will not be disturbing. Some are very noisy. Most of these units are equipped with a built-in thermostat with a sensor mounted in the intake air stream. Some types are available for mounting on high walls or even ceilings, but the additional force required to move the air to a usable area produces even more noise.

Floor insert convection heaters: Floor insert convection heaters require no wall space, as they fit into the floor. They are best suited for placement beneath conventional or sliding glass doors to form an effective draft barrier. All are equipped with safety devices, such as a thermal cutout to disconnect the heating element automatically in the event that normal operating temperatures are exceeded.

Floor insert convector heaters may be installed in both old and new homes by cutting through the floor, inserting the metal housing and wiring, according to the manufacturer's instructions. A heavy-gauge floor grille then fits over the entire unit.

Electric kick-space heaters: Modern kitchens contain so many appliances and so much cabinet space for the convenience of the owner that there often is no room to install electric heaters except on the ceiling. Therefore, a kick-space heater was added to the lines of electric heating manufacturers to overcome this problem.

For the most comfort, kick-space heaters should not be installed in such a manner that warm air blows directly on the occupant's feet. Ideally, the air discharge should be directed along the outside wall adjacent to normal working areas, not directly under the sink.

Radiant heating cable: Radiant heating cable provides an enormous heating surface over the ceiling or concrete floor so that the system need not be raised to a high temperature. Rather, gentle warmth radiates downward (in the case of ceiling-mount cable) or upward (in the case of floor-mounted cable), heating the entire room or area evenly.

There is virtually no maintenance with a radiant heating system, as there are no moving parts and the entire heating system is invisible — except for the thermostat.

Combination heating and cooling units: One way to have individual control of each room or area in the home, as far as heating and cooling are concerned, is to install through-wall heating and cooling units. Such a system gives the occupants complete control of their environment with a room-by-room choice of either heating or cooling at any time of year at any temperature they desire. Operating costs are lower than for many other systems due to the high efficiency of room-by-room control. Another advantage is that if a unit should fail, the defective chassis can be replaced immediately or taken to a shop for repair without shutting down the remaining units in the building.

When selecting any electric heating units, obtain plenty of literature from suppliers and manufacturers before settling on any one type. In most cases you are going to get what you pay for, but most contractors and their personnel shop around at different suppliers before ordering the equipment. Delivery of any of these units may take some time, so once the brand, size, and supplier have been selected, the order should be placed well before the unit is actually needed.

Electric furnaces: Electric furnaces are becoming more popular, although they are somewhat surpassed by the all-electric heat pump. Most are very compact, versatile units designed for either wall, ceiling, or closet mounting. The vertical model can be flush-mounted in a wall or shelf mounted in a closet; the horizontal design (Figure 12-13) can be fitted into a ceiling (flush or recessed).

Central heating systems of the electrically energized type distribute heat from a centrally located source by means of circulating air or water. Compact electric boilers can be mounted on the wall of a basement, utility room, or closet with the necessary control and circuit protection, and will furnish hot water to convectors or to embedded pipes. Immersion heaters may be stepped in one at a time to provide heat capacity to match heat loss.

The majority of electric furnaces are commonly available in sizes up to 24 kW for residential use. The larger boilers with proper controls can take advantage of lower off-peak electricity rates, where they prevail, by heating water during off-peak periods, storing it in insulated tanks, and circulating it to convectors or radiators to provide heat as needed.

Electric hot-water systems: A zone hydronic (hot-water) system permits selection of different temperatures in each zone of the home. Baseboard heaters located along the outer walls of rooms provide a blanket of warmth from floor to ceiling; the heating unit also supplies domestic hot water simultaneously, through separate circuits. A special attachment coupled to the hot-water unit can be used to melt snow and ice on walkways and driveways in winter, and a similar attachment can be used to heat, say, a swimming pool during the spring and fall seasons.

A typical hot-water system operating diagram is shown in Figure 12-14, and is explained as follows: When a zone thermostat calls for heat, the

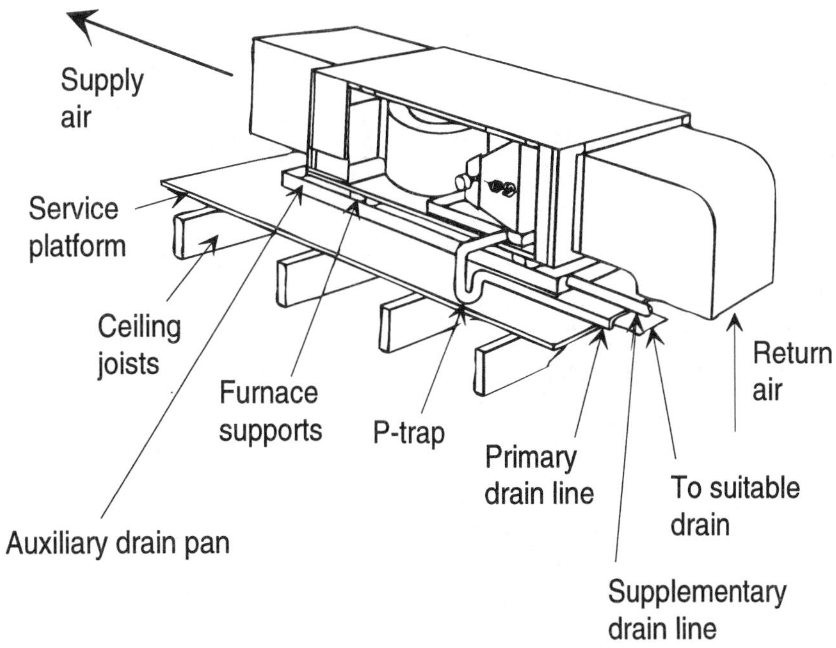
Supply
air
Service
platform
Ceiling
joists
Furnace
supports
P-trap
Primary
drain line
Return
air
To suitable
drain
Auxiliary drain pan
Supplementary
drain line

Figure 12-13: Horizontal application of an electric furnace

appropriate zone valve motor begins to run, opening the valve slowly; when the valve is fully opened, the valve motor stops. At that time, the operating relay in the hydrostat is energized, closing contacts to the burner and the circulator circuits. The high-limit control contacts (a safety device) are normally closed so the burner will now fire and operate. If the boiler water temperature exceeds the high-limit setting, the high-limit contacts will open and the burner will stop, but the circulator will continue to run as long as the thermostat continues to call for heat. If the call for heat continues, the resultant drop in boiler water temperature — below the high-limit setting — will bring the burner back on. Thus, the burner will cycle until the thermostat is satisfied; then both the burner and circulator will shut off.

Hot-water boilers for the home are normally manufactured for use with oil, gas, or electricity. While a zoned hot-water system is comparatively costly to install, the cost is still competitive with the better hot-air systems. The chief disadvantage of hot-water systems is that they don't use ducts. If a central air-conditioning system is wanted, a separate duct system must be installed along with the hot-water system.

Chillers, or refrigerated water, are sometimes used with commercial and industrial systems when hot water is used for heating. The cycle is reversed in warmer months to provide cold water through the system to help cool the conditioned areas.

Heat pumps: The term heat pump, as applied to a year-round air-conditioning system, commonly denotes a system in which refrigeration equipment is used in such a manner that heat is taken from a heat source and transferred to the conditioned space when heating is desired; heat is removed from the space and discharged to a heat sink when cooling and dehumidification are desired. Therefore, the heat pump is essentially a heat-transfer refrigeration device that puts the heat rejected by the refrigeration process to good use. A heat pump can do the following:

- Provide either heating or cooling.
- Change from one to the other automatically as needed.
- Supply both simultaneously if so desired.

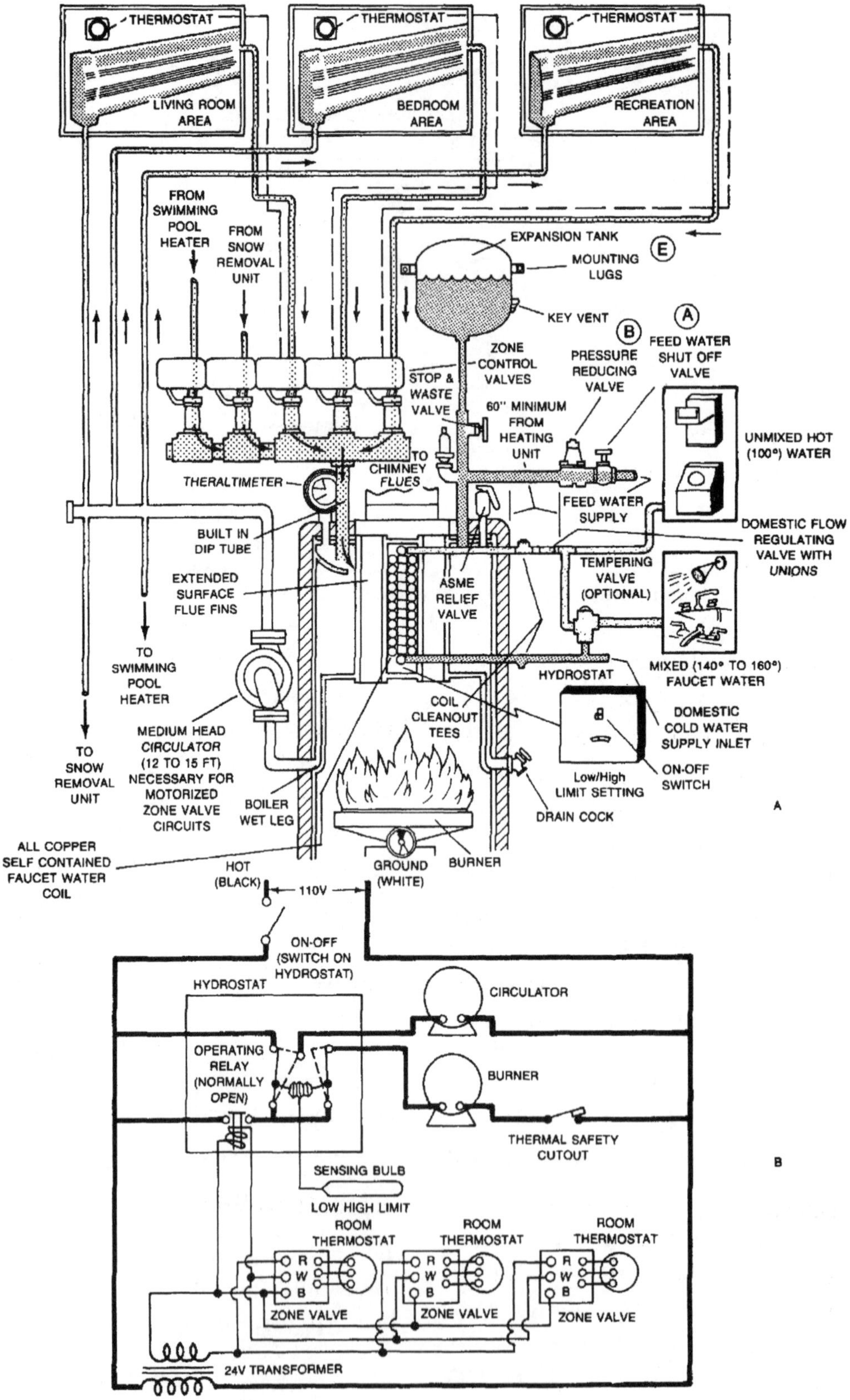
THERMOSTAT
LIVING ROOM AREA
BEDROOM AREA
RECREATION AREA
FROM SWIMMING POOL HEATER
FROM SNOW REMOVAL UNIT
EXPANSION TANK
MOUNTING LUGS
KEY VENT
ZONE CONTROL VALVES
PRESSURE REDUCING VALVE
FEED WATER SHUT OFF VALVE
STOP & WASTE VALVE
60" MINIMUM FROM HEATING UNIT
UNMIXED HOT (100°) WATER
TO CHIMNEY FLUES
THERALTIMETER
FEED WATER SUPPLY
DOMESTIC FLOW REGULATING VALVE WITH UNIONS
BUILT IN DIP TUBE
TEMPERING VALVE (OPTIONAL)
EXTENDED SURFACE FLUE FINS
ASME RELIEF VALVE
TO SWIMMING POOL HEATER
HYDROSTAT
MIXED (140° TO 160°) FAUCET WATER
COIL CLEANOUT TEES
DOMESTIC COLD WATER SUPPLY INLET
TO SNOW REMOVAL UNIT
MEDIUM HEAD CIRCULATOR (12 TO 15 FT) NECESSARY FOR MOTORIZED ZONE VALVE CIRCUITS
BOILER WET LEG
Low/High LIMIT SETTING
ON-OFF SWITCH
DRAIN COCK
A
ALL COPPER SELF CONTAINED FAUCET WATER COIL
HOT (BLACK)
GROUND (WHITE)
BURNER
110V
ON-OFF (SWITCH ON HYDROSTAT)
HYDROSTAT
CIRCULATOR
OPERATING RELAY (NORMALLY OPEN)
BURNER
THERMAL SAFETY CUTOUT
B
SENSING BULB
LOW HIGH LIMIT
ROOM THERMOSTAT
ROOM THERMOSTAT
ROOM THERMOSTAT
ZONE VALVE
ZONE VALVE
ZONE VALVE
24V TRANSFORMER

Figure 12-14: Typical hot-water system operating diagram with control-wiring diagram

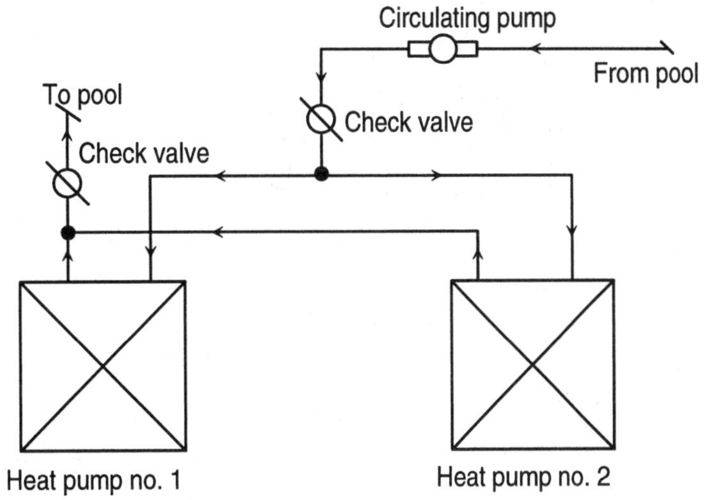
Circulating pump
From pool
To pool
Check valve
Check valve
Heat pump no. 1
Heat pump no. 2

Figure 12-15: Diagram of water-to-air heat pumps

A heat pump has the unique ability to furnish more energy than it consumes. This uniqueness is due to the fact that electric energy is required only to move the heat absorbed by the refrigerant. Thus, a heat pump attains a heating efficiency of two or more to one; that is, it will put out an equivalent of 2 or 3 watts of heat for every watt consumed. For this reason, its use is highly desirable for the conservation of energy.

Although air-to-air heat pumps are the most popular, there are other types available. The water-to-air heat pump, for example, is more efficient than the air-to-air heat pump. Furthermore, water-to-air heat pumps have the following additional advantages:

- The pumps can be located anywhere in the building since no outside air is connected to them.
- The outside air temperature does not affect the performance of the heat pump, as it does in an air-to-air type of heat pump.
- Since the water source of the water-to-air heat pump will seldom vary more than a few degrees in temperature, a more consistent performance can be expected.

The schematic drawing in Figure 12-15 shows how two heat pumps were connected (in parallel) for a large residence with an indoor swimming pool; the swimming-pool water was used as a means of heat exchange, and since this water was preheated to approximately 78°F, the efficiency of the heat pumps utilizing this same water approached the maximum.

During the warm months the heat pumps were reversed (cooling cycle) to cool the area. Again, the pool water was used as the heat exchange, this time acting as a "cooling tower." The air from these heat pumps was distributed by means of underfloor transit ducts with supply-air diffusers and return-air grilles raised above the floor level and mounted in a wainscot around the perimeter of the pool area. This was done to prevent water from entering the openings in the ductwork.

ELECTRIC BASEBOARD HEATERS

All requirements of the *NE Code* apply for the installation of electric baseboard heaters, especially *NE Code* Article 424 — *Fixed Electric Space Heating Equipment.* In general, electric baseboard heaters must not be used where they will be exposed to severe physical damage unless they are adequately protected from such possible damage. Heaters and related equipment installed in

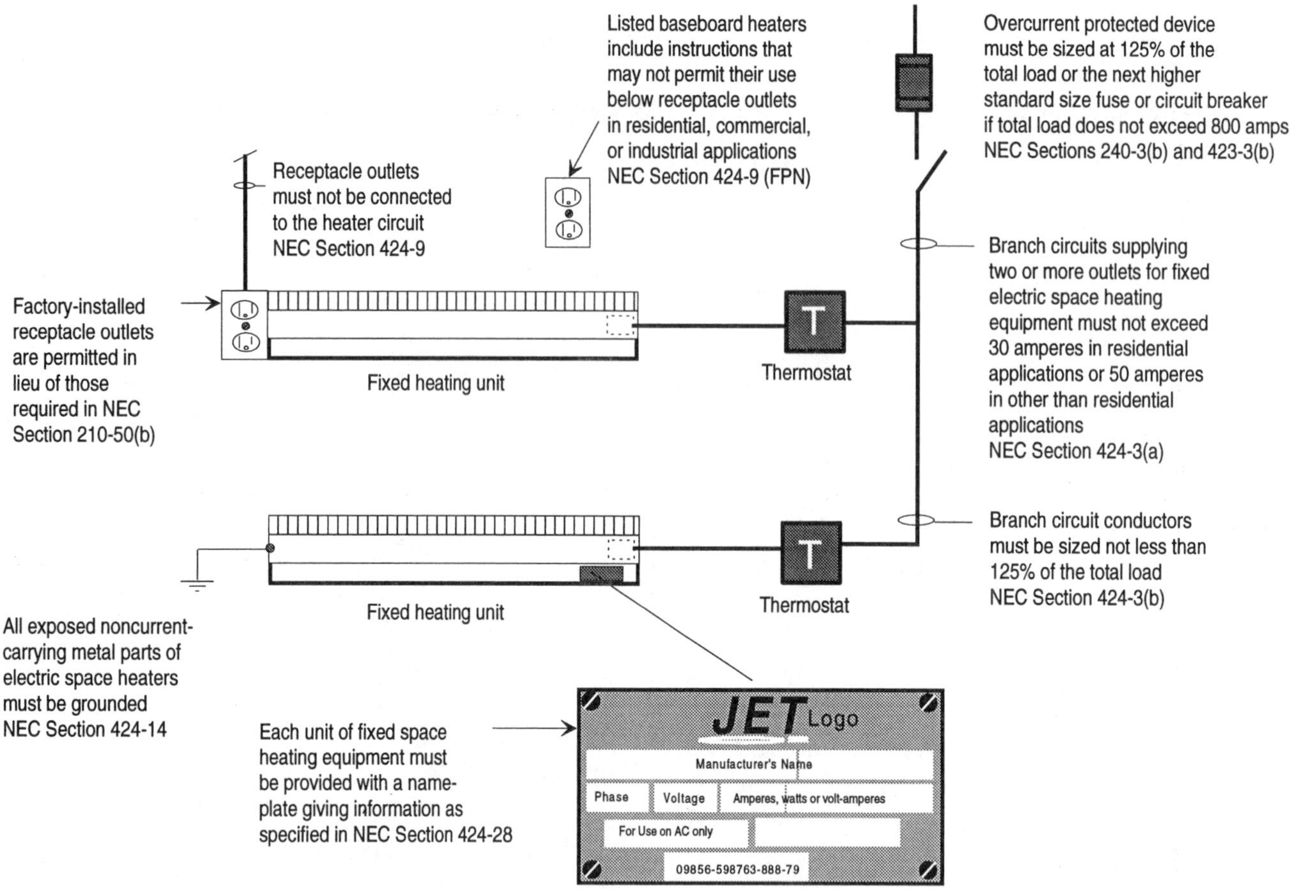
Listed baseboard heaters include instructions that may not permit their use below receptacle outlets in residential, commercial, or industrial applications NEC Section 424-9 (FPN)
Overcurrent protected device must be sized at 125% of the total load or the next higher standard size fuse or circuit breaker if total load does not exceed 800 amps NEC Sections 240-3(b) and 423-3(b)
Receptacle outlets must not be connected to the heater circuit NEC Section 424-9
Branch circuits supplying two or more outlets for fixed electric space heating equipment must not exceed 30 amperes in residential applications or 50 amperes in other than residential applications NEC Section 424-3(a)
Factory-installed receptacle outlets are permitted in lieu of those required in NEC Section 210-50(b)
T
Thermostat
Fixed heating unit
Branch circuit conductors must be sized not less than 125% of the total load NEC Section 424-3(b)
T
Thermostat
Fixed heating unit
All exposed noncurrent-carrying metal parts of electric space heaters must be grounded NEC Section 424-14
Each unit of fixed space heating equipment must be provided with a name-plate giving information as specified in NEC Section 424-28
JET Logo
Manufacturer's Name
Phase
Voltage
Amperes, watts or volt-amperes
For Use on AC only
09856-598763-888-79

Figure 12-16: *NE Code* installation rules for electric baseboard heaters

damp or wet locations must be approved for such locations and must be constructed and installed so that water cannot enter or accumulate in or on wired sections, electrical components, or duct work.

Baseboard heaters must be installed to provide the required spacing between the equipment and adjacent combustible material and each unit must be adequately grounded in accordance with *NE Code* Section 424-14 and *NE Code* Article 250.

Figure 12-16 summarizes the *NE Code* regulations governing the installation of electric baseboard heaters, while Figure 12-18 shows a residential floor plan layout for electric heat.

Compare the floor plan in Figure 12-17 with the heat-loss calculations that were described earlier in this chapter. Note that the closest standard size heaters were used to match the heat loss of each area.

ELECTRIC SPACE-HEATING CABLES

Radiant ceiling heat is acknowledged to be one of the greatest advances in structural heating since the Franklin stove, or so say the manufacturers, and thousands of homeowners all over the country who have chosen this type of heat for their homes.

The enormous heating surface precludes the necessity of raising air temperatures to a high degree. Rather, gentle warmth flows downward (or upward in the case of cable embedded in concrete floors) from the surfaces, heating the entire room or area evenly, and usually leaving no cold spots or drafts.

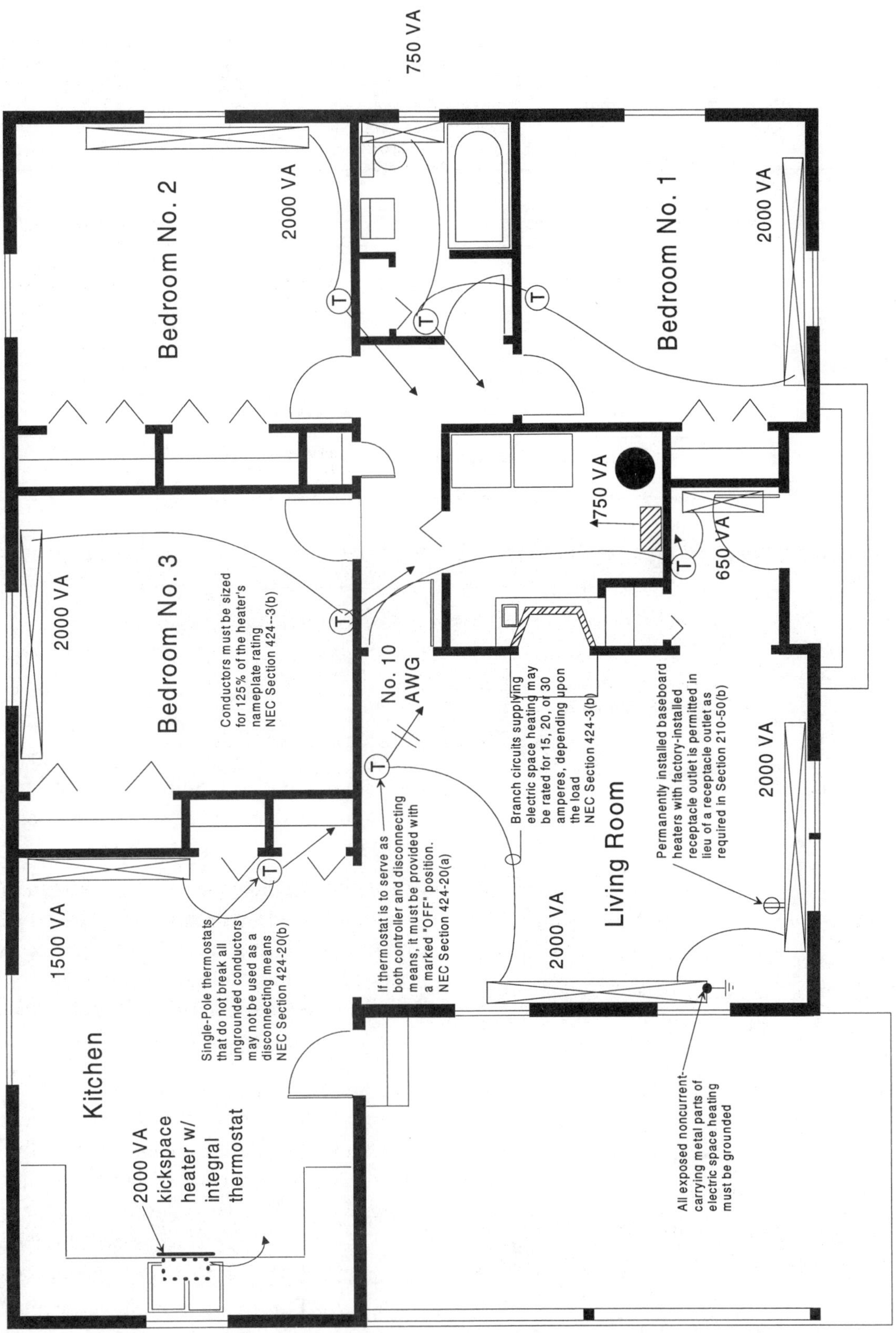
750 VA
Bedroom No. 2
2000 VA
Bedroom No. 1
2000 VA
Bedroom No. 3
2000 VA
Conductors must be sized for 125% of the heater's nameplate rating NEC Section 424--3(b)
No. 10 AWG
750 VA
650 VA
Branch circuits supplying electric space heating may be rated for 15, 20, or 30 amperes, depending upon the load NEC Section 424-3(b)
Permanently installed baseboard heaters with factory-installed receptacle outlet is permitted in lieu of a receptacle outlet as required in Section 210-50(b)
2000 VA
Living Room
2000 VA
If thermostat is to serve as both controller and disconnecting means, it must be provided with a marked "OFF" position. NEC Section 424-20(a)
Single-Pole thermostats that do not break all ungrounded conductors may not be used as a disconnecting means NEC Section 424-20(b)
1500 VA
Kitchen
2000 VA kickspace heater w/ integral thermostat
All exposed noncurrent-carrying metal parts of electric space heating must be grounded

Figure 12-17: Floor-plan layout for a residential electric-heating application

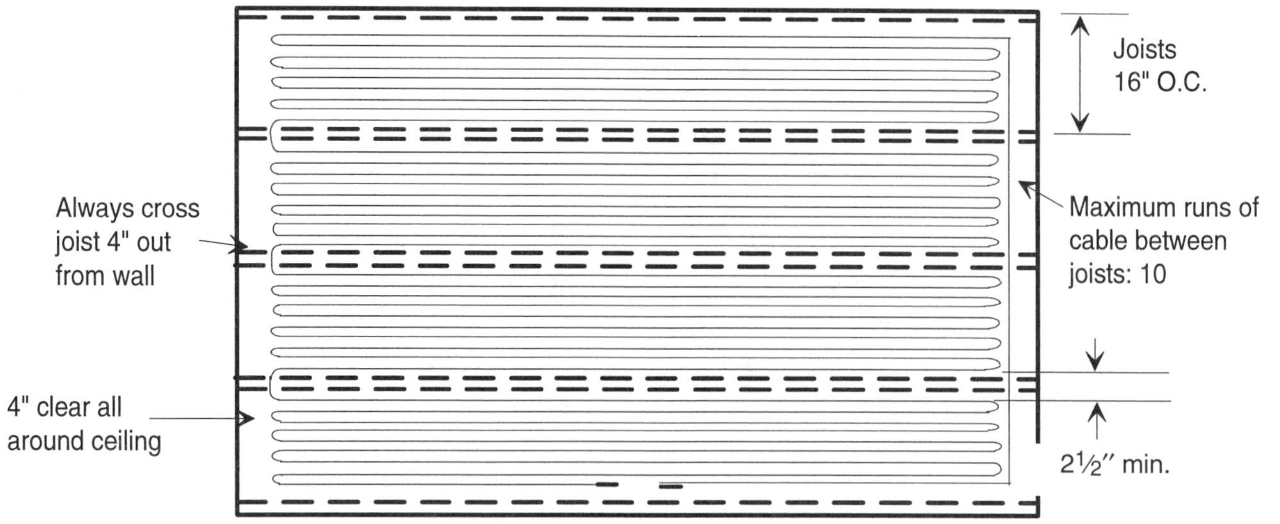
Joists
16" O.C.
Always cross
joist 4" out
from wall
Maximum runs of
cable between
joists: 10
4" clear all
around ceiling
2½" min.

Figure 12-18: Recommendations for heat cable layout

There is no maintenance with a radiant heating system as there are no moving parts, nothing to get clogged up, nothing to clean, oil, or grease, and nothing to wear out.

The installation of this system is within reach of even the smallest electrical contractor. The most difficult part of the entire project is the layout of the system; that is, how far apart to string the cable on the ceiling or in a concrete slab.

An ideal application of electric radiant heating cable would be during the renovation of an area within an existing residence, where the ceiling plaster is beginning to crack and this ceiling will be recovered with drywall or other type of plaster board. Or, perhaps the basement floor needs repair and three inches of additional concrete will be poured over the existing floor. These are ideal locations to install radiant heating cable.

Installation in Plaster Ceilings

To determine the spacing of the cable on a given ceiling, deduct 1 foot from the room length and 1 foot from the room width and multiply this new length by the new width, which will give the usable ceiling area in square feet. Multiply the square feet of the ceiling by 12 to get the ceiling area in inches before dividing by the length of the heating cable. The result will be the number of inches apart to space the cable.

For example, assume that a room 14 × 12 feet has a calculated heat loss of 2,000 watts. Therefore, (14 ft. - 1 ft.) (12 ft. - 1 ft.) × 12 in. = 13 × 11 × 12 = 1,716. We then look at manufacturers' tables and see that a 2,000-watt heating cable is 728 feet in length. Dividing the usable area (1,716 sq. in.) by this length (728), we find that the cable should be spaced 2.3 inches apart.

The drawing in Figure 12-18 shows a floor plan of a typical heating cable installation as suggested by one manufacturer. However, nearly every brand of heating cable will be installed in exactly the same way, and the procedure is as follows:

Step 1. Nail outlet box on inside wall approximately 5 feet above the finished floor for your thermostat location.

Step 2. Drill two holes in wall plate above this junction box location.

Step 3. Drill two holes through ceiling lath above thermostat junction box location.

Step 4. Put the spool of heat cable on a nail, screwdriver, or any type of shaft you have at hand for unwinding the cable from the spool.

Step 5. Cover the accessible end of the 8-feet nonheating lead wire with nonmetallic loom. Loom should be long enough so at least 2 inches will go on ceiling surface and reach to the thermostat outlet box, but leave 6 inches of lead wire inside of this junction box for viewing the identification tags. Never, for any reason, remove those tags. Also do not cut or shorten the nonheating leads.

Step 6. Run the accessible end (loom and all) of your cable through one of the holes in the ceiling, down the wall, through the plate, into the thermostat outlet or junction box.

Step 7. Pull slack out of your nonheating leads and staple securely to the ceiling. Any excess nonheating lead should be covered with plaster the same as the heat cable. Do not staple or bend the cable.

Step 8. Your next step will be to mark the ceiling. First mark a line all the way around the room 6 inches away from each wall; this accounts for the 1 foot you deducted from the width and length. A chalk line is best for this. Now take a yardstick, often given to regular customers at the local hardware store, and notch it for your calculated spacing. Run the cable along the ceiling 6 inches out from the wall to an outside wall, where you attach it in parallel spacing. In this example the spacing is 2½ inches. Always keep the cable at least 2 inches from metal corner lath, or other metal reinforcing.

Step 9. When you are down to the return lead wire, you are back to the starting wall. Cover this lead with the same length of loom as you did the starting lead. Then staple this return lead securely to the ceiling, run through the other hole in the ceiling, down the wall and into the thermostat outlet box.

Step 10. Connect the thermostat, which should be fed by a circuit of proper wire size according to the current, in amperes, drawn by the heating cable. Divide the wattage by the rated voltage. Your answer will be the load in amperes. Then use the following table to size your wire.

AMPS	TYPE TW WIRE
15 or lower	10 AWG
15-20	8 AWG
20-30	6 AWG
30-50	4 AWG

Always make certain that the heating cable is connected to the proper voltage. A 120-volt cable connected to a 240-volt circuit will melt the cable, while a 240-volt heating cable connected to 120 volts will produce only 25% of the rated wattage of the cable. See Figure 12-19 for a summary of *NE Code* requirements governing the installation of electric space-heating cable.

Installing Concrete Cable

The heat loss is calculated in the same manner as for any other area. For best results, the heating cable should never be spaced less than 2½ inches apart when installing in concrete floor except around outside walls, which may be spaced on 1½-inch minimum centers for the first 2 feet out from the wall. The concrete thickness above the cable should be from ½ to 1 inch.

The spacing of the cable is found exactly as previously shown for the spacing of the cable on the ceiling, and then the installation procedure is as follows:

Step 1. Secure junction box on an inside wall approximately 5 feet from the finished floor to house the thermostat.

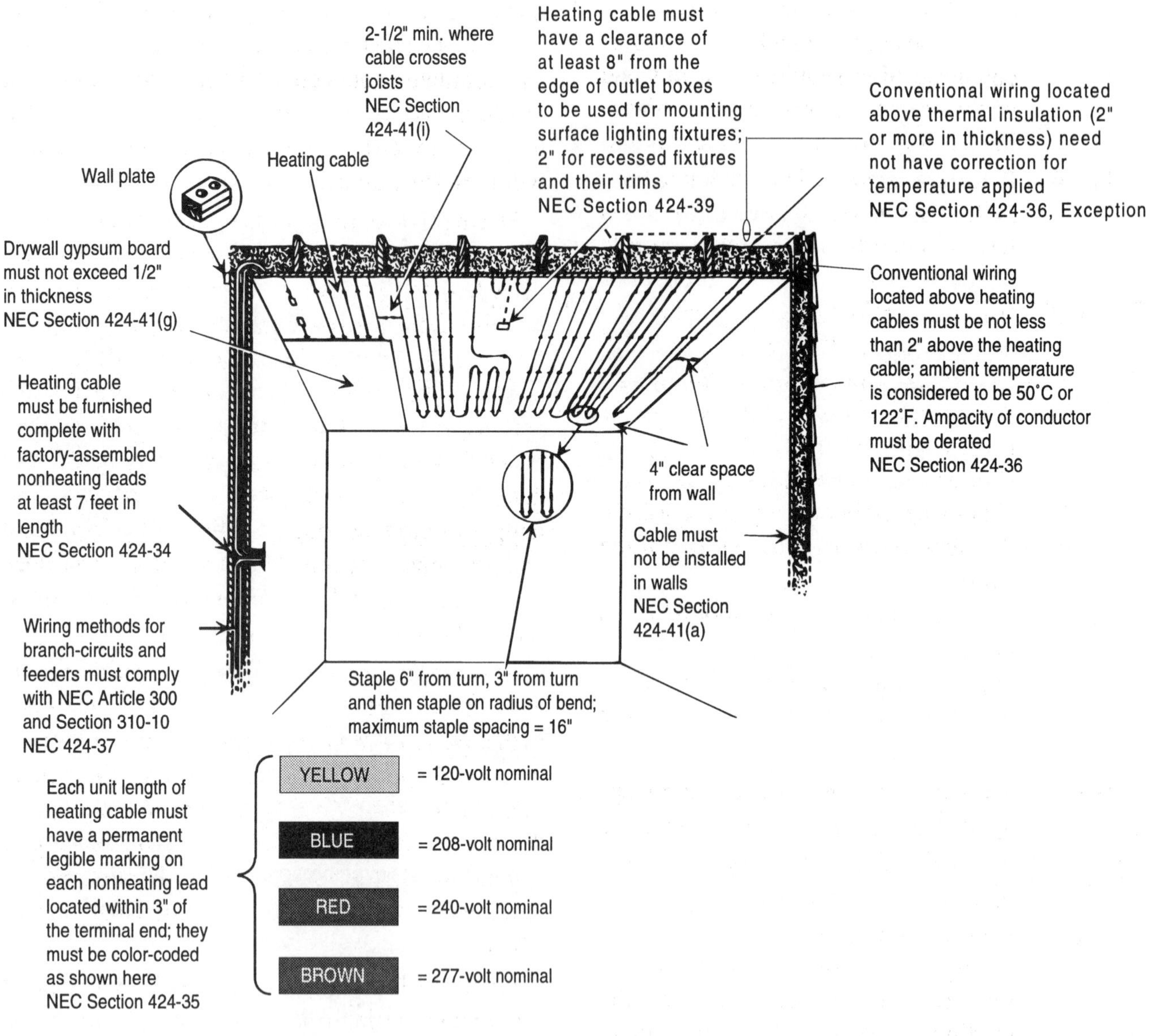
2-1/2" min. where cable crosses joists NEC Section 424-41(i)
Heating cable must have a clearance of at least 8" from the edge of outlet boxes to be used for mounting surface lighting fixtures; 2" for recessed fixtures and their trims NEC Section 424-39
Conventional wiring located above thermal insulation (2" or more in thickness) need not have correction for temperature applied NEC Section 424-36, Exception
Heating cable
Wall plate
Drywall gypsum board must not exceed 1/2" in thickness NEC Section 424-41(g)
Conventional wiring located above heating cables must be not less than 2" above the heating cable; ambient temperature is considered to be 50°C or 122°F. Ampacity of conductor must be derated NEC Section 424-36
Heating cable must be furnished complete with factory-assembled nonheating leads at least 7 feet in length NEC Section 424-34
4" clear space from wall
Cable must not be installed in walls NEC Section 424-41(a)
Wiring methods for branch-circuits and feeders must comply with NEC Article 300 and Section 310-10 NEC 424-37
Staple 6" from turn, 3" from turn and then staple on radius of bend; maximum staple spacing = 16"
Each unit length of heating cable must have a permanent legible marking on each nonheating lead located within 3" of the terminal end; they must be color-coded as shown here NEC Section 424-35
YELLOW
= 120-volt nominal
BLUE
= 208-volt nominal
RED
= 240-volt nominal
BROWN
= 277-volt nominal

Figure 12-19: *NE Code* regulations governing the installation of electric space-heating equipment

Step 2. Install a piece of rigid conduit from the switch or thermostat junction box to house the nonheating leads, between the concrete slab and the switch or thermostat outlet box.

Step 3. Approximately 6 inches of the lower end of the conduit should be embedded in the concrete and a smooth porcelain bushing should be on this end of the conduit to protect the nonheating leads where they leave the conduit.

Step 4. Place the spool of heating cable on a nail, screwdriver, or other shaft you have at hand for unwinding the cable from the spool.

Step 5. Run the accessible end of the 8-foot nonheating lead through the conduit to the thermostat junction box, leaving 6 inches extending out of the box. Never remove the identification tags or shorten the nonheating leads. They should be embedded in the concrete the same as the heating portion of the cable.

Step 6. Run the cable along the floor 6 inches out from the wall to the outside or exposed wall, fastening the cable to the floor either with staples or masking tape.

Step 7. The cable usually is spaced 1½ inches apart for the first 2 feet around the exposed wall and never less than 2½ inches apart for the remaining area.

Step 8. Run the return nonheating lead wire through the conduit to the thermostat junction box in the same manner as the starting nonheating lead was run in Step 5.

In general, the concrete slab should be prepared by applying a vapor barrier over 4 to 6 inches of gravel. Then pour 4 inches of vermiculite or other insulating concrete over the gravel after the outside edges of the slab have been insulated, in accordance with good building practice. The heating cable is then installed as described previously.

At this time, workers should inspect and test the cable before the final layer of concrete is installed, because once the concrete is poured, it's an expensive matter to repair. First, visually inspect the cable for any possible damage to the insulation during the application. Then, with a suitable ohmmeter, check for continuity and capacity of the cable. Concealed breaks may be found by leaving an ohmmeter connected to the cable, and then brushing the cable lightly with the bristles of a broom. Any erratic movement of the meter dial will indicate a fault.

During the pouring of the final coat of concrete, it is recommended that the ohmmeter be left connected to the heating cable leads to detect any possible damage to the cable during the pouring of the finish layer of ordinary concrete (do not use insulating concrete). If an ohmmeter is not available at the time, a 100-watt lamp may be connected in series with the cable to immediately detect any damage during the installation of the concrete. The lamp will glow as long as the circuit is complete and no damage occurs. However, if a break does occur, the lamp will go out and the break can be repaired before the concrete hardens.

Make repairs to a broken cable by stripping the ends of the broken cable and rejoining the ends with a No. 14 AWG pressure-type connector provided and approved for this purpose. The splice must be insulated with thermoplastic tape to a thickness equal to the insulation of the cable. Use any thermoplastic tape listed by the Underwriters' Laboratories as suitable for temperatures up to 176°F.

Once the finish layer of concrete sets, asphalt tile, linoleum tile, or linoleum can be laid on the concrete in the normal manner.

Besides the heating of interior spaces, electric-heating cable also has many other uses. It can be embedded in concrete or asphalt surfaces for the removal of ice and snow, provide water pipes exposed to cold weather with freeze protection, deice roofs and gutters, and heat soil in a hotbed or window box — keeping the temperature at a constant 70°F. The cost of heat cable is relatively inexpensive and the installation goes fast.

FORCED-AIR SYSTEMS

Electric forced-air heating systems are usually of three types:

- Heat pumps
- Electric furnaces
- Air conditioners with duct heaters

In the majority of installations utilizing central, forced-air systems, the system is usually designed by mechanical engineers and installed by mechanical contractors. Branch circuits, feeders, motor starters, and some control wiring are frequently installed by electrical workers. Consequently, every electrician should have a basic working knowledge of forced-air systems, along with applicable *NE Code* installation requirements.

Duct Heaters

Duct heater is a term applied to any heater mounted in the air stream of a forced-air system where the air-moving (fan-coil) unit is not provided as an integral part of the equipment. Duct heaters are used in electric furnaces, combination electric heating/cooling systems, and most of the time in heat pumps to offer auxiliary heat when the pumps themselves cannot supply the demand.

In general, heaters installed in an air duct must be identified as suitable for the installation, and some means must be provided to assure uniform and adequate airflow over the face of the heater in accordance with the manufacturer's instructions. This latter requirement is normally accomplished by airflow controls and other components involving turning vanes, pressure plates, or other devices on the inlet side of the duct heater to assure an even distribution of air over the face of the heater.

Duct heaters installed closer than 4 feet to a heat pump or air conditioner must have both the duct heater and heat pump or air conditioner identified as suitable for such installation and must be so marked.

Duct heaters intended for use with elevated inlet air temperature must be identified as suitable for use at the elevated temperatures. Furthermore, duct heaters used with air conditioners or other air-cooling equipment that may result in condensation of moisture must be identified as suitable for use with air conditioners.

The *NE Code* requires that all duct heaters are installed according to the manufacturer's instructions. Furthermore, duct heaters must be located with respect to building construction and other equipment so as to permit access to the heater. Sufficient clearance must be maintained to permit replacement of controls and heating elements and for adjusting and cleaning of controls and other parts requiring such maintenance. See Figure 12-20.

Control requirements — including disconnecting means — are specified in *NE Code* Sections 424-63, 424-64, and 424-65. Complete coverage of these requirements, along with HVAC controls in general, is presented in Chapter 13 — *HVAC Controls*. But in general, a fan-circuit interlock is one of the requirements. Such a control ensures that the fan circuit is energized when any heater circuit is energized. It would be a waste of energy, and perhaps also be a hazard, if the duct heaters became energized and no air flowed over or through them. However, the *NE Code* permits a slight time- and temperature-delay before the fan may be energized. This prevents the system from blowing cold air into the conditioned space. In other words, such a control gives the duct heaters time to "warm-up" before air is induced in the system.

Furthermore, each duct heater must be provided with an automatic-reset limit control to deenergize the duct-heater circuit or circuits in case overheating or other faults occur. In addition, an integral independent supplementary control or controller shall be provided in each duct heater that will disconnect a sufficient number of conductors to

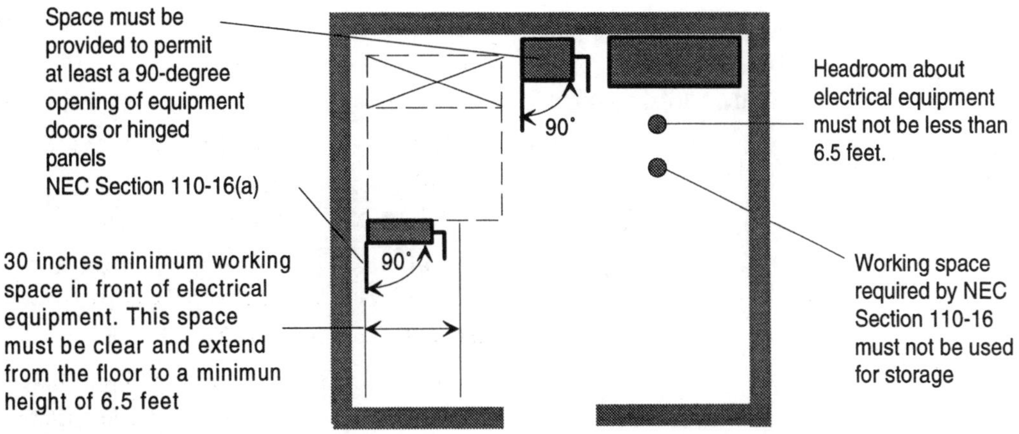
Space must be provided to permit at least a 90-degree opening of equipment doors or hinged panels NEC Section 110-16(a)
90°
90°
Headroom about electrical equipment must not be less than 6.5 feet.
30 inches minimum working space in front of electrical equipment. This space must be clear and extend from the floor to a minimun height of 6.5 feet
Working space required by NEC Section 110-16 must not be used for storage

Figure 12-20: Sufficient clearance must be provided for HVAC equipment

interrupt current flow. This device must be manually resettable or replaceable.

The disconnect for duct heaters — the same as for other types of HVAC equipment — must be accessible and within sight of the controller. All control equipment must also be accessible.

ROOM AIR CONDITIONERS

There are millions of room air conditioners in use throughout the United States and Canada. Consequently, the *NE Code* deemed it necessary to provide Article 440 G, beginning with *NE Code* Section 440-60, to ensure that such equipment will be installed so as not to provide a hazard to life or property. These *NE Code* requirements apply to electrically energized room air conditioners that control temperature and humidity. In general, this section of the *NE Code* considers room air conditioners (with or without provisions for heating) to be an alternating-current appliance of the air-cooled window, console, or in-wall (through-wall) type that is installed in a conditioned room or space and that incorporates a hermetic refrigerant motor-compressor(s). Furthermore, this *NE Code* provision covers only equipment rated at 250 volts or less, single-phase, and such equipment is permitted to be cord- and attachment plug-connected.

Three-phase room air conditioners, or those rated at over 250 volts, are not covered under *NE Code* Article 440 G. This type of equipment must be directly connected to a wiring method as described in *NE Code* Chapter 3.

The majority of room air conditioners covered under *NE Code* 440 G are connected by cord- and attachment-plug to receptacle outlets of general-purpose branch circuits. The rating of any such unit must not exceed 80% of the branch-circuit rating if connected to a 15-, 20-, or 30-ampere general-purpose branch circuit. The rating of cord-and-plug connected room air conditioners must not exceed 50% of the branch-circuit rating if lighting units and other appliances are also supplied.

Figures 12-21 and 12-22 depict the *NE Code* application rules for room air conditioners. Note that the attachment plug and receptacle are allowed to serve as the disconnecting means. In some cases, the attachment plug and receptacle may also serve as the controller, or the controller may be a switch that is an integral part of the unit. The required overload protective device may be supplied as an integral part of the appliance and need not be included in the branch-circuit calculations.

Equipment grounding, as required by *NE Code* Section 440-61, may be handled by the grounded receptacle; that is, the grounding terminal of the cord connects to the grounding terminal of the receptacle.

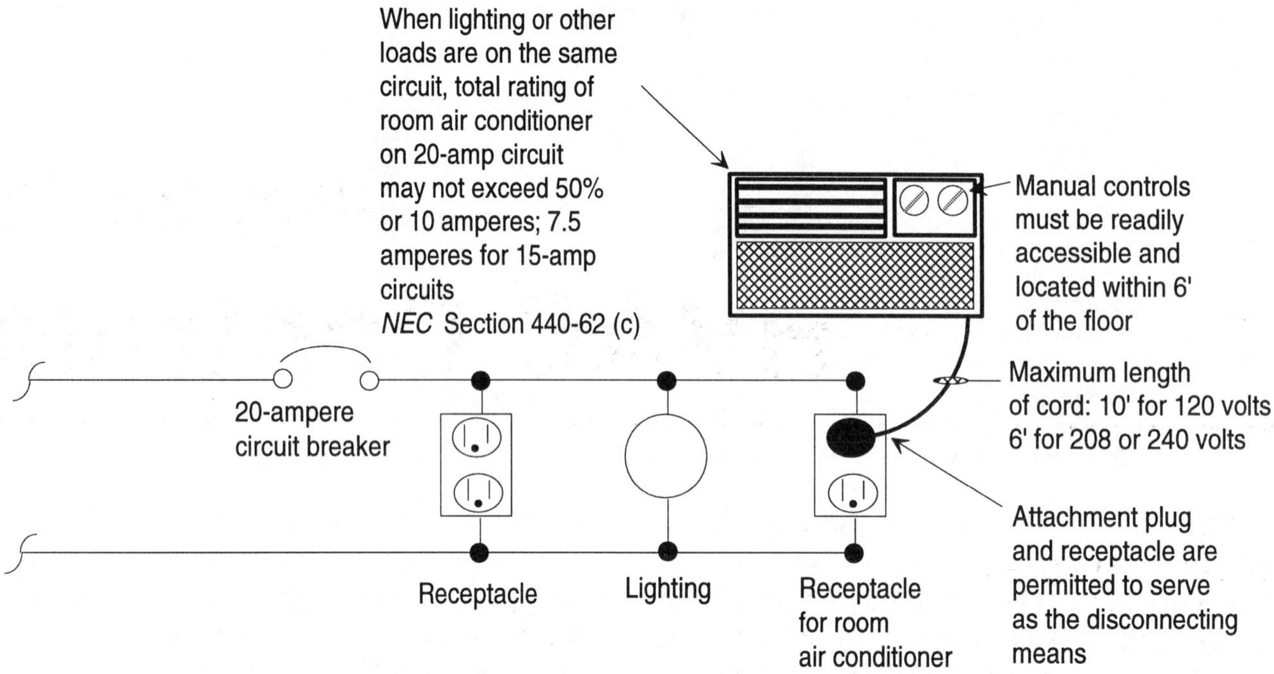
When lighting or other loads are on the same circuit, total rating of room air conditioner on 20-amp circuit may not exceed 50% or 10 amperes; 7.5 amperes for 15-amp circuits *NEC* Section 440-62 (c)
Manual controls must be readily accessible and located within 6' of the floor
Maximum length of cord: 10' for 120 volts 6' for 208 or 240 volts
20-ampere circuit breaker
Attachment plug and receptacle are permitted to serve as the disconnecting means
Receptacle
Lighting
Receptacle for room air conditioner

Figure 12-21: Branch circuits for room air conditioners

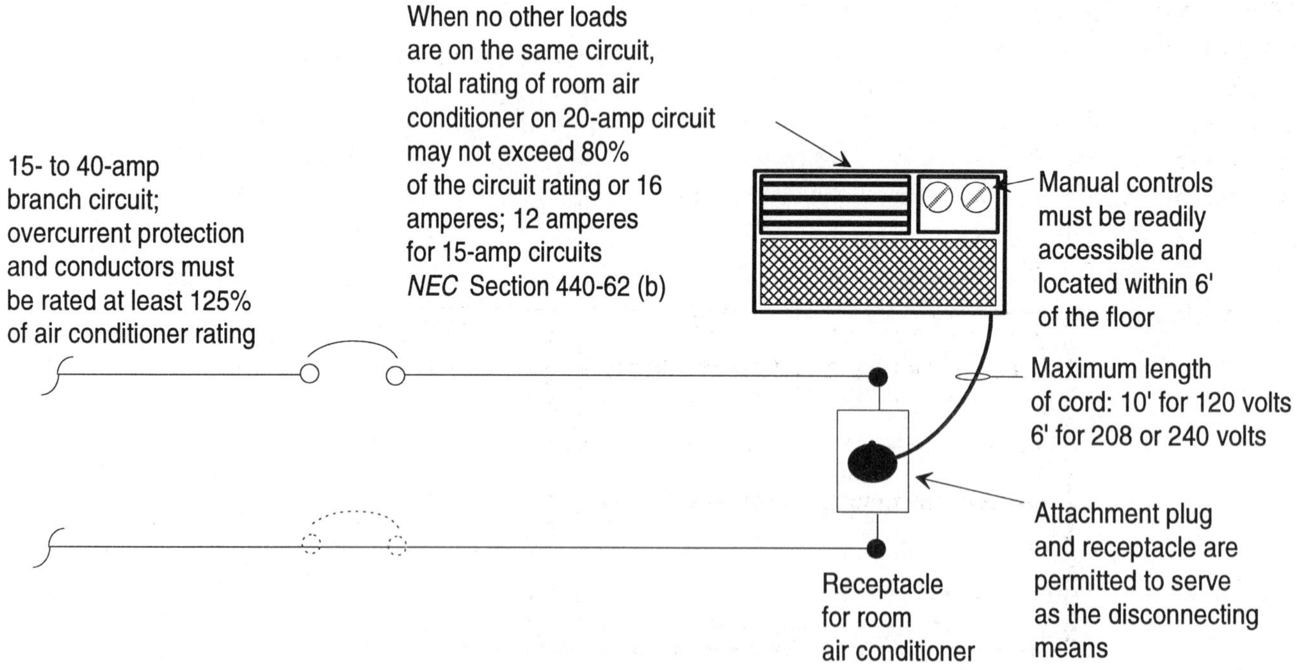
When no other loads are on the same circuit, total rating of room air conditioner on 20-amp circuit may not exceed 80% of the circuit rating or 16 amperes; 12 amperes for 15-amp circuits *NEC* Section 440-62 (b)
15- to 40-amp branch circuit; overcurrent protection and conductors must be rated at least 125% of air conditioner rating
Manual controls must be readily accessible and located within 6' of the floor
Maximum length of cord: 10' for 120 volts 6' for 208 or 240 volts
Attachment plug and receptacle are permitted to serve as the disconnecting means
Receptacle for room air conditioner

Figure 12-22: Fixed air conditioner connected to branch circuit

Chapter 13

HVAC Controls

All of the HVAC systems described in Chapter 12 would not be able to operate without some means of control, if only to stop and start a system. For example, without a controlling mechanism, an air-conditioning system would be turned on and run year-round. Even when the space became cool or hot enough, the system would continue to run and make the area more uncomfortable than if it had not been installed in the first place. Therefore, control is an important area of comfort conditioning for buildings.

Electronic control circuits are the principal controls used in HVAC systems of any consequence at the present time. Not too long ago, their main use was in highly sophisticated commercial and industry applications, but they are now used in virtually all HVAC applications — from residential, upward. Electronic controls provide quick response to temperature changes, and temperature averaging is easily accomplished.

Although solid-state controls dominate the field, some of the earlier controls will also be covered, because many are still in use and will remain in use for some time to come.

INDOOR THERMOSTAT

HVAC controls can be electric, electronic or pneumatic. Almost limitless combinations of each type are possible. Therefore, it is not practical to

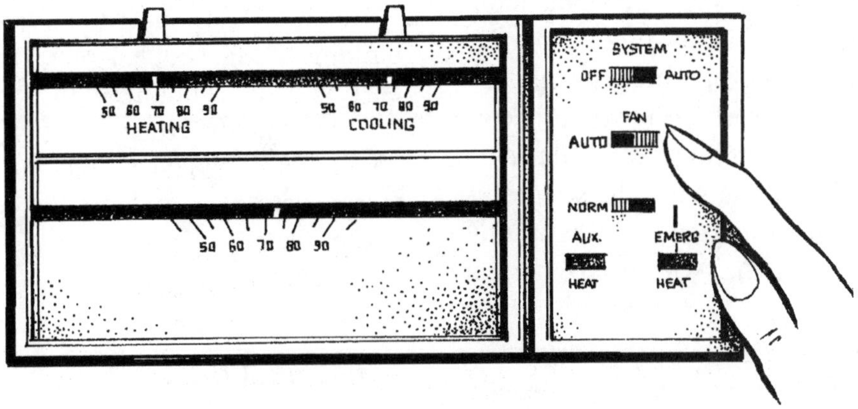
50 60 70 80 90
HEATING
50 60 70 80 90
COOLING
50 60 70 80 90
SYSTEM
OFF
AUTO
FAN
AUTO
NORM
AUX.
EMERG
HEAT
HEAT

Figure 13-1: Typical indoor thermostat

attempt to describe all of the possible combinations in this chapter. Rather, fundamental terms, functions, operations, and simple maintenance and troubleshooting techniques are presented.

The indoor thermostat (Figure 13-1) is the most easily recognized control device. Such controls can be found in practically every motel room in the United States. It is also a familiar ornament in every home utilizing central heating or cooling systems. Here's an overview of a thermostat used to control an electric heat pump.

A two-stage indoor changeover thermostat is the most popular type used to control heat pumps during the heating season. During the cooling season, it functions as a conventional air conditioning thermostat. The first stage of this thermostat controls the heat pump. The second stage (usually preset 1°F to 2°F below the first stage) allows the supplementary heat to be energized.

By placing the selector/switch on AUTO, the automatic changeover thermostat switches the system from heating to cooling or cooling to heating automatically. A manual thermostat also is available to offer the option of selecting manually the mode of operation for heating or cooling.

Emergency heat switch (EM.HT.): The indoor thermostat should have an emergency heat selector (EM.HT.) with a light to indicate full use of the supplementary heat when the heat pump is not in operation. If the heat pump should become inoperative, this switch enables bypassing the normal operation of the heat pump and heats the area by the supplementary heat until the problem is corrected. An indicator light, usually red, mounted on the thermostat, will come on when the selector is in the emergency heat position. As soon as the unit has been repaired, return the switch to the normal operating position.

Supplementary heat light: Some thermostats have a light which indicates the supplementary heat is on to assist the heat pump in normal operation. When the outdoor temperature falls below the balance point of the heat pump and energizes the supplementary heat, the light will cycle on and off intermittently as the supplementary heat is activated.

The temperature range at which the light comes on will vary depending on the balance point.

Temperature Setting

Heating: The recommended setting for the heating cycle is 68°F. Once the thermostat is set, the best policy is to leave it alone. Raising the thermostat as little as 2°F may cause the supplementary heat to energize, thereby increasing the energy usage.

Night setback: Although night setback is recommended during the winter for most types of heating systems — to save energy and reduce costs — it is not generally recommended for a heat pump. When using a heat pump to raise the room temperature in the morning, the supplementary heat may come on, using more energy than was saved during the night. However, thermostat temperature settings for weekend trips or vacations during the heating season should be reduced to save energy. The use of a standard automatic nighttime setback control with a heat pump system is not recommended.

Cooling: A setting of 78°F or higher is recommended for cooling. For each degree the temperature is set below 78°F, the cooling energy usage will increase approximately 5%.

Raising the temperature when the room or building is unoccupied is recommended to save energy. If the building will not be occupied for several days, the cooling system can be turned off altogether. However, frequent changes of the thermostat setting reduces the economical operation of the heat pump and tends to shorten the life of the compressor.

Fan Operation

Operation of the fan is the same for the automatic or manual changeover thermostat. The fan switch set in the AUTO position gives fan operation only when the unit is actually heating or

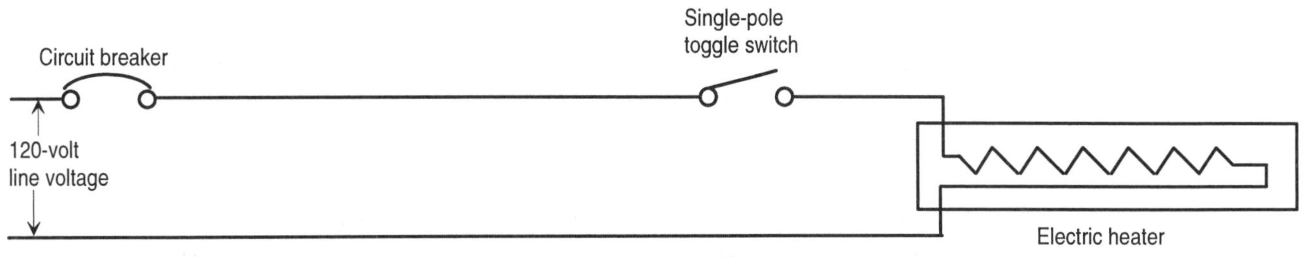
Circuit breaker
Single-pole
toggle switch
120-volt
line voltage
Electric heater

Figure 13-2: Simple heater control circuit

cooling. In the ON position, the indoor fan will run continuously, which may cost a little more. The ON position is recommended to obtain a more even temperature throughout the house. However, operating the fan in the ON position may allow the thermostat to be set at a lower temperature, or reduce the operating time of the compressor, which may offset the increased cost. It should also increase occupants' comfort.

Thermostat Location

The indoor thermostat should be located on an interior wall in the central portion of the home, approximately 5 feet above the floor. It must be installed level. The location should be free from drafts, vibrations and any interior heat sources such as a lamp or a television set. Care should be taken to seal behind the wall where the thermostat wire or mounting box penetrates the wall.

BASIC PRINCIPLES OF CONTROL

The circuit in Figure 13-2 shows a simple "control" for an electric resistance-type heater, consisting of a 120-volt, 2-wire circuit feeding the heater and a conventional single-pole toggle switch to interrupt the power supply. A 1-pole, 15-ampere circuit breaker is used to protect the circuit against short circuits and ground faults. When the toggle switch is in the ON position, the heater becomes energized; when in the OFF position, the heater is de-energized. Obviously, this is a manual control and leaves much to be desired. For example, once the space to be heated reaches the desired temperature, the toggle switch cannot sense the difference and continues to stay in the ON position until it is manually shut off. If the switch is left on, the space will continue to receive heat, even after the desired temperature has been reached.

A better solution is to replace the single-pole toggle switch with a thermostat control. A thermostat utilizing bimetal elements is shown in Figure 13-3. The bimetal elements consist of thin strips of two different metals securely attached to each other. Since different metals expand and contract at different rates, the thin strips of metal actually curve toward or away from a given point when there is a change in temperature. In this way, the thermostat "makes" or "breaks" its contacts when the temperature changes — providing automatic control of the heater in question.

A floor plan of a typical year-round HVAC system is shown in Figure 13-4, including the probable controls that will exist within the system. To get an overview of this system, let's take a look at the various controls and their relationship to the entire system.

The system is started by closing the fan starter switch. At the same time, an electric circuit provides the energy for operation of the controls. Notice that the outdoor and return air dampers are connected by a mechanical linkage so that the opening of one will close the other. Similarly, the

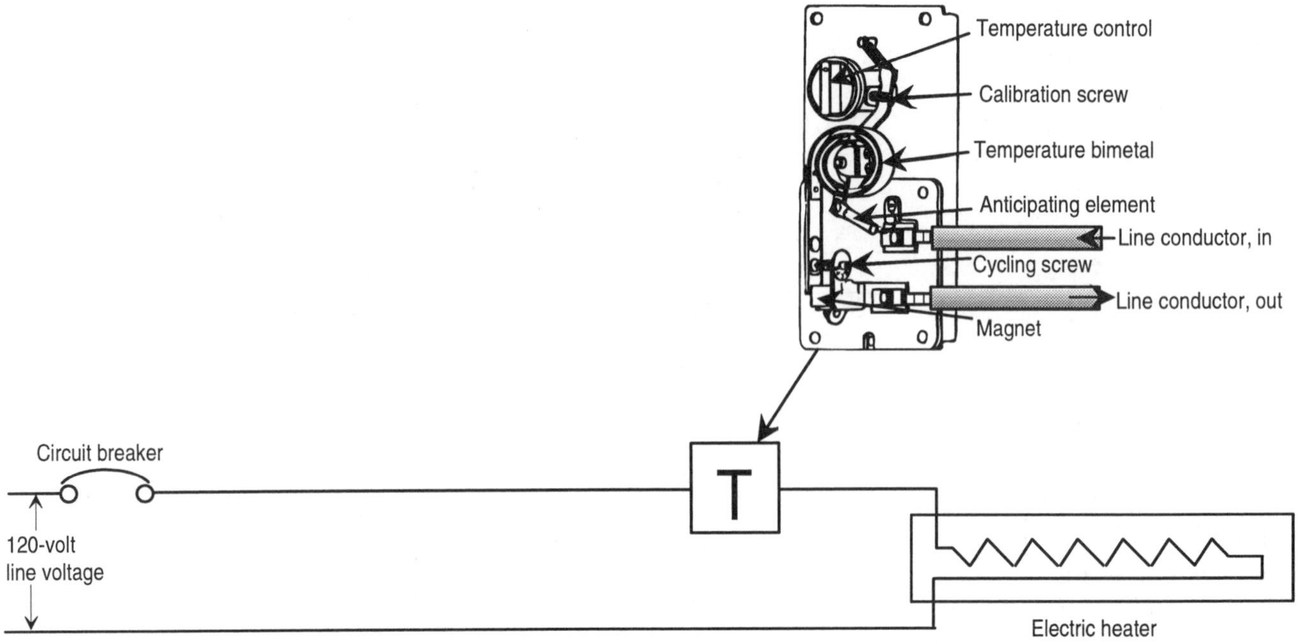
Temperature control
Calibration screw
Temperature bimetal
Anticipating element
Line conductor, in
Cycling screw
Line conductor, out
Magnet
Circuit breaker
120-volt
line voltage
T
Electric heater

Figure 13-2: Simple heater thermostat control circuit

face and bypass dampers for the cooling coil are interconnected.

When de-energized, the damper motors, D-1 and D-2, hold the outdoor air and exhaust dampers in the closed positions.

During the heating cycle, the space thermostat, T-1, operates the hot-water heating coil valve, V-1, to maintain room-air temperature. The insertion thermostat, T-2, positions the damper motor, D-1. The minimum quantity of outdoor air is determined by the switch, S. As the temperature of the air from the outside rises, T-2 moves D-1 so that more outside air and less return air is included in the mixture to be heated until 100% outside air is admitted at the top setting for T-2. Simultaneously, damper motor, D-2, must open the exhaust damper. The thermostat, T-4, is a low-limit controller to protect coils from freezing. It prevents the admission of too much cold outside air. A low-limit controller, T-6, in the supply duct is interconnected with the heating-coil valve control line to keep the supply temperature above a minimum value. During the warm season, the outdoor air controller, T-5, reacts to the high-temperature outdoor air to deactivate T-6 and the heating valve.

For cooling, switch S is positioned for minimum outside air. When the outdoor air temperature reaches the setting of the summer controller, T-3, damper motor, D-1, positions the damper to admit the minimum outdoor air quantity. The space humidity controller, H-1, opens the cooling valve, V-2, whenever dehumidification is required. The cooling coil may be a direct-expansion or chilled-water coil. The controller, H-1, also sets the position of damper motor, D-3, so that the face and bypass dampers for the cooling coil allow the proper portion of air to pass over the coil to satisfy the dehumidification load. The space thermostat, T-1, can call for reheating when necessary by operating V-1. It can also control V-2 and D-3 if their operation for humidity control is inadequate for cooling air to maintain the desired space temperature.

Obviously, other controls are required for the operation of central heating and cooling equipment. A description of the more popular HVAC controls follows.

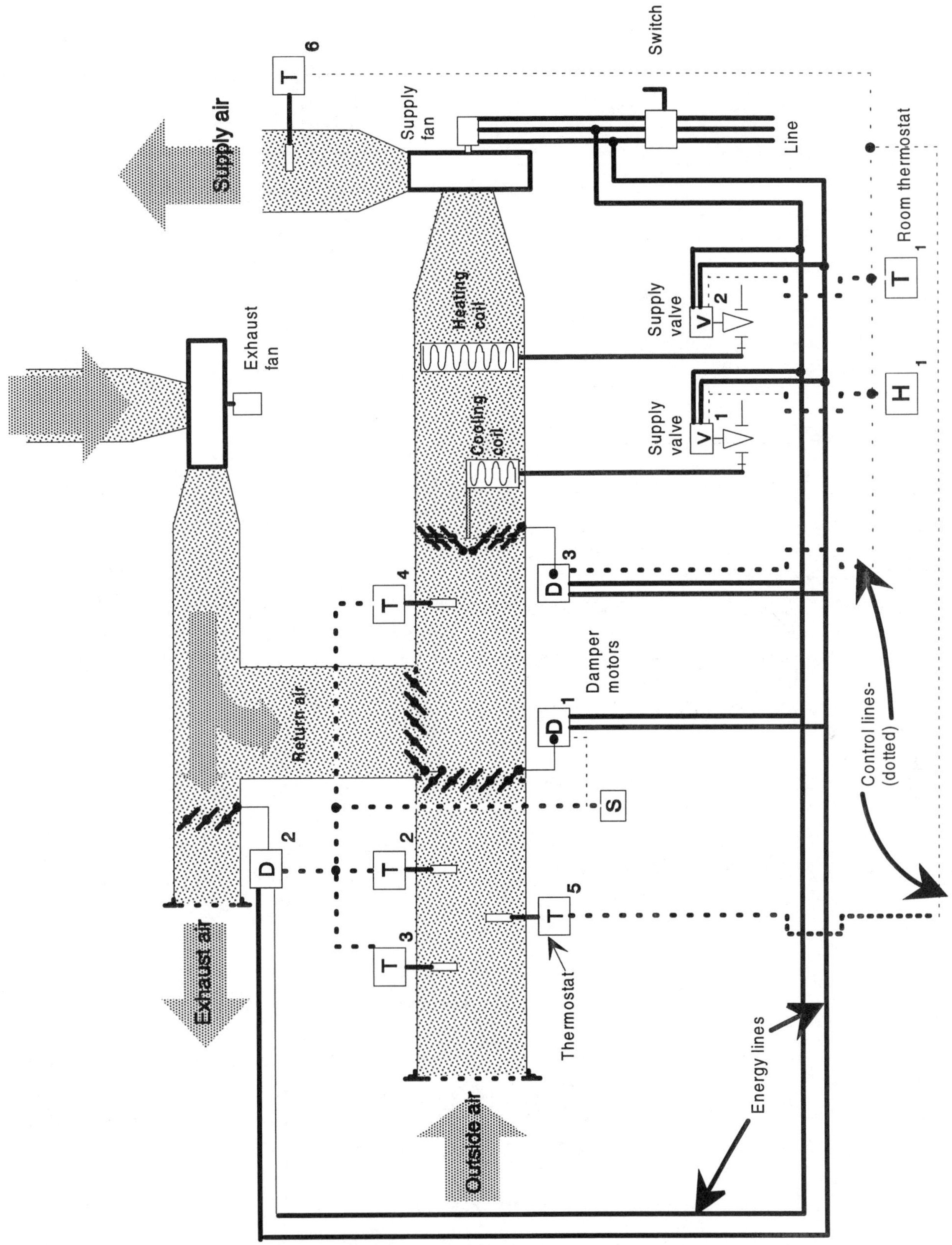
Supply air
Exhaust air
Outside air
Return air
Supply fan
Exhaust fan
Heating coil
Cooling coil
Supply valve
Supply valve
Damper motors
Thermostat
Energy lines
Control lines-
(dotted)
Room thermostat
Switch
Line

Figure 13-4: Typical HVAC control system

BASIC ELECTRONIC CONTROLS

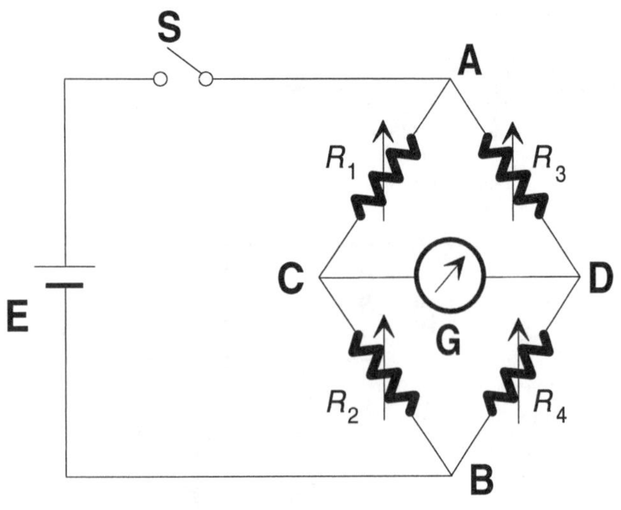
S
A
R_1
R_3
C
D
E
G
R_2
R_4
B

Figure 13-5: Basic Wheatstone-bridge circuit

The Wheatstone-bridge circuit is the basis of many electronic circuits installed in the past couple of decades, and many are still in use. The bridge consists of two sets of two series-wired resistors, connected in parallel across a dc voltage source. One set of series-wired resistors is R_1 and R_2: the second set is R_3 and R_4. See Figure 13-5. The voltage source, E, is between points A and B. A sensitive electric current indicator (galvanometer), G, is connected across the parallel sets of resistors at points C and D.

When switch S is closed, the voltage from E flows through both sets of resistors. If the potential at point C is the same as at point D, the galvanometer reads zero. This means that there is no potential difference between the two points, and the bridge is balanced.

When any of the four resistors has a different voltage, the galvanometer registers a value other than zero. This indicates that there is a current between points C and D. When this occurs, the bridge is unbalanced. Some control manufacturers have modified this basic bridge circuit and put it to work in electronic circuits. Figure 13-6 shows a typical modification.

The dc voltage has been replaced by an ac power supply. Also, the switch is eliminated and the galvanometer is replaced by an amplifier-switching relay unit. Resistor R_2 is replaced by a temperature-sensing element, T_1 — making a sturdier, trouble-free component.

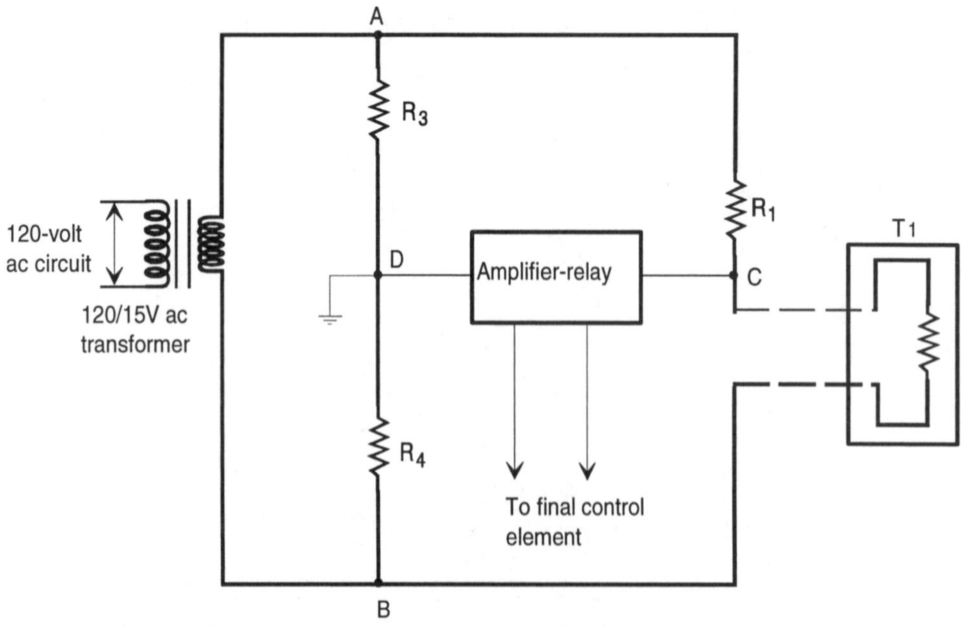
A
R_3
R_1
T_1
120-volt
ac circuit
D
Amplifier-relay
C
120/15V ac
transformer
R_4
To final control
element
B

Figure 13-6: Modified Wheatstone-bridge circuit

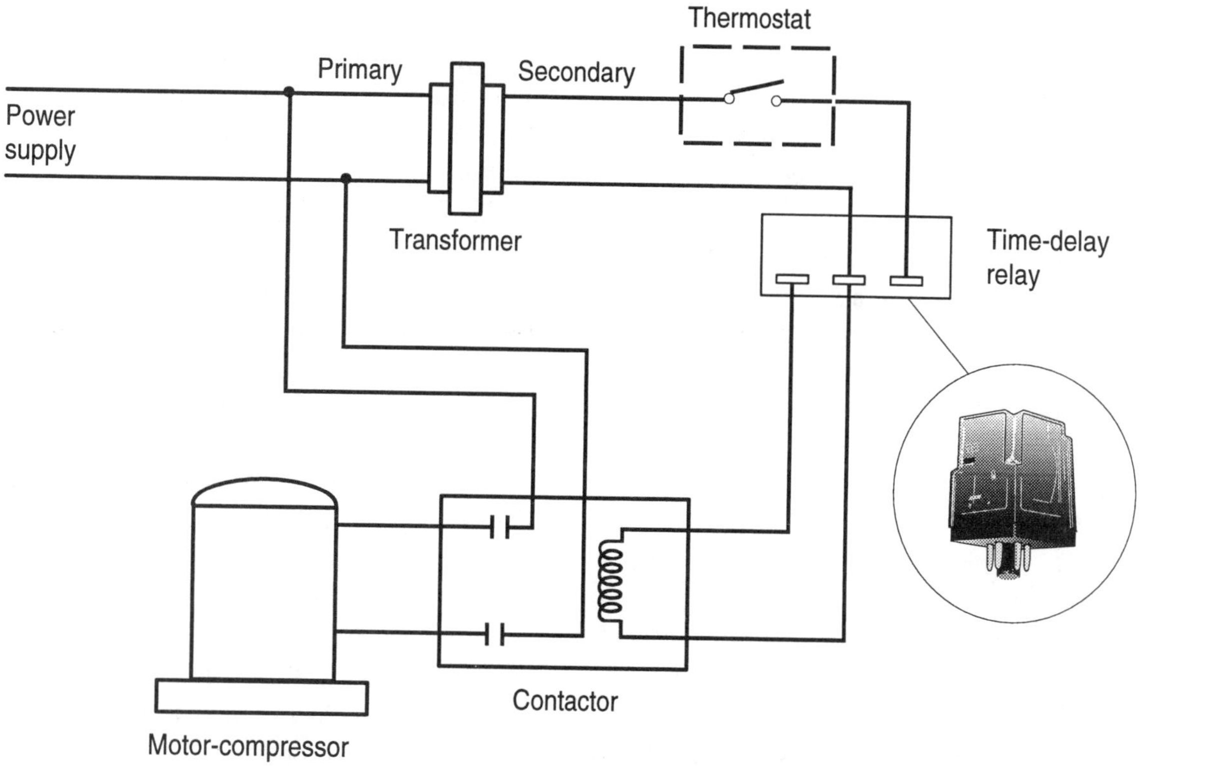
Thermostat
Primary
Secondary
Power
supply
Transformer
Time-delay
relay
Contactor
Motor-compressor

Figure 13-7: Solid-state time-delay relay circuit

In this modification, a resistance of 1,000 ohms is assigned to each of the three fixed resistors (R1, R2, R4) and to the thermostat element, T1. When conditions in the area to be conditioned are satisfied, the resistance in T1 does not change, and the voltage across the amplifier relay is zero. Because there is no voltage across this amplifier relay, the final control element (motor, valve, and the like) cannot be energized. The bridge circuit is balanced.

When the air temperature in the space changes, the thermostat element senses the change. This causes a corresponding change in the resistance at T1, and the bridge becomes unbalanced. Voltage now flows through the amplifier-relay to the final control element. In a heating application, a drop in temperature in the space causes a decrease in resistance at T1. This results in voltage relationships in the bridge circuit that cause the amplifier relay to increase the voltage to the final control element, and heat is added to the space. As heat is added, resistance at T1 increases, and the resulting voltage change in the bridge circuit inactivates the heating action (shuts off the burner).

This modified bridge circuit can also be applied to cooling applications. Depending on whether the bridge voltage fed to the amplifier-relay is in phase or out of phase with the supply voltage, the final control element is opened or closed.

TIME-DELAY CIRCUITS

The thermostat or other control for HVAC systems must respond to a gradual or average change in the space to be conditioned. An average change is produced by adding a tiny heater (timing device) near the thermostat temperature-sensing element in electrical controls. See Fig. 13-7. As long as the air temperature is maintained within certain limits, the thermal heater element cools to its "off" point and deenergizes the heating element. Then it heats to its "on" point and energizes the heating element.

It cools to its "off" point again and repeats the cycle.

The timer sequence for cooling the space is the opposite of the sequence for heating. Consequently, the temperature differential is much less than if the timer is not used.

The timer or thermal-heater principle reduces the swing (differential) in air temperature in the room to a point at which the temperature remains almost constant. The differential is much greater when the timer is not used.

BASIC CONTROL COMPONENTS

Thermostat: In general, low-voltage room thermostats should be used for the best temperature control. The low-voltage thermostats respond much faster to temperature changes than the greater-mass line-voltage devices.

From a cost standpoint, the less-expensive installation of low-voltage wiring more than offsets the extra cost of the transformer. Also low-voltage thermostats are safer.

The room thermostat is provided with a heat anticipator connected in series with the rest of the control circuit. These anticipators are made of a resistance-type material that produces heat in accordance with the current drawn through it. Heat anticipators are adjustable and are normally set to correspond with the current rating of the main gas valve, motor starter, or relay. The purpose of these devices is to make the room temperature more stable.

In operation, when the thermostat is calling for heat, the anticipator is also producing heat to the thermostat. This heating action causes the thermostat to become satisfied before the room actually reaches the set point of the thermostat. Thus, the thermostat stops the flame or de-energizes the motor starter, relay, etc., and the room temperature will not overshoot, or go too high.

Transformer: The transformer is a device used to reduce line voltage to a useable control voltage, usually 24 V. Transformers must be sized to provide sufficient power to operate the control circuit. Most are oversized sufficiently to provide enough power to operate an air-conditioning control circuit also. This rating is usually 40 VA for small systems; greater for larger systems.

Fan control: The fan control is a temperature-actuated control that, when heated, will close a set of contacts to start the indoor fan motor. The sensing element of the fan control is positioned inside the heat exchanger where the temperature is the highest.

This control is actuated by a bimetal element that opens or closes the contacts on temperature change. The fan control is usually set to bring the fan on at about 100°F in the heating mode.

In operation, the burner provides heat to the heat exchanger for a few seconds to warm the furnace before the fan is started. This operation is to prevent blowing cold air into the room on furnace start-up. When the thermostat is satisfied, the main burner stops providing heat, but the fan continues to operate until the temperature in the furnace has been reduced, thus removing any excess heat in the furnace. See Figures 13-8 and 13-9.

Limit control: The limit control is also a heat-actuated switch with a bimetal sensing element positioned inside the heat exchanger. This is a safety control that is wired into the primary side of the transformer. If the temperature inside the furnace reaches approximately 200°F, the power to the transformer will be shut off, which also stops all power in the temperature-control circuit.

Main gas valve: The main gas valve is the device that acts on demand from the thermostat to either admit gas to the main burners or to stop the gas supply. This valve has many functions. It has a gas pressure regulator, a pilot safety, a main gas cock, a pilot gas cock, and the main gas solenoid all in a single unit — the combination gas control.

As the thermostat calls for heat, energizing the solenoid coil, the valve lever opens the cycling valve. The inlet gas now flows through the control orifice past the cycling valve — all of which are monitored by the various control devices. At this point, gas flow is in two directions.

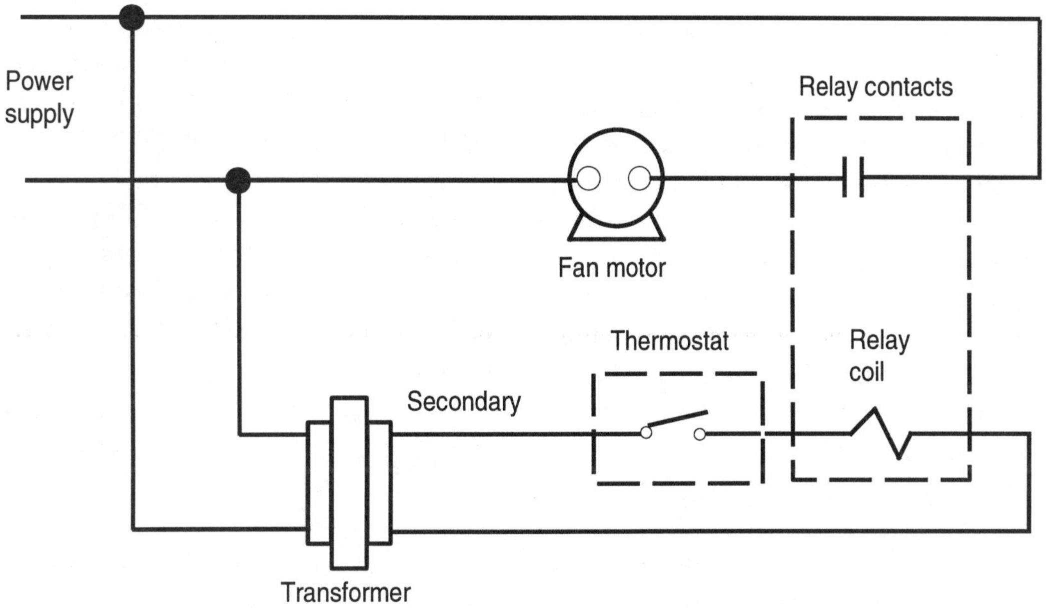
Power
supply
Relay contacts
Fan motor
Thermostat
Relay
coil
Secondary
Transformer

Figure 13-8: Wiring for a fan relay

1. Part of the flow is to the back of the diaphragm by means of internal passageways. The resulting increase in pressure pushes the main valve to the open position, compressing the diaphragm springs lightly.
2. Part of the flow is through the seat of the master regulator into the valve outlet by means of internal passageways. This causes the master regulator to begin its function.

The gas valve remains in this position and the master regulator continues to regulate until the relay coil is de-energized, at which time the cycling valve seals off.

As the cycling valve closes, the regulator spring causes the seat of the master regulator to close off. The function of the bypass orifice is to permit gas in the passageways to escape into the outlet of the valve, thereby causing the main gas valve to close.

Pilot safety: There are two types of pilot safety controls, 90% safe, and 100% safe. These two names refer to the amount of gas cut off when the pilot light is unsafe. The 100% safe device is incorporated in the combination main gas valve. The 90% safe device incorporates a set of contacts

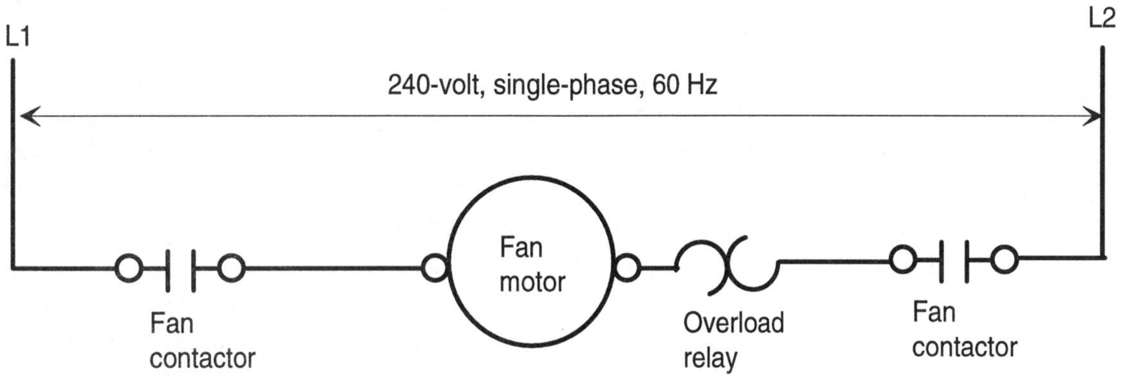
L1
L2
240-volt, single-phase, 60 Hz
Fan
motor
Fan
contactor
Overload
relay
Fan
contactor

Figure 13-9: Fan control circuit

that open the control circuit during an unsafe condition.

These units are used in conjunction with a thermocouple to keep the control contacts closed or the valve open during normal operation. If, at any time, the control "drops out" (the contacts open), the reset button must be manually reset before operation of the furnace can be resumed.

Thermocouple: The thermocouple is a device that uses the difference in metals to provide electron flow. The hot junction of the thermocouple is put in the pilot flame where the dissimilar metals are heated. When heat is applied to the welded junction, a small voltage is produced. This small voltage is measured in millivolts (mV) and is the power used to operate the pilot safety control. The output of a thermocouple is approximately 30 mV. This simple device can cause many problems if the connections are not kept clean and tight.

The fire-stat: The fire-stat is a safety device mounted in the fan compartment to stop the fan if the return air temperature reaches about 160° F. It is a bimetal-actuated, normally closed switch that must be manually reset before the fan can operate.

The reason for stopping the fan when the high return-air temperatures exist is to prevent agitation of any open flame in the house, thus helping to prevent the spread of any fire that may be present in the building.

Furnace wiring: There are three different circuits and three different voltages in a modern furnace 24-volt control system. The following diagrams will illustrate each of these circuits:

- The fan or circulator circuit (Figure 13-9)
- The temperature control circuit (Figure 13-10)
- The pilot safety circuit, 30 mV (Figure 13-11)

When all three of these circuits are connected together (Figure 13-12), we have a modern furnace 24V control system.

The control of electric furnaces is much the same as that just described; however, there are no pilot safety devices, and the main gas valve is replaced with relays that actuate to complete the electrical circuit to the heating elements.

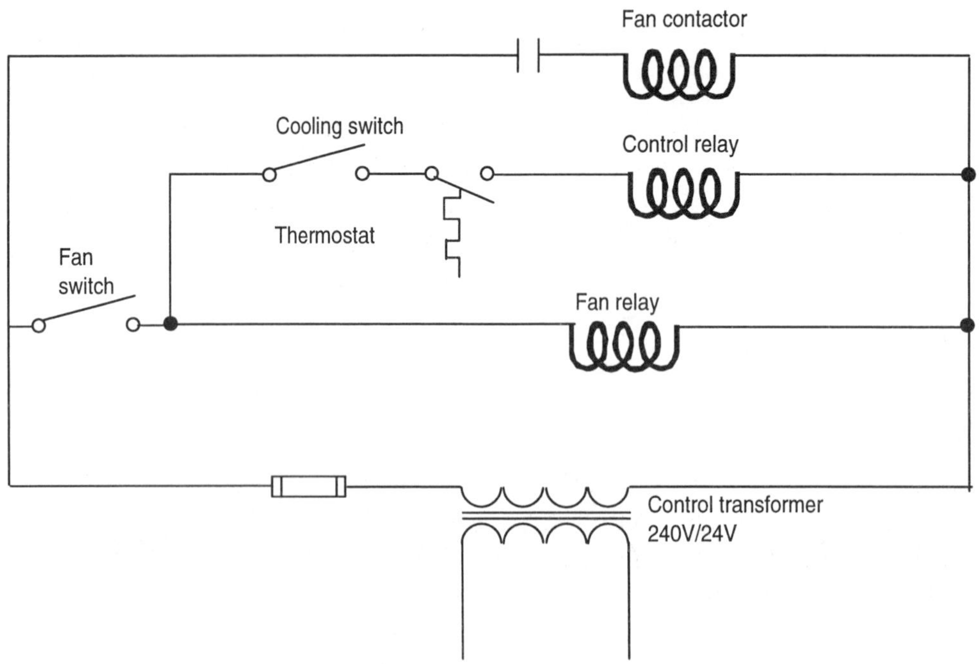
Fan contactor
Cooling switch
Control relay
Thermostat
Fan
switch
Fan relay
Control transformer
240V/24V

Figure 13-10: Temperature control circuit

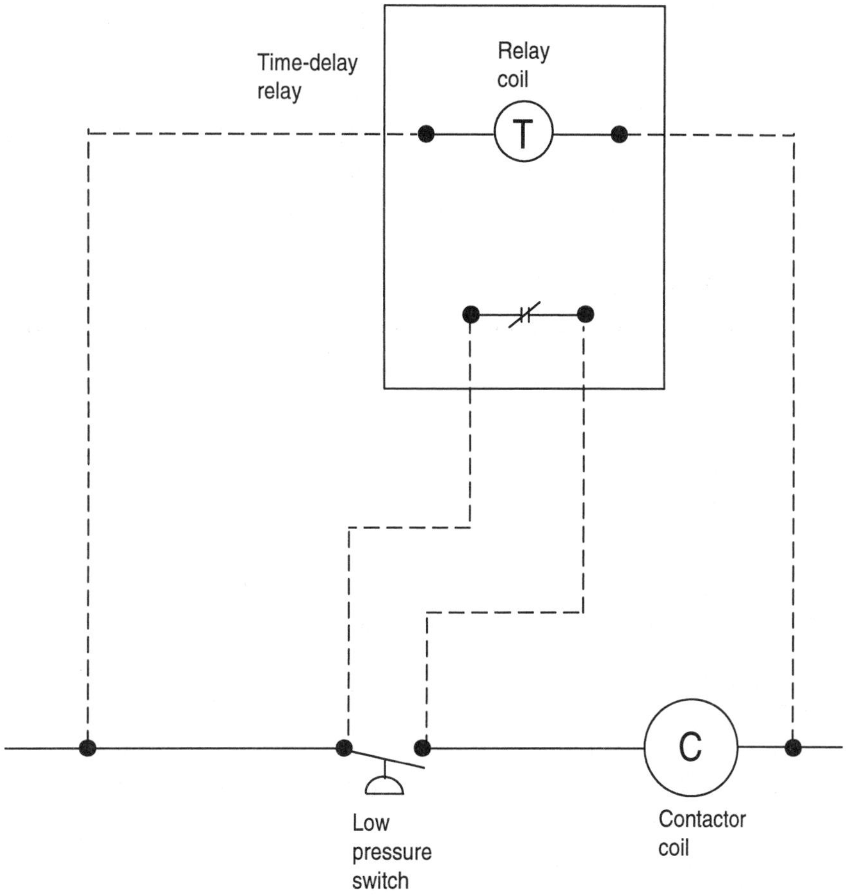
Time-delay relay
Relay coil
T
Low pressure switch
C
Contactor coil

Figure 13-11: Pilot safety circuit

The variety of functions performed by a heating system is limited only by the use of controls. The more a service technician knows about controls, the easier his or her job will be in servicing such equipment.

NE CODE REQUIREMENTS FOR HVAC CONTROLS

Several *NE Code* sections cover the requirements for HVAC controls. For example, Part C of *NE Code* Article 424 deals with the control and protection of fixed electric space heating equipment (*NE Code* Sections 424-19 through 424-22); Part E of *NE Code* Article 440 covers requirements for motor-compressor controllers, while Part F of *NE Code* Article 440 deals with motor-compressor and branch-circuit overload protection.

In general, a means must be provided to disconnect heating equipment, including motor controllers and supplementary overcurrent protective devices, from all ungrounded conductors. The disconnecting means may be a switch, circuit breaker, unit switch, or a thermostatically-controlled switching device. The selection and use of a disconnecting device are governed by the type of overcurrent protection and the rating of any motors that are part of the HVAC equipment.

In certain heating units, supplementary overcurrent protective devices other than the branch-circuit overcurrent protection are required. These supplementary overcurrent devices are usually used when heating elements rated more than 48 amperes are supplied as a subdivided load. In this case, the disconnecting means must be on the supply side of the supplementary overcurrent protective

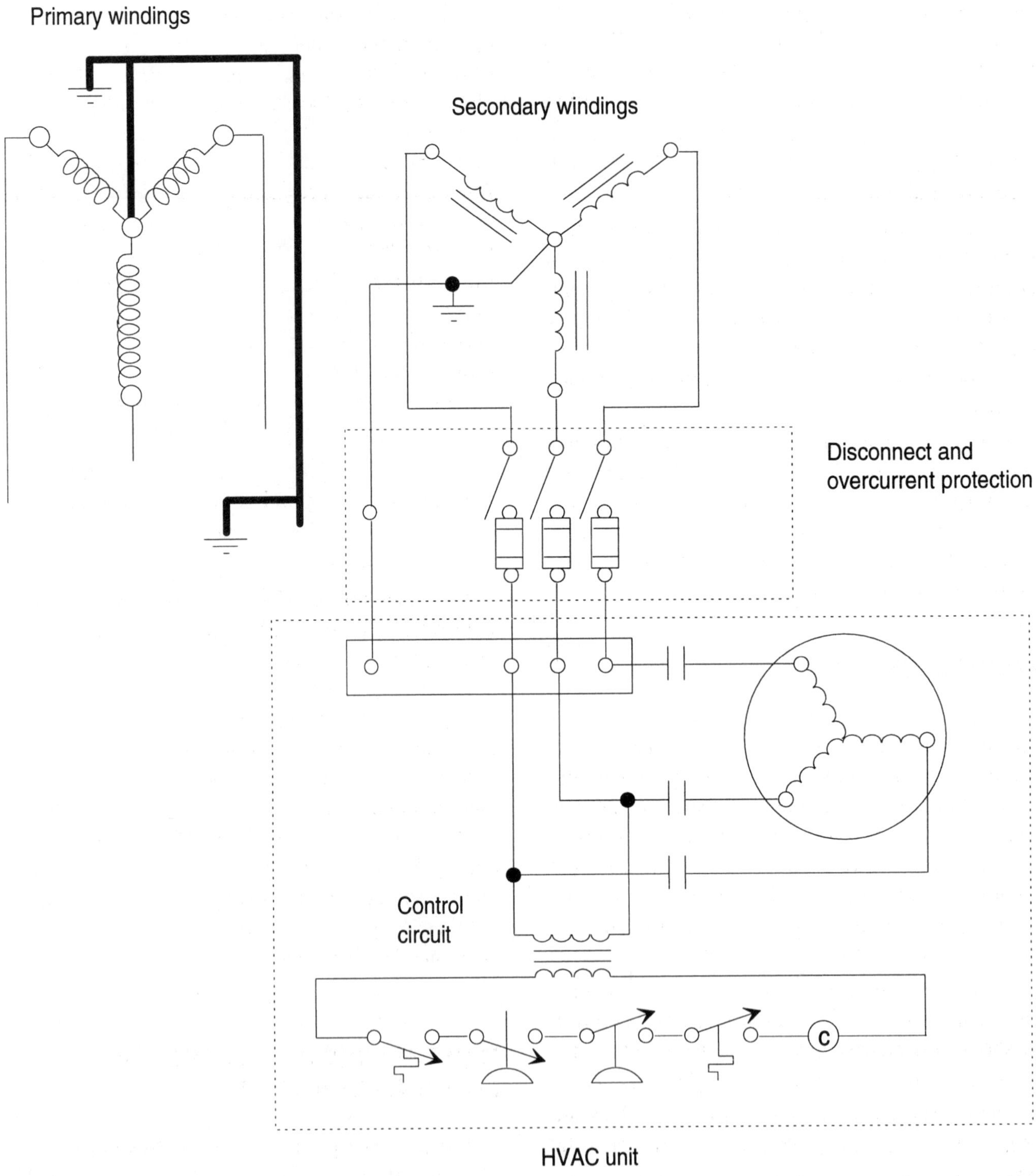
Primary windings
Secondary windings
Disconnect and
overcurrent protection
Control
circuit
C
HVAC unit

Figure 13-12: Major components for an HVAC control circuit

device and within sight of it. This disconnecting means may also serve to disconnect the heater and any motor controllers, provided the disconnecting means is within sight from the controller and heater, or it can be locked in the open position. If the motor is rated over $1/8$ HP, a disconnecting means must comply with the rules for motor disconnecting means unless a unit switch is used to disconnect all ungrounded conductors.

A heater without supplementary overcurrent protection must have a disconnecting means that complies with rules similar to those for permanently connected appliances. A unit switch may be the disconnecting means in certain occupancies when other means of disconnection are provided as specified in the *NE Code*.

Figure 13-13 summarizes some of the *NE Code* requirements for HVAC controls, including thermostats, motor controllers, disconnects, and overcurrent protection. Refer to *NE Code* Sections 424-19 through 424-22 as well as *NE Code* Sections 440-41 through 440-55.

TROUBLESHOOTING

Troubleshooting heating and cooling systems covers a wide range of electrical and mechanical problems, from finding a short circuit in the power supply line, through adjusting a pulley on a motor shaft, to tracing loose connections in complex control circuits. However, in nearly all cases, the technician can determine the cause of the trouble by using a systematic approach, checking one part of the system at a time in the right order.

Every heating and cooling system's problems can be solved, and it is the purpose of this section to show you exactly how to go about solving the more conventional ones in a safe and logical manner.

The troubleshooting charts in Figure 13-14 are arranged so that the problem is listed first. The possible causes of this problem are listed in the order in which they should be checked. Finally, solutions to the various problems are given, including step-by-step procedures where it is felt that they are necessary.

To better illustrate the use of these solutions to heating and cooling equipment problems, assume that an air-conditioner fan or blower motor is operating, but the compressor motor is not. Glance down the left-hand column in the troubleshooting charts until you locate the problem titled "Compressor motor and/or condenser motor will not start, but blower motor operates." Begin with the first item under "Probable Cause" which tells you to check the thermostat system switch to make sure it is set to "Cool." Finding that the switch is set in the proper position, you continue on to the next item; that is, "Check the thermostat . . . ." You may find that the temperature setting is above the room temperature so the system is not calling for cooling. Set the thermostat below room temperature, and the cooling unit will function.

This example is, of course, very simple, but most of the heating and cooling problems can be just as simple if a systematic approach to troubleshooting is used.

Manufacturers of HVAC equipment also provide troubleshooting and maintenance manuals for their equipment. These manuals can be one of the most helpful "tools" imaginable for troubleshooting specific HVAC equipment. When unpacking equipment, controls, and other components for the system, always save any manuals or instructions that accompany the items. File them in a safe place so that you and other maintenance personnel can readily find them. Many electricians like to secure these manuals on the inside of a cabinet door within the equipment. This way, they will always be available when needed.

MAINTENANCE OF HEATING AND COOLING EQUIPMENT

The old saying, "an ounce of prevention . . ." certainly holds true for heating and cooling systems. A correctly-installed system that is maintained according to the manufacturer's recommendations will give years of trouble-free service at minimum cost.

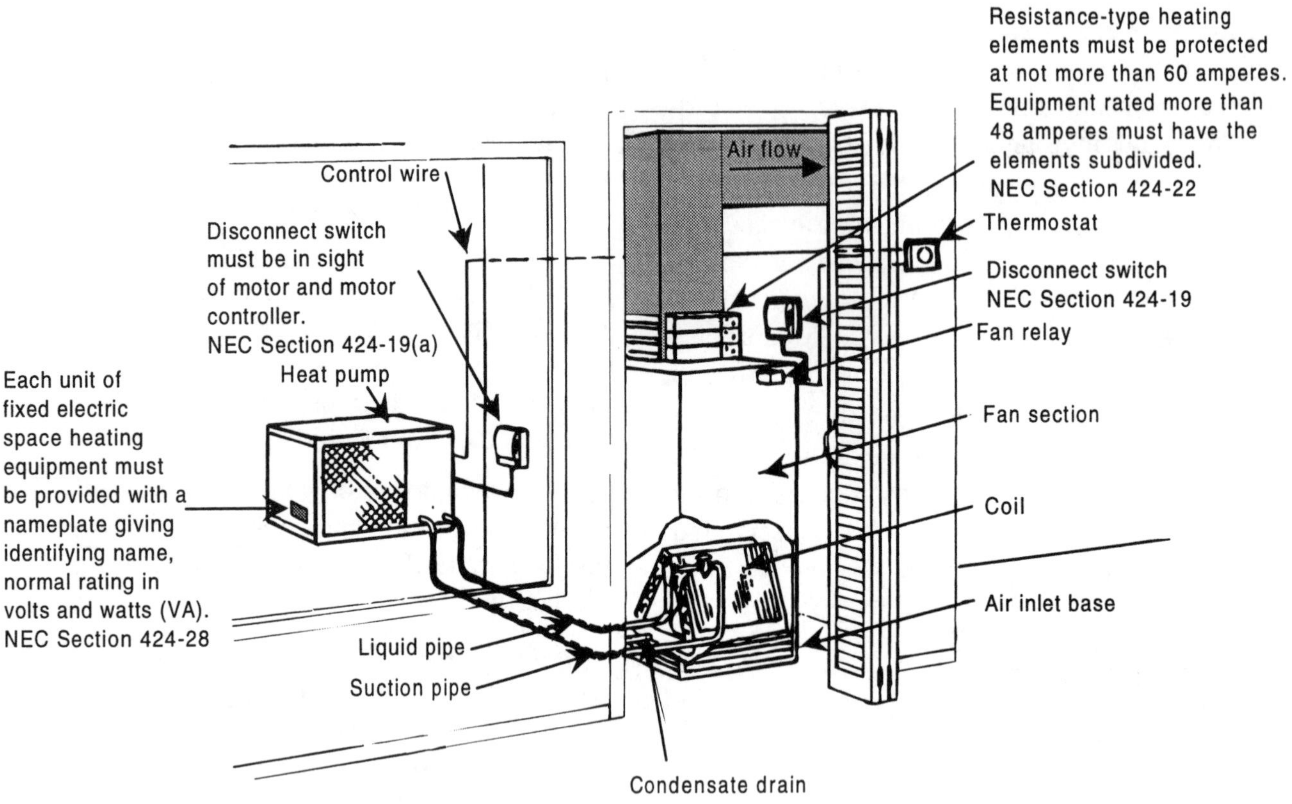
Resistance-type heating elements must be protected at not more than 60 amperes. Equipment rated more than 48 amperes must have the elements subdivided. NEC Section 424-22
Air flow
Control wire
Disconnect switch must be in sight of motor and motor controller. NEC Section 424-19(a)
Heat pump
Thermostat
Disconnect switch NEC Section 424-19
Fan relay
Fan section
Each unit of fixed electric space heating equipment must be provided with a nameplate giving identifying name, normal rating in volts and watts (VA). NEC Section 424-28
Coil
Air inlet base
Liquid pipe
Suction pipe
Condensate drain

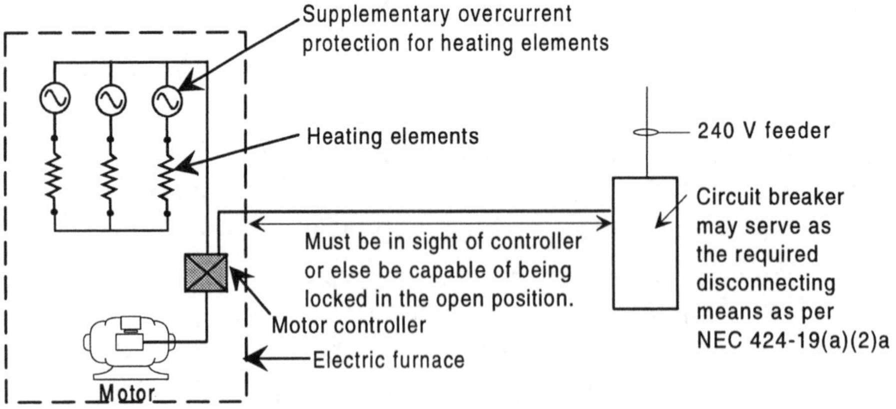
Supplementary overcurrent protection for heating elements
Heating elements
240 V feeder
Circuit breaker may serve as the required disconnecting means as per NEC 424-19(a)(2)a
Must be in sight of controller or else be capable of being locked in the open position.
Motor controller
Electric furnace
Motor

Figure 13-13: Summary of *NE Code* requirements for HVAC controls

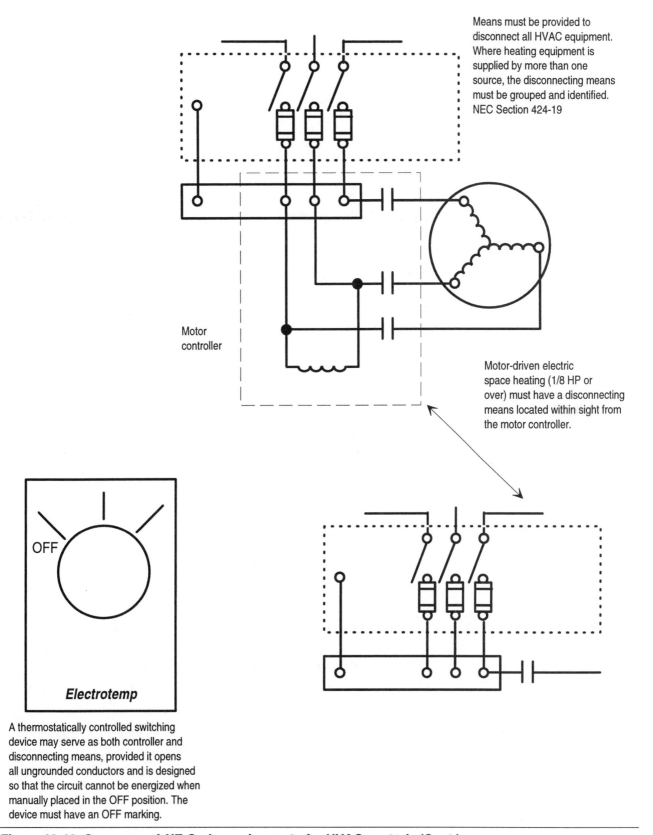

Means must be provided to disconnect all HVAC equipment. Where heating equipment is supplied by more than one source, the disconnecting means must be grouped and identified. NEC Section 424-19

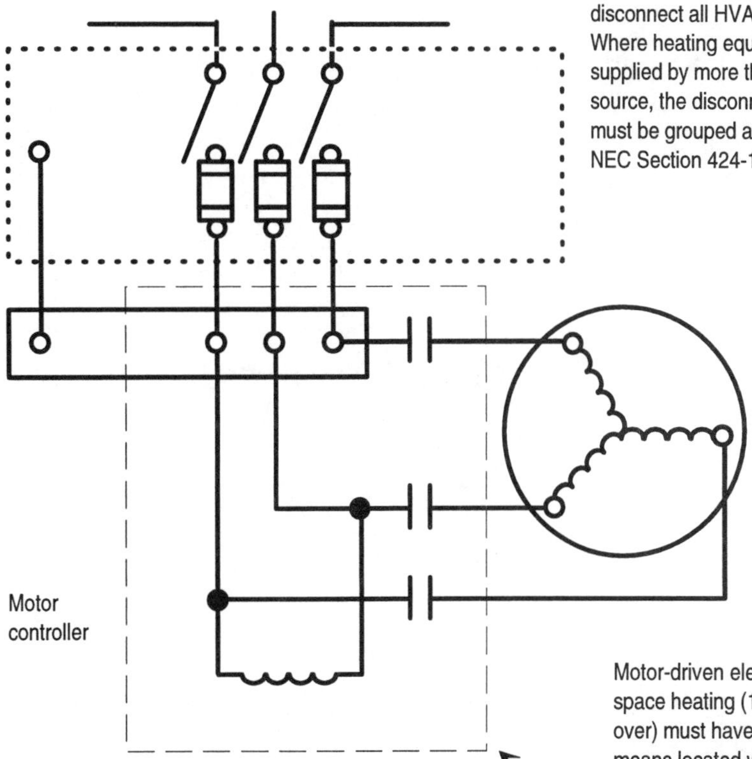
Motor
controller

Motor-driven electric space heating (1/8 HP or over) must have a disconnecting means located within sight from the motor controller.

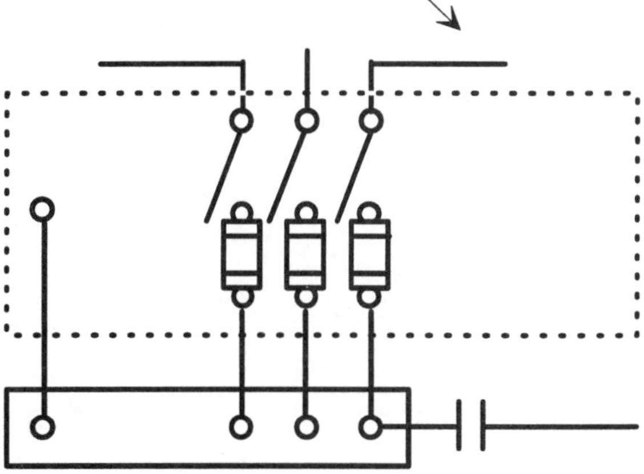

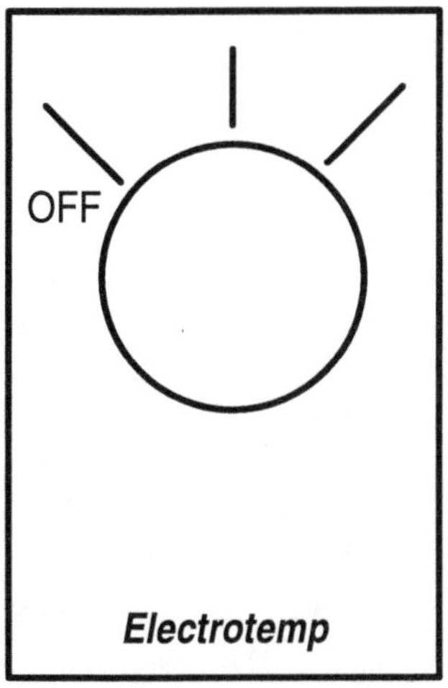
OFF
Electrotemp

A thermostatically controlled switching device may serve as both controller and disconnecting means, provided it opens all ungrounded conductors and is designed so that the circuit cannot be energized when manually placed in the OFF position. The device must have an OFF marking.

Figure 13-13: Summary of *NE Code* requirements for HVAC controls (Cont.)

Malfunction	Probable Cause	Corrective Action
Compressor motor and condenser motor will not start, but fan/coil unit (blower motor) operates normally	Check the thermostat system switch to ascertain that it is set to "COOL."	Make necessary adjustments to settings.
	Check the thermostat to make sure that it is set below room temperature.	Make necessary adjustments.
	Check the thermostat to see if it is level. Most thermostats must be mounted level; any deviation will ruin their calibration.	Remove cover plate, place a spirit level on top of the thermostat base, loosen the mounting screws, and adjust the base until it is level; then tighten the mounting screws.
	Check all low-voltage connections for tightness.	Tighten.
	Make a low-voltage check with a voltmeter on the condensate float switch; the condensate may not be draining.	The float switch is normally found in the fan/coil unit. Repair or replace.
	Low air flow could be causing the trouble, so check the air filters.	Clean or replace.
	Make a low-voltage check of the antifrost control.	Replace if defective.
	Check all duct connections to the fan-coil unit.	Repair if necessary.
Compressor, condenser, and fan/coil unit motors will not start	Check the thermostat system switch setting to ascertain that it is set to "COOL."	Adjust as necessary.
	Check thermostat setting to make sure it is below room temperature.	Adjust as necessary.

Figure 13-14: HVAC troubleshooting chart

Malfunction	Probable Cause	Corrective Action
Compressor, condenser, and fan/coil unit motors will not start	Check thermostat to make sure it is level.	Correct as required.
	Check all low-voltage connections for tightness.	Tighten.
	Check for a blown fuse or tripped circuit breaker.	Determine the cause of the open circuit and then replace the fuses or reset the circuit breaker.
	Make a voltage check of low-voltage transformer.	Replace if defective.
	Check the electrical service against minimum requirements; that is, for correct voltage, amperage, etc.	Update as necessary.
Condensing unit cycles too frequently, contactor opens and closes on each cycle and blower motor operates	Check condensate drain.	Repair or replace.
	Check all low-voltage wiring connections for tightness.	Tighten.

Figure 13-14: HVAC troubleshooting chart (Cont.)

Malfunction	Probable Cause	Corrective Action
Condensing unit cycles too frequently, contactor opens and closes on each cycle and blower motor operates	Defective blower motor.	Test amperage reading while motor is running. Do not confuse the full-load (starting) amperes shown on the motor nameplate with the actual running amperes. The latter should be about 25% less. If the amperage varies considerably from that on the nameplate, check the motor for bad bearings, defective winding insulation, etc.
	Low voltage.	Test circuit for proper voltage.
Inadequate cooling with condensing unit and blower running continuously	Check all low-voltage connections (control wiring) against the wiring diagram furnished with the system.	Correct if necessary. Then check for leaks in the refrigerant lines.
	Check all joints in the supply and return ductwork.	Make all joints tight.
	The equipment could be undersized. Check heat gain calculations against the output of the unit.	Correct structural deficiencies with insulation, awnings, etc., or install properly sized equipment.
Condensing unit cycles but blower motor does not run	Check all low-voltage connections against the wiring diagram furnished with the system.	Correct if necessary.
	Check all low-voltage connections for tightness.	Tighten.

Figure 13-14: HVAC troubleshooting chart (Cont.)

Malfunction	Probable Cause	Corrective Action
Condensing unit cycles but blower motor does not run	Make a voltage check on the blower relay.	Replace if necessary.
	Make electrical and mechanical checks on blower motor. Check for correct voltage at motor terminals. Mechanical problems could be bad bearings or a loose blower wheel. Bearing trouble can be detected by turning the blower wheel by hand (with current off), and checking for excessive wear, roughness, or seizure.	Repair or replace defective components.
Continuous short cycling of blower coil unit and insufficient cooling	Make electrical and mechanical checks.	Repair or replace motor if necessary.
Sweating at blower coil output or at electric duct heater outlet	Check to see if the insulation is installed properly.	Insulate properly.
	Inspect the joints at the duct heater or blower coil receiving collar.	Seal properly.
Thermostat calls for heat but blower motor will not operate	Check all low-voltage connections against the wiring diagram furnished with the system.	Correct if necessary.
	Check all low-voltage connections for tightness.	Tighten.
	Check all low-voltage against the unit's nameplate.	Correct if necessary.
	Check all line-voltage connections for tightness.	Tighten.

Figure 13-14: HVAC troubleshooting chart (Cont.)

Malfunction	Probable Cause	Corrective Action
Thermostat calls for heat but blower motor will not operate	Check for blown fuses or a tripped circuit breaker in the line.	Determine the reason for the open circuit and replace the fuses or reset circuit breaker.
	Check the low-voltage transformer.	Replace if defective.
	Make a low-voltage check on the magnetic relay.	Repair or replace if necessary.
	Make electrical and mechanical checks on the blower motor.	Repair or replace the motor if defective.
Thermostat calls for heat, blower motor operates but delivers cold air	Make a visual and electrical check on the heating elements.	If not operating, continue on to next step.
	Make an electrical check on the heater limit switch — first disconnecting all power to the unit — using an ohmmeter to check continuity between the two terminals of the switch.	If the limit switch is open, repair or replace.
	Make an electrical check on the time-delay relay. Most are rated at 24 volts and have one set of normally-open auxiliary contacts for pilot duty.	If the relay heater coil is open or grounded, repair or replace.
	Make an electrical check on the magnetic relay.	Repair or replace if defective.
	Check the electric service-entrance and related circuits against the minimum recommendations.	Upgrade if necessary.

Figure 13-14: HVAC troubleshooting chart (Cont.)

Malfunction	Probable Cause	Corrective Action
Thermostat calls for heat and blower motor operates continuously, system delivers warm air but the thermostat is not satisfied	Check all joints in the ductwork for air leaks.	Make all defective joints tight.
	Check all duct joints and blower outlets for tightness.	Seal where necessary.
	Make a visual and electrical check of the electric heating element.	Repair or replace if necessary.
	Make an electrical check on the heater limit switch.	Repair or replace.
	Check the heating element against the blower unit for the possibility of a mismatch.	Replace if incompatible.
	Check your heat loss calculations. The equipment could be undersized.	If so, correct structural deficiencies by installing more insulation, storm windows and doors, etc., or install properly sized equipment.
Blower unit operates properly and delivers air but thermostat is not satisfied	Check all joints in the ductwork for air leaks.	Repair if necessary.
	Check the air filter.	Clean or replace if necessary. Also check the number of air outlets for adequacy and make sure they are balanced.

Figure 13-14: HVAC troubleshooting chart (Cont.)

Malfunction	Probable Cause	Corrective Action
Blower unit operates properly and delivers air but thermostat is not satisfied	Check for undersized equipment.	Correct structural deficiencies or install properly sized equipment.
Electric heater cycles on limit switches but blower motor does not operate	Make an electrical check on the magnetic relay.	Repair or replace if defective.
	Make electrical and mechanical checks on the blower motor.	Repair or replace if defective.
	Check the line connection against the wiring diagram furnished with the system.	Make any necessary changes.
Excessive air noise at terminator	Duct or outlet undersized; air velocity too great.	Increase size of duct and/or outlet.
	Make external static pressure check.	Correct restrictions in system if necessary.
	Check for properly balanced system.	Make corrections if necessary.
Excessive noise at return air grille	Check the return duct to make sure it has a 90 degree bend.	Correct if necessary.
	Make a visual check of the blower unit to ascertain that all shipping blocks and angles have been removed.	Remove if necessary.
	Check blower motor assembly suspension and fasteners.	Tighten if necessary.
Excessive vibration at blower unit	Visually check for vibration isolators (which isolate the blower coil from the structure).	If missing, install as recommended by the manufacturer.
	Visually check to ascertain that shipping blocks and angles have been removed from the blower unit.	Remove if necessary.

Figure 13-14: HVAC troubleshooting chart (Cont.)

Maintenance procedures and frequency will depend on the type of system, but the information given here covers most residential and commercial HVAC systems in general. For further information, consult the Owner's Handbook accompanying the equipment. If none is available, obtain a copy from the equipment supplier or write directly to the manufacturer.

Besides saving on repairs, good maintenance of heating and cooling equipment will ensure that the equipment operates at maximum efficiency, which will save fuel and reduce other operating expenses.

Cleaning: Heating and cooling equipment should be cleaned at regular intervals in order to maintain operating efficiency, lengthen the life of the equipment, and minimize energy consumption and operating costs.

Begin by removing lint and dust with a cloth and brush from all finned convector-type heaters, and then vacuum them. This includes electric and hot-water baseboard heaters, wall, and unit heaters. Air conditioning evaporator and condenser coils should be vacuumed, scrubbed with a liquid solvent or detergent, and flushed.

Air filters in forced-air systems collect dust, as is their purpose. Periodic inspections will tell you how often they should be cleaned or replaced. Permanent metal mesh and electrostatic air filters should be washed and treated; throw-away filters should be replaced with the same size and type.

Vacuum or wipe off lint and dust from all supply and return grilles, diffusers, and registers, using a detergent solution if necessary. While you're doing this, also remove any dirt on the dampers and check the levers for proper operation.

Motors should also be vacuumed to remove dust and lint, but you should first turn off the power supply. Once free of dust and lint, wipe their exterior surfaces clean with a rag and reconnect the power.

Propeller fans and blower wheels are especially susceptible to dust deposits and should be cleaned often. Again make certain that the power is turned off — to prevent losing a finger — and remove the dirt deposits with a liquid solvent.

Every periodic inspection of any HVAC system should include a check of the condensate drain. Wash the drain pan with a mild detergent and flush out the drain line. All loose particles of dirt should be brushed from the evaporator coil and a fin comb should be used to open all clogged air passages in the coil. If the coil is extremely dirty, a small pressurized sprayer may be used with a strong dishwasher detergent to flush the coil. Always rinse with clean water after using detergent.

Lubrication: Adequate, regular lubrication ensures efficient operation, long equipment life, and minimum maintenance cost, but never over-lubricate! Observe the manufacturer's instructions when lubricating all bearings, rotary seals, and movable linkages of:

- Motors — direct or belt-drive type.
- Shafts — fans, blower wheel, and damper.
- Pumps — water circulating.
- Motor Controllers — sequential and damper operators.

Periodic inspection: This checklist should be followed during periodic inspections of most heating and cooling systems in order to minimize maintenance expense and to save as much fuel as possible. This checklist may vary with different types of systems.

When checking the electrical system, examine all components for evidence of overheating and insulation deterioration.

HVAC Inspection Checklist

1. Drive Belts

Examine for:

- ☐ Proper tension and alignment
- ☐ Sidewall wear
- ☐ Deterioration, cracks
- ☐ Greasy surfaces
- ☐ Safety guards in position and secure

2. "V" Pulleys

Examine for:

- ☐ Alignment
- ☐ Wear of "V" wall
- ☐ Tightness of pulley and set screws

3. Fan Blade and Blower Wheel Assemblies

Examine for:

- ☐ Metal fatigue cracks
- ☐ Tightness of hubs and set screws
- ☐ Balance
- ☐ Safety guards in position and secure

4. Electrical Components

Examine, burnish, or replace electrical contacts of:

- ☐ Magnetic contactors and relays
- ☐ Control switches, thermostats, timers, etc.

Examine wire and terminals for:

- ☐ Corrosion and looseness at switches, relays, thermostats, controllers fuse clips, capacitors, etc.

Examine motor capacitors for:

- ☐ Case swelling
- ☐ Electrolyte leakage

Appendix I

Electrical Specifications

The basic factors influencing the choice of an electrical system in a modern residence are:

- Adequacy
- Safety
- Reliability
- Performance
- Convenience

In order for the electrical designer to achieve high-quality results in a residential electrical system, he must make certain that his design is clearly conveyed to the workmen on the job. This is usually achieved by plans (working drawings) and written specifications, the latter of which will be described in this appendix.

As mentioned earlier in this book, the electrical plans usually will show the layout of all outlets for lighting, power, signal and communication systems, and related electrical circuits. Also, they will include the location of the service-entrance equipment, and related diagrams and schedules.

The specifications for a building or project give a written description of what will be required by the contractor. Together with the working drawings, they form the basis of the contract requirements for the construction of the electrical system. The specifications should clearly identify:

- The type and quality of equipment and materials selected by the designer, architect, and/or owners.
- Local code requirements relating to approved materials and wiring methods.
- The scope or summary of the work required of the electrical contractor.
- A summary of certain types of work not required of the electrical contractor but which involve electrical work to be done by others and covered in other divisions of the written specifications.

The length and detail of the specifications will, of course, depend upon the size and complexity of the job, but nearly all specifications will contain basically the same number of sections within a division. These sections may consist of only a few concise paragraphs for smaller projects or of several pages of detailed instructions for larger projects.

In the following paragraphs, a set of written specifications (patterned after the format used by the Construction Specification Institute) is presented for the residence described in this book. This set of specifications, and the accompanying explanation, will act as a guide for the reader when you write specifications for projects now or in the future. All you need do is delete items or add different items in place of those that appear in this book. In fact, the set of specifications presented here has been used for dozens of residences over the past few years with only minor changes for each — with very satisfactory results. However, since the written specifications normally take precedence over the drawings, use caution when writting or providing any set of specifications for residential projects.

General Provisions

The General Provisions section is usually the first section in the Electrical Division of the specifications. It covers such items as scope of the work, cutting and patching, permits, etc. A specification for a typical residence may begin as follows:

DIVISION 16 ELECTRICAL

Section 16010—General Provisions

(A) The "Instructions to Bidders," "General Conditions," "Supplementary General Conditions," and "Special Conditions" of the Architectural Specifications govern work under this Section.

(B) It is understood and agreed that the Electrical Contractor has, by careful examination of the working drawings and these Specifications, and the job site, where appropriate, has satisfied himself as to the nature and location of the work and all conditions that must be met in order to carry out the work under this Section of the Contract.

(C) The Drawings are diagrammatic and indicate generally the location of material and equipment. These Drawings shall be followed as closely as possible. The Electrical Contractor shall coordinate the work under this Section with the architectural, structural, heating, ventilation, air conditioning and plumbing drawings, and the drawings of other trades for exact dimensions, clearances, and roughing-in locations. This Electrical Contractor shall cooperate with all other trades in order to make minor field adjustments to accommodate the work of others.

(D) The Drawings and Specifications are complementary, each to the other, and the work required by either shall be included in the Contract as if called for by both.

(E) The work under this Section includes the furnishing of all labor, materials, equipment, and incidentals necessary for the installation of the complete electrical system as shown on the Drawings and as called for in the Specifications.

(F) All work under this Section shall conform, in all respects, to the *National Electrical Code*, requirements of the utility company, and state, city, and local ordinances. Where Plans and Specifications indicate work in excess of the above minimum requirements, the Plans and Specifications shall be followed.

(G) The Electrical Contractor shall pay for all permits, inspection fees, and other installation fees required to complete the work under this Section of the Contract.

(H) Upon completion of the project, the Electrical Contractor shall provide the Owners with a certificate of final inspection and approval of all work in this Section, by the electrical inspector having jurisdiction, if applicable.

(I) The Electrical Contractor shall be responsible for notifying the Electrical Inspector at each inspection stage, and no work shall progress until the inspection has been completed and the work approved.

(J) The Electrical Contractor shall do all cutting and patching of work as necessary to complete the work under this Section of the Contract. All finished work damaged in connection with this work shall be repaired to perfectly match adjoining finished work. The Electrical Contractor shall employ tradesmen skilled in carrying out this particular phase of the work.

(K) The Electrical Contractor shall furnish and present five (5) copies of shop drawings or brochures for all lighting fixtures, equipment, and accessories to the Architect/Engineer for approval. No equipment shall be installed prior to approval.

(L) The naming of a certain brand or manufacturer in the Specifications is to establish a quality standard for the article desired. The Contractor is not restricted to the use of the specific brand of the manufacturer named unless so indicated in the Specifications. However, where a substitute is requested, it will be permitted only with the written approval of the Architect/Engineer. The Contractor shall assume all responsibility for additional expenses as required in any way to meet changes from the original material or equipment specified.

(M) Should the Electrical Contractor perform any work that does not comply with the requirements of the applicable building codes, state laws, local ordinances, and utility company regulations, he shall bear the cost of correcting any such deficiency.

Before continuing onto Section 16100, "Basic Materials and Methods," it would be appropriate to analyze some of the subparagraphs in the preceding section.

These subparagraphs will apply to nearly all residential electrical systems without any change, and while some of the paragraphs may seem too obvious, there is a definite reason for each.

Subparagraph (A) tells the electrical contractor that he must comply with the architectural conditions set forth in other divisions. Subparagraph (B) leaves little doubt that the electrical contractor must examine the plans and specifications very carefully prior to bidding so that there will be no disputes later on concerning interpretation of the documents. In other words, if the contractor does not understand certain parts of the construction documents, he must ask the architect/engineer to clarify these prior to bidding.

Section 16100, "Basic Materials and Methods," encompasses products, assemblies, and methods of installation for such items as raceways, conductors and cable, outlet boxes, panelboards and load centers, switches and receptacles, and motor and motor controls, including overload and overcurrent protective devices.

Section 16100—Basic Materials and Methods

(A) All materials and equipment shall be new and undamaged, and shall bear the approval label of Underwriters' Laboratories, Inc., and shall be listed for use in each specific location, unless approval does not apply.

Section 16110—Raceways:

(A) Rigid steel conduit shall be used for all service-entrance raceways unless specifically excepted on the Plans and in the Specifications. Rigid steel conduits shall be low-carbon, hot-dipped galvanized both inside and outside, with threaded joints. Other finishes may be substituted only with the approval of the Architect/Engineer.

(B) Conduit fittings shall be cast aluminum alloy or cast ferrous alloy, galvanized, and shall be UL approved.

(C) Special conduit fittings shall be T & B, O. Z., or approved equal; style shall be appropriate for each application.

(D) Conduit system shall be installed to provide a continuous bond throughout the entire raceway system.

(E) Junction boxes and pull boxes shall be furnished and installed where indicated on the Drawings or where required by the *NE Code* or where necessary to facilitate pulling wire and cable without damage.

(F) Exposed wiring in basement and garage area shall be installed in electrical metallic tubing (EMT). The EMT shall be UL approved and galvanized inside and outside. Fittings of the same type material and finish shall be used.

Section 16120—Conductors and Cables:

(A) Service-entrance conductors shall be THW and shall be as manufactured by the Triangle Conduit & Cable Company, Inc., or approved equal.

(B) Branch-circuit conductors installed in conduit shall be TW. All other wiring for branch circuits shall be nonmetallic-sheathed cable or armored cable, adequately sized and installed according to the *NE Code,* except where prohibited by local ordinances.

Section 16130—Outlet Boxes:

(A) Outlet boxes for switch and receptacle outlets used with concealed wiring shall be sheet steel, galvanized or cadmium plated, as manufactured by Steel City.

(B) Masonry boxes shall be used in all exposed masonry walls.

(C) Boxes for switch and receptacle outlets used with exposed wiring shall be a type approved for surface mounting. "Utility" or "Handy" boxes may be used for single-gang switches and receptacles on masonry walls.

(D) Exposed outlet boxes for outside wiring shall be solid cast aluminum and shall be approved for weatherproof use.

(E) Throughout the entire installation, all metal boxes shall be grounded in accordance with the methods set forth in the *NE Code.*

Section 16134—Panelboards and Load Centers:

(A) Panelboards and load centers shall be the dead-front safety type, with thermal magnetic, quick-break, trip-free, plug-in-type molded-case circuit breakers as indicated on the Drawings and in the Panelboard Schedule.

(B) All panel directories shall be typed, and the terminology approved by the Owners.

Section 16140—Switches and Receptacles:

(A) All flush switches shall be the quiet ac-rated toggle type and shall be as manufactured by Leviton Cat. No. 53501(I). Three- and four-way switches shall be of the same series. All wiring devices shall be Ivory unless otherwise indicated.

(B) Receptacles shall be equal to Leviton Cat. No. 5014(I) and shall be mounted 16 inches to center of box from finished floor except where otherwise noted.

(C) All wiring devices shall be trimmed with Ivory wall plates except in the kitchen and bathrooms where stainless-steel, satin-finish plates shall be used.

(D) Special wiring devices shall be as indicated on the Drawings and in the Electrical Symbol List.

Section 16150—Motors and Motor Controls:

(A) Motor controllers or starters will be furnished as part of the equipment on which they are used. These controls shall be furnished by the respective subcontractors and shall be mounted and connected by the Electrical Contractor.

Service and Distribution

Service-entrance capacity can be computed by methods outlined in Articles 220, 230, and Chapter 9 of the *National Electrical Code* as well as Chapter 5 of this book. While the results obtained will be safe, they may not always be adequate from the standpoint of performance. The designer should always check with the owners (where possible) to determine their requirements as far as electrical use is concerned.

This division of the specifications should supplement the drawings so that together they will provide the owners of the house with an adequate, efficient, and convenient electrical distribution system.

The panelboard schedules and power-riser diagrams usually appear on the drawings giving the:

- Type and size of service.
- Manufacturer and size of panelboard.
- Number and size of circuit breakers to be contained within the panelboard.
- Wire and conduit size.

Therefore, all that is required in the specifications is the following:

Section 16400—Service and Distribution

Section 16410—Electric Service:

(A) The Electrical Contractor shall furnish and install service-entrance conductors, panelboards, load centers, and all necessary fittings and equipment to complete the service entrance, including the meter socket. The local utility company will furnish the meter base unless otherwise indicated on the drawings.

(B) The service entrance shall be installed as indicated on the Drawings and as shown in the Power-Riser Diagram.

(C) The Electrical Contractor shall bond and ground all the service-entrance, equipment in accordance with the *NE Code* and local ordinances.

(D) The Electrical Contractor shall pay all costs incurred in the installation of the electric service, including connection fees of the local utility company.

(E) It shall be the responsibility of the Electrical Contractor to connect the various loads in such a way that there is minimum phase unbalance throughout the building. The Electrical Contractor shall operate the building under full heating and other load conditions, including full lighting, and shall provide a record of the amperes per phase of each feeder installed to the main distribution center.

Lighting Equipment and Fixtures

Residential lighting fixtures are sometimes selected by the owner with the help of the architect, or the architect/designer may provide a fixture allowance in the base bid so that the owner may select fixtures of his choice. If the cost of the fixtures selected by the owner exceeds the allowance, the owner must pay extra. For this reason, it is usually best to have the exact lighting fixtures selected and listed by manufacturers' names, catalog numbers, etc., prior to bidding the job. All of this data usually appears in a lighting-fixture schedule and is shown on the drawings.

The specifications include supplemental data that are not practical to include in the form of the schedule.

Section 16500—Equipment and Methods

(A) Lighting fixtures are to be as shown on the Drawings and as indicated in the Lighting-Fixture Schedule.

(B) Before installation of lighting fixtures, the Electrical Contractor shall verify all mounting heights and the exact location of each fixture with the Architect/Engineer.

(C) All lighting fixtures shall be lamped as indicated in the Lighting-Fixture Schedule and shall be those manufactured by General Electric, Westinghouse, or Sylvania.

(D) Where the finish of a lighting fixture is damaged during installation, all such damage shall be touched up with a matching finish before acceptance of the building.

(E) At the time of acceptance of the building, all lamps shall be operating; all lighting controls shall be functional, and the entire lighting system must be operating according the the plans and written specifications.

Appendix II
Trade Organizations

ASA
Acoustical Society of America
500 Sunnyside Blvd.
Woodberry, NY 11797
(516) 349-7800

ASC
Adhesive and Sealant Council, Inc.
627 K St. NW
Washington, DC 20001
(202) 452-1500

ARI
Air-Conditioning and Refrigeration Institute
1501 Wilson Blvd.
Arlington, VA 22209
(703) 524-8800

ACCA
Air Conditioning Contractors of America
1513 16th St.
Washington, DC 20036
(202) 583-9370

ACEA
Allied Construction Employers Association
180 N. Executive Drive
Brookfield, WI 53008
(414) 785-1430

AA
Aluminum Association
900 19th ST., NW
Washington, DC 20006
(202) 862-5100

AAA
American Arbitration Association
140 W. 51st St.
New York, NY 10020
(212) 484-4000

American Association of State Highway and Transportation Officials
444 N. Capitol St., NW, Suite 225
Washington, DC 20001
(202) 624-5800

ABCA
American Building Contractors Association
11100 Valley Blvd., Suite 120
El Monte, CA 91731
(818) 401-0071

ACI
American Concrete Institute
22400 W. Seven Mile Rd.
Detroit, MI 48219
(313) 532-2600

ACPA
American Concrete Pavement Association
3800 N. Wilke Rd., Suite 490
Arlington Heights, IL 60004
(708) 394-5577

ACPA
American Concrete Pipe Association
8300 Boone Blvd.
Vienna, VA 22180
(703) 821-1990

ACEC
American Consulting Engineers Council
1015 15th St., NW, Suite 802
Washington, DC 20005
(202) 347-7474

AGA
American Gas Association, Inc.
1515 Wilson Blvd.
Arlington, VA 22209
(703) 841-8400

AHA
American Hardboard Association
520 N. Hicks Rd.
Palatine, IL 60067

AHMA
American Hardware Manufacturers Association
931 N. Plum Grove Rd.
Schaumburg, IL 60173
(708) 605-1025

AIA
American Institute of Architects
1735 New York Ave., NW
Washington, DC 20006
(202) 626-7300

ASID
American Society of Interior Designers
200 Lexington Ave.
New York, NY 10016

AISC
American Institute of Steel Construction, Inc.
400 N. Michigan Ave.
Chicago, IL 60611
(312) 670-2400

AITC
Amerian Institute of Timber Construction
11818 S.E. Mill Plain Blvd.
Vancouver, WA 98684
(206) 254-9132

AISI
American Iron and Steel Institute
1133 15th St. NW, Suite 300
Washington, DC 20005
(202) 452-7100

ALSC
American Lumber Standards Committee
P.O. Box 210
Germantown, MD 20874
(301) 972-1700

ANSI
American National Standards Institute
1430 Broadway
New York, NY 10018
(212) 642-4900

APFA
American Pipe Fitting Association
6203 Old Keene Mill Ct.
Springfield, VA 22152
(703) 644-0001

APA
American Plywood Association
PO Box 11700
Tacoma, WA 98411
(509) 565-6600

ARTBA
American Road and Transportation Builders Association
525 School St. SW
Washington, DC 20024
(202) 488-2722

ASTM
American Society for Testing and Materials
1916 Race St.
Philadelphia, PA 19103
(215) 299-5400

ASCE
American Society of Civil Engineers
345 E. 47th St.
New York, NY 10017
(212) 705-7496

ASCC
American Society of Concrete Construction
426 S. Westgate
Addison, IL 60101
(708) 543-0870

ASHRAE
American Society of Heating, Refrigerating, and Air-Conditioning Engineers, Inc.
1791 Tullie Circle, NE
Atlanta, GA 30329
(404) 636-8400

ASHI
American Society of Home Inspectors
1735 N. Lynn St., Suite 950
Arlington, VA 22209-2022
(703) 524-2008

ASID
American Society of Interior Designers
1430 Broadway
New York, NY 10018
(212) 685-3480

ASME
American Society of Mechanical Engineers
345 E. 47th St.
New York, NY 10017
(212) 705-7800

ASSE
American Society of Sanitary Engineers
PO Box 40362
Bay Village, OH 44140
(216) 835-3040

ASA
American Subcontractors Association
1004 Duke St.
Alexandria, VA 22314
(703) 684-3450

AWS
Amerian Welding Society, Inc.
550 N.W. LeJeune Rd.
Miami, FL 33126
(305) 443-9353

APA
Architectural Precast Association
825 E. 64th St.
Indianapolis, IN 46220
(317) 251-1214

AWI
Architectural Woodwork Institute
2310 S. Walter Reed Dr.
Arlington, VA 22206
(703) 671-9100

AIA/NA
Asbestos Information Association/North America
1745 Jefferson Davis Hwy., Suite 509
Arlington, VA 22202
(703) 979-1150

AI
Asphalt Institute
Asphalt Institute Building
College Park, MD 20740
(301) 779-9354

ABC
Associated Builders and Contractors, Inc.
1300 North 17th Street
Rosslyn, Virginia 22209
(703) 812-2000

AGC
Associated General Contractors of America
1957 East St., NW
Washington, DC 20005
(202) 393-2040

SMACNA
Associated Sheet Metal Contractors, Inc.
3121 W. Hallandale Beach Blvd., Suite 114
Hallandale, FL 33009
(305) 961-0440

ABC
Association of Bituminous Contractors
2020 K St. NW, Suite 800
Washington, DC 20006

AWCI
Association of the Wall and Ceiling Industries International
1600 Cameron St.
Alexandria, VA 22314
(703) 684-2924

BIA
Brick Institute of America
11490 Commerce Park Dr., Suite 300
Reston, VA 22091
(703) 620-0010

BHMA
Builder's Hardware Manufacturers Association, Inc.
60 E. 42nd St., Rm. 511
New York, NY 10165
(212) 661-4261

BRB
Building Research Board
2101 Constitution Ave., NW
Washington, DC 20418

BSC
Building Systems Contractors
6710 Persimmon Tree Rd.
Bethesda, MD 20817
(303) 320-2505

CRA
California Redwood Association
405 Enfrente Dr., Suite 200
Novalto, CA 94949
(415) 382-0662

CRI
Carpet and Rug Institute
PO Box 2048
Dalton, GA 30722-2048
(404) 278-3176

CISCA
Ceilings and Interior Systems Construction Association
104 Wilmot, Suite 201
Deerfield, IL 60015
(708) 940-8800

CTI
Ceramic Tile Institute
700 N. Virgil Ave.
Los Angeles, CA 90029
(213) 660-1911

CRSI
Concrete Reinforcing Steel Institute
933 N. Plum Grove Rd.
Schaumburg, IL 60195
(708) 517-1200

CIEA
Construction Industry Employers Association
625 Ensminger Rd.
Tonawanda, NY 14150
(716) 875-4744

CCIC
Construction Consultants International Corp.
8133 Leesburg Pike
Vienna, VA 22180
(703) 734-2393

CCC
Construction Contractors Council
6120 Brandon Ave.
Springfield, VA 22150
(703) 644-2215

CC&M
Construction Costs and Management
6575 Edsall Rd.
Sterling, VA 22170
(703) 354-7991

CEMC
Construction Education Management Corporation.
8133 Leesburg Pike
Vienna, VA 22180
(703) 734-2399

CEI
Construction Environment Inc.
5655-D General Washington Dr.
Alexandria, VA 22312
(703) 750-0525

CIMA
Construction Industry Manufacturers Association
111 E. Wisconsin Ave., Suite 940
Milwaukee, WI 53202-4879
(414) 272-0943

CMC
Construction Management Collaborative
901 N. Pitt St.
Alexandria, VA 22314
(703) 836-3344

CM&C
Construction Managers and Consultants, Inc.
12107 Stallion Ct.
Woodbridge, VA 22192
(703) 690-1635

CSI
Construction Specifications Institute
601 Madison St.
Alexandria, VA 22314
(703) 684-0300

Corps of Engineers/U.S. Department of the Army
20 Massachuesetts Ave., NW
Washington, DC 20314
(202) 272-0660

CABO
Council of American Building Officials
5205 Leesburg Pike, Suite 708
Falls Church, VA 22041
(703) 931-4533

DHI
Door and Hardware Institute
7711 Old Springhouse Rd.
McLean, VA 22102-3474
(703) 556-3990

DIPRA
Ductile Iron Pipe Research Association
245 Riverchase Parkway E., Suite 0
Birmingham, AL 35244
(205) 988-9870

EPA
Environmental Protection Agency
401 M St., SW
Washington, DC 20460
(202) 382-2090

FPRS
Forest Products Research Society
2801 Marshall Ct.
Madison, WI 53705
(608) 231-1361

GBCA
General Building Contractors Association
36 S. 18th St.
PO Box 15959
Philadelphia, PA 19103
(215) 568-7015

HPMA
Hardwood Plywood Manufacturers Association
PO Box 2789
Reston, VA 22090
(703) 435-2900

IESNA
Illuminating Engineering Society of North America
345 E. 47th St.
New York, NY 10017
(212) 705-7926

ILIA
Indiana Limestone Institute of America
Stone City Bank Building, Suite 400
Bedford, IN 47421
(812) 275-4426

IDSA
Industrial Designers Society of America
1142 E. Walker St.
Great Falls, VA 22066
(703) 759-0100

IFI
Industrial Fasteners Institute
1505 E. Ohio Building
Cleveland, OH 44114
(216) 241-1482

IHEA
Industrial Heating Equipment Association
1901 N. Moore St.
Arlington, VA 22209
(703) 525-2513

ISEA
Industrial Safety Equipment Association
1901 N. Moore St.
Arlington, VA 22209
(703) 525-1695

IEEE
Institute of Electrical / Electronics Engineers
345 E. 47th St.
New York, NY 10017
(212) 705-7900

ICAA
Insulation Contractors Association of America
15819 Crabbs Branch Way
Rockville, MD 20855
(301) 590-0030

International Association of Bridge, Structural and Ornamental Iron Workers
1750 New York Ave., NW, Suite 400
Washington, DC 20006
(202) 383-4800

IALD
International Association of Lighting Designers
18 E. 16th St., Suite 208
New York, NY 10003
(212) 206-1281

IAPMO
International Association of Plumbing and Mechanical Officials
20001 Walnut Dr., S
Walnut, CA 91789
(714) 595-8449

IBB
International Brotherhood of Boilermakers, Iron Ship Builders, Blacksmiths, Forgers and Helpers
753 State Avenue, Suite 565
Kansas City, KS 66101
(913) 371-2640

IBEW
International Brotherhood of Electrical Workers
1125 15th St. NW
Washington, DC 20005
(202) 833-7000

IBPAT
International Brotherhood of Painters and Allied Trades
1750 New York Avenue, NW
Washington, DC 20005
(202) 637-0700

IMI
International Masonry Institute
823 15th St. NW, Suite 1001
Washington, DC 20005
(202) 783-3908

IRF
International Road Federation
525 School St. SW
Washington, DC 20024
(202) 554-2106

IUBAC
International Union of Bricklayers and Allied Craftsmen
Bowen Building
815 15th St. NW
Washington, DC 20005
(202) 783-3788

IUOE
International Union of Operating Engineers
1125 17th St., NW
Washington, DC 20036
(202) 429-9100

LIUNA
Laborers' International Union of North America
905 16th St., NW
Washington, DC 20006-1765
(202) 737-8320

Manufacturers Standardization Society of the Valve and Fittings Industry
127 Park St., NE
Vienna, VA 22180
(703) 281-6613

MFMA
Maple Flooring Manufacturers Association
60 Revere Dr., Suite 500
Northbrook, IL 60062
(708) 480-9080

MIA
Marble Institute of America
33505 State St.
Farmington, MI 48024
(313) 476-5558

MCAA
Mason Contractors Association of America
17W 601 14th St.
Oakbrook Terrace, IL 60181
(708) 620-6767

MCAA
Mechanical Contractors Association of America
1385 Piccard Dr.
Rockville, MD 20850
(301) 869-5800

MBMA
Metal Building Manufacturers Association
1230 Keith Building
Cleveland, OH 44115
(216) 241-7333

MLSFA
Metal Lath/Steel Framing Association
600 S. Federal, Suite 400
Chicago, IL 60605
(312) 922-6222

NAPA
National Asphalt Pavement Association
6811 Kenilworth Ave., Suite 620
PO Box 517
Riverdale, MD 20737
(301) 779-4880

NABD
National Association of Brick Distributors
212 S. Henry St.
Alexandria, VA 22314
(703) 549-2555

NADC
National Association of Demolition Contractors
4415 W. Harrison St.
Hillside, IL 60162
(708) 449-5959

NADCO
National Association of Development Companies
1730 Rhode Island Ave., NW
Washington, DC 20036
(202) 785-8484

NADO
National Association of Development Organizations
400 N. Capitol St.
Washington, DC 20001
(202) 624-7806

NADC
National Association of Dredging Contractors
1733 King St.
Alexandria, VA 22314
(703) 548-8300

NAEC
National Association of Elevator Contractors
4053 LaVista Rd., Suite 120
Tucker, GA 30084
(404) 496-1270

NAHB
National Association of Home Builders
15th and M St., NW
Washington, DC 20005
(202) 822-0200

NAPHCC
National Association of Plumbing, Heating, and Cooling Contractors
PO Box 6808
Falls Church, VA 22046
(703) 237-8100

NARSC
National Association of Reinforcing Steel Contractors
PO Box 225
Fairfax, VA 22030
(703) 591-1870

NAWIC
National Association of Women in Construction
327 S. Adams St.
Fort Worth, TX 76104
(817) 877-5551

NBMDA
National Building Material Distributors Association
1417 Lake Cook Rd.
Deerfield, IL 60015
(708) 945-7201

NCMA
National Concrete Masonry Association
PO Box 781
Herndon, VA 22070
(703) 435-4900

NCSBCS
National Conference of States on Building Codes and Standards
505 Huntmar Park Dr.
Herndon, VA 22070
(703) 437-0100

NCA
National Constructors Association
1730 M St. NW
Washington, DC 20036

NCIC
National Construction Industry Council
1919 Pennsylvania Ave. NW
Washington, DC 20006
(202) 887-1494

NCRP
National Council on Radiation Protection and Measurement
7910 Woodmont Ave., Suite 800
Bethesda, MD 20814
(301) 657-2625

NECA
National Electrical Contractors Association
7315 Wisconsin Ave.
13th Floor, West Building
Bethesda, MD 20814
(301) 657-3110

NEMA
National Electrical Manufacturers Association
2101 L St., NW, Suite 300
Washington, DC 20037
(202) 457-8400

NFPA
National Fire Protection Association
Batterymarch Park
Quincy, MA 02269
(617) 770-3000

NFPA
National Forest Products Association
1250 Connecticut Ave., NW, Suite 200
Washington, DC 20036
(202) 463-2700

NGA
National Glass Association
8200 Greensboro Dr., Suite 302
McLean, VA 22102
(703) 442-4890

NHRA
National Housing Rehabilitation Association
1726 18th St. NW
Washington, DC 20009
(202) 328-9171

NKCA
National Kitchen Cabinet Association
6711 Lee Highway
Arlington, VA 22205
(703) 237-7580

NLA
National Lime Association
3601 N. Fairfax Dr.
Arlington, VA 22201
(703) 243-5463

NLBMDA
National Lumber and Building Material Dealers Association
40 Ivy St., SE
Washington, DC 20003
(202) 547-2230

NOFMA
National Oak Flooring Manufacturers Association
PO Box 3009
Memphis, TN 38173-0009
(901) 526-5016

NPCA
National Paint and Coatings Association
1500 Rhode Island Ave., NW
Washington, DC 20005
(202) 462-6272

NPA
National Particleboard Association
18928 Premiere Ct.
Gaithersburg, MD 20879
(301) 690-0604

NPCA
National Precast Concrete Association
825 E. 64th St.
Indianapolis, IN 46220
(317) 253-0486

NRMCA
National Ready Mixed Concrete Association
900 Spring St.
Silver Spring, MD 20910
(301) 587-1400

NRCA
National Roofing Contractors Association
1 O'Hare Center
6250 River Rd.
Rosemont, IL 60018
(708) 299-9070

NSPE
National Society of Professional Engineers
1420 King St.
Alexandria, VA 22314
(703) 684-2800

NSA
National Stone Association
1415 Elliot Pl., NW
Washington, DC 20007
(202) 342-1100

NTMA
National Terrazzo and Mosaic Association
3166 Des Plaines Ave., Suite 132
Des Plaines, IL 60018
(708) 635-7744

NWWDA
National Wood Window and Door Association
1400 E. Touhey Ave.
Des Plaines, IL 60018
708) 299-5200

OPCMIA
Operative Plasterers' and Cement Masons' International Association of the United States and Canada
1125 17th St., NW, 6th Floor
Washington, DC 20036
(202) 393-6569

PDCA
Painting and Decorating Contractors of America
3913 Old Lee Hwy.
Fairfax, VA 22030
(703) 359-0826

PPI
Plastics Pipe Institute
355 Lexington Ave.
New York, NY 10017
(212) 351-5420

PHCIB
Plumbing-Heating-Cooling Information Bureau
303 E. Wacker Dr., Suite 711
Chicago, IL 60601
(312) 372-7331

PMI
Plumbing Manufacturers Institute
800 Roosevelt Rd., Building C, Suite 20
Glen Ellyn, IL 60137
(708) 858-9172

PTI
Post-Tensioning Institute
1717 W. Northern Ave., Suite 218
Phoenix, AZ 85021
(602) 870-7540

PCA
Portland Cement Association
5420 Old Orchard Rd.
Skokie, IL 60077
(708) 966-6200

PCI
Prestressed Concrete Institute
175 W. Jackson Blvd., Suite 1859
Chicago, IL 60604
(312) 786-0300

RFCI
Resilient Floor Covering Institute
966 Hungerford Dr., Suite 12B
Rockville, MD 20850
(301) 340-8580

SSFI
Scaffolding, Shoring, and Forming Institute, Inc.
1230 Keith Building
Cleveland, OH 44115
(216) 241-7333

SMA
Screen Manufacturers Association
655 Irving Park, Suite 201
Chicago, IL 60613-3198
(312) 525-2644

SWI
Sealant and Waterproofers Institute
3101 Broadway, Suite 300
Kansas City, MO 64111
(816) 561-8230

SIGMA
Sealed Insulating Glass Manufacturers Association
111 E. Wacker Dr., Suite 600
Chicago, IL 60601
(312) 644-6610

SMACNA
Sheet Metal and Air Conditioning Contractors National Association, Inc.
4201 Lafayette Center Dr.
Chantilly, VA 22021
(703) 803-2980

SMWIA
Sheet Metal Workers International Association
1750 New York Ave., NW
Washington, DC 20006
(202) 783-5880

SBCCI
Southern Building Code Congress International, Inc.
900 Montclair Rd.
Birmingham, AL 35213
(205) 591-1853

SDI
Steel Door Institute
712 Lakewood Center N
14600 Detroit Ave.
Cleveland, OH 44107
(216) 899-0010

SJI
Steel Joist Institute
1205 48th Ave., N, Suite A
Myrtle Beach, SC 29577
(803) 449-0487

SSPC
Steel Structures Painting Council
4400 5th Ave.
Pittsburgh, PA 15213
(412) 268-3327

SWI
Steel Window Institute
1230 Keith Building
Cleveland, OH 44115
(216) 241-7333

SBA
Systems Builders Association
PO Box 117
West Milton, OH 45383
(513) 698-4127

TCAA
Tile Contractors Association of America, Inc.
112 N. Alfred St.
Alexandria, VA 22314
(703) 836-5995

TCA
Tile Council of America
PO Box 2222
Princeton, NJ 08542
(609) 921-7050

UL
Underwriters' Laboratories, Inc.
333 Pfingsten Rd.
Northbrook, IL 60062
(708) 272-8800

USJ&P
United Association of Journeymen and Apprentices of the Plumbing and Pipe Fitting Industry of the United States and Canada
901 Massachusetts Ave., NW
Washington, DC 20001
(202) 628-5823

UBC
United Brotherhood of Carpenters and Joiners of America
101 Constitution Ave., NW
Washington, DC 20001
(202) 546-6206

US/OSHA
U.S. Department of Labor/Occupational Safety and Health Administration
200 Constitution Ave., NW
Washington, DC 20210
(202) 523-8148

U.S. Forest Products Laboratory
One Gifford Pinchot Dr.
Madison, WI 53705-2398
(608) 231-9200

UURWAW
United Union of Roofers, Waterproofers and Allied Workers
1125 17th St. NW, 5th Floor
Washington, DC 20036
(202) 638-3228

VMA
Valve Manufacturers Association of America
1050 17th St., NW Suite 701
Washington, DC 20036
(202) 331-8105

Index

D

E

F

G

H

I

J, K

L

M

N

O

P

Q, R

U

V

W

X, Y, Z

Other Practical References

National Construction Estimator

1994
FREE ESTIMATING DISK ENCLOSED
NATIONAL CONSTRUCTION ESTIMATOR

Current building costs for residential, commercial, and industrial construction. Estimated prices for every common building material. Manhours, recommended crew, and labor cost for installation. *Includes Estimate Writer, an electronic version of the book on computer disk, with a stand-alone estimating program — **FREE** on 5¼" high density (1.2Mb) disk.* (If your computer can't use high density disks, add $10 for *Estimate Writer* on extra 5¼" 360K disks or 3½" 720K double density disks.) **592 pages, 8½ x 11, $31.50. Revised annually**

Residential Wiring Revised

Shows how to install rough and finish wiring in new construction, alterations, and additions. Complete instructions on troubleshooting and repairs. Every subject is referenced to the most recent *National Electrical Code*, and there's over 24 pages of the most-needed *NEC* tables to help make your wiring pass inspection — the first time. **352 pages, 5½ x 8½, $19.75**

Blueprint Reading for the Building Trades

How to read and understand construction documents, blueprints, and schedules. Includes layouts of structural, mechanical, HVAC and electrical drawings. Shows how to interpret sectional views, follow diagrams and schematics, and covers common problems with construction specifications. **192 pages, 5½ x 8½, $11.25**

National Electrical Estimator

This year's prices for installation of all common electrical work: conduit, wire, boxes, fixtures, switches, outlets, loadcenters, panelboards, raceway, duct, signal systems, and more. Provides material costs, manhours per unit, and total installed cost. Explains what you should know to estimate each part of an electrical system. *Includes Electrical Estimate Writer **FREE** with the book on a 5¼" high density (1.2 Mb) disk.* (If your computer can't use high density disks, add $10 for extra 5¼" 360K disks or 3½" 720K double density disks.) **464 pages, 8½ x 11, $31.75. Revised annually**

Contractor's Survival Manual

How to survive hard times and succeed during the up cycles. Shows what to do when the bills can't be paid, finding money and buying time, transferring debt, and all the alternatives to bankruptcy. Explains how to build profits, avoid problems in zoning and permits, taxes, time-keeping, and payroll. Unconventional advice on how to invest in inflation, get high appraisals, trade and postpone income, and stay hip-deep in profitable work. **160 pages, 8½ x 11, $16.75**

Audiotapes: Electrician's Exam Preparation Guide

These tapes are made to order for the busy electrician looking for a better-paying career as a licensed apprentice, journeyman, or master electrician. This two-audiotape set asks you over 150 often-used exam questions, waits for your answer, then gives you the correct answer and an explanation. This is the easiest way to study for the exam. **Two 50-minute audiotapes, $26.50**

Electrical Blueprint Reading Revised

Shows how to read and interpret electrical drawings, wiring diagrams, and specifications for constructing electrical systems. Shows how a typical lighting and power layout would appear on a plan, and explains what to do to execute the plan. Describes how to use a panelboard or heating schedule, and includes typical electrical specifications. **208 pages, 8½ x 11, $18.00**

Audiotapes: Estimating Electrical Work

Listen to Trade Service's two-day seminar and study electrical estimating at your own speed for a fraction of the cost of attending the actual seminar. You'll learn what to expect from specifications, how to adjust labor units from a price book to your job, how to make an accurate take-off from the plans, and how to spot hidden costs that other estimators may miss. *Includes six 30-minute tapes, a workbook that includes price sheets, specification sheet, bid summary, estimate recap sheet, blueprints used in the actual seminar, and blank forms for your own use.* **$65**

Construction Estimating Reference Data

Provides the 300 most useful manhour tables for practically every item of construction. Labor requirements are listed for sitework, concrete work, masonry, steel, carpentry, thermal and moisture protection, door and windows, finishes, mechanical and electrical. Each section details the work being estimated and gives appropriate crew size and equipment needed. This new edition contains ***DataEst***, a computer estimating program on a high density disk. This fast, powerful program and complete instructions are yours free when you buy the book. **432 pages, 11 x 8½, $39.50**

Planning Drain, Waste & Vent Systems

How to design plumbing systems in residential, commercial, and industrial buildings. Covers designing systems that meet code requirements for homes, commercial buildings, private sewage disposal systems, and even mobile home parks. Includes relevant code sections and many illustrations to guide you through what the code requires in designing drainage, waste, and vent systems. **192 pages, 8½ x 11, $19.25**

Estimating Electrical Construction

Like taking a class in how to estimate materials and labor for residential and commercial electrical construction. Written by an A.S.P.E. National Estimator of the Year, it teaches you how to use labor units, the plan take-off, and the bid summary to make an accurate estimate, how to deal with suppliers, use pricing sheets, and modify labor units. Provides extensive labor unit tables and blank forms for your next electrical job. **272 pages, 8½ x 11, $19.00**

Electrician's Exam Preparation Guide

Need help in passing the apprentice, journeyman, or master electrician's exam? This is a book of questions and answers based on actual electrician's exams over the last few years. Almost a thousand multiple-choice questions — exactly the type you'll find on the exam — cover every area of electrical installation: electrical drawings, services and systems, transformers, capacitors, distribution equipment, branch circuits, feeders, calculations, measuring and testing, and more. It gives you the correct answer, an explanation, and where to find it in the code. Also tells how to apply for the test, how best to study, and what to expect on examination day. **320 pages, 8½ x 11, $23.00**

NO POSTAGE
NECESSARY
IF MAILED
IN THE
UNITED STATES

BUSINESS REPLY MAIL
FIRST CLASS MAIL PERMIT NO. 271 CARLSBAD CA

POSTAGE WILL BE PAID BY ADDRESSEE

Craftsman Book Company
6058 Corte del Cedro
P. O. Box 6500
Carlsbad, CA 92018-9974

NO POSTAGE
NECESSARY
IF MAILED
IN THE
UNITED STATES

BUSINESS REPLY MAIL
FIRST CLASS MAIL PERMIT NO. 271 CARLSBAD CA

POSTAGE WILL BE PAID BY ADDRESSEE

Craftsman Book Company
6058 Corte del Cedro
P. O. Box 6500
Carlsbad, CA 92018-9974